PARTICLE
PHYSICS

PARTICLE PHYSICS

An Introduction

ROBERT PURDY, PHD

MERCURY LEARNING AND INFORMATION

Dulles, Virginia
Boston, Massachusetts
New Delhi

Original title and copyright: *The Fundamentals of Particle Physics*
by Robert Purdy. Copyright ©2017 Pantaneto Press. All rights reserved.

Publisher: David Pallai
MERCURY LEARNING AND INFORMATION
22841 Quicksilver Drive
Dulles, VA 20166
info@merclearning.com
www.merclearning.com
1-(800)-232-0223

This book is printed on acid-free paper.

R. Purdy. *Particle Physics: An Introduction.*
ISBN: 978-1-683921-42-4

The publisher recognizes and respects all marks used by companies, manufacturers, and developers as a means to distinguish their products. All brand names and product names mentioned in this book are trademarks or service marks of their respective companies. Any omission or misuse (of any kind) of service marks or trademarks, etc. is not an attempt to infringe on the property of others.

Library of Congress Control Number: 2017960711

171819321 Printed in the United States of America

Our titles are available for adoption, license, or bulk purchase by institutions, corporations, etc. For additional information, please contact the Customer Service Dept. at 1-(800)-232-0223.

The sole obligation of MERCURY LEARNING AND INFORMATION to the purchaser is to replace the book, based on defective materials or faulty workmanship, but not based on the operation or functionality of the product.

To my parents, who have always listened to my ramblings

CONTENTS

INTRODUCTION   *xiii*

1. A HISTORY OF PARTICLE PHYSICS   1
 1.1 Atomic Theory   1
 1.2 Atomic Structure   4
 1.3 Forces and Interactions   10
 1.4 Strange and Unexpected Developments   16
 1.5 Quarks and Symmetries   21
 1.6 The Standard Model of Particle Physics   28
 1.7 The Current State of the Field   31
   Exercises   32

2. SPECIAL RELATIVITY   35
 2.1 Lorentz Transformations   35
   2.1.1 Scalars, Vectors, and Reference Frames   35
   2.1.2 Special Relativity   40
   2.1.3 Minkowski Space   41
 2.2 Energy and Momentum in Minkowski Space   45
   2.2.1 Invariant Mass   47
   Exercises   48

3. QUANTUM MECHANICS   51
 3.1 States and Operators   51
 3.2 The Schrödinger Equation   54
 3.3 Probability Current   56
 3.4 Angular Momentum and Spin   58
 3.5 Spin $\frac{1}{2}$ Particles and the Pauli Matrices   62
 3.6 The Hamiltonian   64
   3.6.1 The Lagrangian   66

3.7 Quantum Mechanics and Electromagnetism:
      The Schrödinger Approach 67
3.8 Quantum Mechanics and Electromagnetism:
      the Pauli Equation 70
      Exercises 73

4. SYMMETRIES AND GROUPS   75
4.1 The Importance of Symmetry in Physics 75
4.2 Discrete Symmetries 76
      4.2.1 Mathematical Structure of
              Discrete Symmetries 76
      4.2.2 Discrete Symmetries in Particle Physics . . . 77
4.3 Continuous Symmetries 85
      4.3.1 Mathematical Structure of Continuous
              Symmetries 85
      4.3.2 Continuous Symmetries in Particle Physics . . 90
      Exercises 100

5 EXPERIMENTAL PARTICLE PHYSICS   103
5.1 Detectors 104
      5.1.1 Interactions of Particles with Matter 104
      5.1.2 Early Detectors 113
      5.1.3 Modern Detectors 116
5.2 Accelerators 123
      5.2.1 Linear Accelerators 125
      5.2.2 Cyclotrons 126
      5.2.3 Synchrotrons 130
5.3 Measurable Quantities in Particle Physics:
      Matching Theory to Experiment 134
      5.3.1 Cross-Sections 135
      5.3.2 Lifetimes 145
      Exercises 148

6. PARTICLE CLASSIFICATION   153
6.1 The Spin-Statistics Theorem 153
6.2 The Strong Force 155
      6.2.1 Isospin 157

6.2.2   Flavor $SU(3)$ 158

6.3   Color 161

6.4   Building Hadrons 164

     6.4.1   Quark Content 165

     6.4.2   Mass 172

     6.4.3   Angular Momentum, Parity, and
              Charge Parity 174

     6.4.4   Larger Flavor Symmetries 178

     6.4.5   Resonances 181

     Exercises 182

7. RELATIVISTIC QUANTUM MECHANICS   **185**

7.1   The Klein-Gordon Equation 185

     7.1.1   A Relativistic Schrödinger Equation 186

     7.1.2   Solutions of the Klein-Gordon Equation . . . 187

     7.1.3   Conserved Current 188

7.2   The Maxwell and Proca Equations 191

     7.2.1   Derivation of the Maxwell Equation 191

     7.2.2   Solutions of the Maxwell Equation 191

     7.2.3   Including Mass: The Proca Equation 194

     7.2.4   Spin of Vector Particles 195

7.3   Combining Equations: How Do Particles
     Interact? 198

     7.3.1   Quantum Field Theory without
              the Math 199

     7.3.2   Feynman Rules 202

     Exercises 206

8. THE DIRAC EQUATION   **209**

8.1   A Linear Relativistic Equation 209

8.2   Representations of the Gamma Matrices 211

     8.2.1   The Dirac Representation 212

     8.2.2   The Weyl Representation 212

8.3   Spinors and Lorentz Transformations 213

8.4   Solutions of the Dirac Equation 216

     8.4.1   Spin 220

8.4.2 Antiparticles 222
8.4.3 Helicity 225
8.4.4 Chirality 226
8.5 Massless Particles 229
8.6 Charge Conjugation 231
8.7 Dirac, Weyl, and Majorana Spinors 232
8.8 Bilinear Covariants 234
Exercises 236

9. QUANTUM ELECTRODYNAMICS   241
9.1 $U(1)$ Symmetry in Wave Equations 241
9.2 Localizing the $U(1)$ Symmetry 243
9.3 The Link with Classical Physics 246
9.4 A Well-Tested Theory 249
9.5 Calculations in QED 250
9.5.1 Feynman Rules for QED 251
9.5.2 Calculating Amplitudes 252
9.5.3 Calculating the Differential Cross-Section . . 260
9.6 Beyond Leading Order: Renormalization 264
9.7 Form Factors and Structure Functions 271
9.7.1 Electromagnetic Form Factors 271
9.7.2 Structure Functions and the Quark Model . . 275
Exercises 276

**10. NON-ABELIAN GAUGE THEORY
AND COLOR   279**
10.1 Non-Abelian Symmetry in the Dirac Equation 280
10.1.1 $SU(3)$ and Color 280
10.1.2 Localizing the $SU(3)$ Symmetry 281
10.2 Gluon Self-Interactions 284
10.3 Strong Force Interactions 285
10.3.1 Quantum Chromodynamics 285
10.3.2 Scale-Dependence 286
10.4 High-Energy QCD 288
10.4.1 Asymptotic Freedom 288
10.4.2 Perturbative QCD 288

10.5 Low-Energy QCD 294
    10.5.1 Quark Confinement 294
    10.5.2 The Residual Nuclear Force 296
    10.5.3 Perturbative and Lattice QCD 298
10.6 Exotic Matter 304
    10.6.1 Pentaquarks and Tetraquarks 304
    10.6.2 Glueballs 306
    10.6.3 Quark-Gluon Plasma 307
    Exercises 308

**11. SYMMETRY BREAKING AND THE HIGGS
MECHANISM**   **311**
11.1 The Weak Force as a Boson-Mediated Interaction . . 311
    11.1.1 $\mathbb{P}$ Violation 312
    11.1.2 $\mathbb{C}$ Violation 313
11.2 Renormalizability and the Need for Symmetry 314
11.3 Hidden Symmetry 315
    11.3.1 Toy Model 1: Z_2 Symmetry Breaking 316
    11.3.2 Toy Model 2: $U(1)$ Symmetry Breaking 318
    11.3.3 The Higgs Mechanism: $SU(2) \otimes U(1)$
        Breaking 324
11.4 Electroweak Interactions 327
    11.4.1 Hypercharge and Weak Isospin 328
    Exercises 330

**12. THE STANDARD MODEL OF PARTICLE
PHYSICS**   **331**
12.1 Putting It All Together 331
12.2 Fermion Masses 332
12.3 Quark Mixing and the CKM Matrix 335
    12.3.1 The Cabibbo Hypothesis 337
    12.3.2 Neutral Mesons 340
    12.3.3 More General Quark Mixing 343
12.4 $\mathbb{CP}$ Violation in the Weak Sector 344
12.5 Successes of the Standard Model 349
    12.5.1 Anomaly Cancellation 349

12.6 Drawbacks of the Standard Model 352

    12.6.1 Baryogenesis 353

    12.6.2 The Hierarchy Problem 354

    12.6.3 The Strong $\mathbb{CP}$ Problem 356

    Exercises 357

13. BEYOND THE STANDARD MODEL   361

13.1 Neutrino Oscillations and the PMNS Matrix 361

13.2 The See-Saw Mechanism 363

13.3 Grand Unification 365

    13.3.1 Magnetic Monopoles 370

13.4 Supersymmetry 371

13.5 Gravitons 374

    13.5.1 Can We Go Further than Spin-2? 377

    13.5.2 Problems with Gravity 378

13.6 Axions 378

13.7 Dark Matter 379

13.8 Dark Energy and Inflation 382

    13.8.1 Inflation 383

    13.8.2 Dark Energy 384

13.9 The Future of Particle Physics 386

    Exercises 390

APPENDIX A   393

APPENDIX B   397

APPENDIX C   403

BIBLIOGRAPHY   405

INDEX   409

INTRODUCTION

WHY STUDY PARTICLE PHYSICS?

Particle physics remains one of the most popular aspects of the study of modern physics. Not only is the subject popular among physics students, it also captures the imagination of the general public. Admittedly, a part of the reason for this may be that the machines constructed for the field's continued development are easily some of the most awe-inspiring engineering feats in human history. However, I do not think that this is the full story. Rather, I believe that the appeal of particle physics is that it addresses some of the key questions that lead people to an interest in physics in the first place. First, particle physics, arguably more than any other discipline, aims to answer the question of what the universe is ultimately made of. In my opinion, the only other discipline to address such weighty problems is cosmology, and the two could not be more different in their approach. While cosmology studies the overall structure and history of the universe on the grandest scales—what we might call the "holistic" approach—particle physics is concerned with building a universe from its simplest constituents. We can think of this as a constructionist or "bottom up" approach to understanding the world. This said, the aim of the two disciplines is ultimately the same, and so despite the many orders of *magnitude* between the scales of their realms of study, there is a surprising amount of overlap between the two. This is a point that we will explore a little in the final chapter of this book.

The second reason I believe particle physics appeals to many is that it is built upon a few guiding principles. Chief among these is the role that symmetry plays in our universe. As we will see throughout this text, symmetry and its implications play a pivotal role in the field of particle physics. As such, there is an elegance and beauty underlying much of modern particle physics. Indeed, it is the desire to produce a more symmetrical theory that has led to many of the developments in the history of particle physics, including as-yet purely hypothetical developments such as supersymmetry and grand unification. The drive to develop theories of the world that are elegant, simple, and symmetrical is not unique to particle physics of course: it is a principle adhered to by all areas of physics, dating back at least as far as William of Ockham and his famous razor in the fourteenth century. However, since particle physics develops this idea to its full potential, this is just another reason that students of physics are commonly drawn to this field.

THE AIM OF THIS BOOK

This book has three main aims. First, I wish to introduce the reader to the concepts of particle physics. As a theorist, my approach to this is mainly to come at things from a theoretical point of view. As such, the emphasis is on developing the ideas of particle physics that are mathematically consistent and then showing that they apply to the real world. This is as opposed to the (arguably more historically accurate) approach of finding and refining theories that fit the experimental observations. In this way, the reader will hopefully develop an appreciation for the elegance of the subject and its reliance on symmetry and simplicity as guiding principles. This said, even as a theorist, I must acknowledge that occasionally it is necessary to observe the real world to ensure that our theories are on the right track. As such, an introduction is given in Chapter 5 to some aspects of experimental physics and of how we may compare theory with experiment.

The crowning achievement of particle physics to date has been the development of the Standard Model of particle physics. The second aim of this book is thus to provide the reader with a solid understanding of the Standard Model. The necessary groundwork is laid throughout the book, and then the Standard Model itself is discussed in Chapter 12, along with some of its properties and consequences.

Despite its successes, however, the Standard Model is known not to be the ultimate theory of fundamental particle interactions, and some of the model's limitations are also discussed. These are further addressed, and some of the extensions of particle physics "beyond the Standard Model" are introduced in Chapter 13.

Finally, the third aim of this book is to provide the reader with the necessary toolkit required for an exploration of particle physics. With this in mind, Chapters 2–4 provide something of a crash course in some of the necessary mathematical background for the subject. Throughout the rest of the book, various additional tools are gradually introduced, such as Dirac spinors, Feynman diagrams and their related calculations, and a brief introduction to the concepts of lattice gauge theory. The hope is that, whatever aspect of particle physics a reader wishes to pursue, this book will provide a useful foundation.

UNITS IN PARTICLE PHYSICS

The *Système International d'Unités* (SI units) is a marvelous achievement of science and diplomacy: a unified, consistent, and logical set of units for the description of any physical quantity (and even some anthropocentric biological ones for good measure). With this system, any two parties anywhere on Earth may know that they are measuring the same quantity in the same way. Despite this, the SI units are not always the most useful or most natural for a given discipline. This is why many scientific disciplines develop their own set of units: solid state physicists commonly use the non-SI Ångstrom, while astrophysics and cosmology routinely express distances in megaparsecs.

The units of choice in particle physics fall into two categories. First, the most important quantity in experimental particle physics is the cross-section for an interaction. This is measured in the same units as area (as it is related to the cross-sectional area of the colliding particles), but the scales involved are of the order of a few femtometers. To avoid a profusion of annoying prefixes, the nuclear physicists involved in the Manhattan Project to construct the first atomic bomb chose a unit of area much more suitable to the kinds of scales being considered. This is the barn (b), equivalent to 100 fm^2, whose name originates from comparison with a barn door, since 100 fm^2 is a relativity large area in nuclear physics terms. Since the barn is so ingrained in experimental particle physics, it is not uncommon to hear it used with SI prefixes. This leads to the rather bizarre "standard" unit for luminosity, or particle collisions per unit cross-section: the inverse femtobarn (fb^{-1}). Similarly, nuclear physicists also introduced the "fermi" as a unit of length, equivalent to 10^{-15} m or one femtometer. This has the fortunate side effect that, when abbreviated, the unit is identical to its SI counterpart: fm. In written form, then, the reader need not consider the fermi as a separate unit. It is worth mentioning here, though, simply because the reader may occasionally come across the term "fermi" in conversation.

In theoretical particle physics, the quantity of interest is generally energy. In a similar vein to the adoption of the barn by experimentalists, theorists have chosen a unit of a suitable scale for the types of quantities they wish to study: the electron volt (eV), equivalent to 1.60217662 × 10^{-19} J, or the amount of energy gained by a single electron when passing through a potential difference of one volt. However, theorists also take this idea much further by introducing the concept of "natural units." By choosing units such that the numerical values of the constants c and $\hbar$ (as well as a handful of other constants) are equal to 1, the need for additional units is negated. In this way, *all* physical quantities may be expressed as powers of eV or, more commonly, mega-, giga- or tera-electron volts (MeV, GeV, or TeV). This idea will be justified further in Chapters 2 and 3. Both experimentalists and theorists alike will, however, also on occasion convert to SI units for clarity or to compare results across disciplines.

1

A HISTORY OF PARTICLE PHYSICS

This chapter is intended to provide the reader with an overview of the history of particle physics, but that is not its only function. Along the way, some of the key concepts in the field will be touched upon in such a way as to prepare the reader for later chapters, when we will revisit these concepts and put them on a firmer footing.

1.1   ATOMIC THEORY

A history of particle physics must start somewhere, and this one starts with Isaac Newton. We could go all the way back to Ancient Greece and talk about the philosophical arguments regarding whether or not matter was infinitely divisible, but we will avoid this for two reasons. First, such matters have been considered in the philosophies of various world cultures, and the author does not wish to be accused of Eurocentrism. Second, and more importantly, the following is a history of the *scientific* study of the particle nature of reality, and science itself only surfaced as a discipline in its modern form much later, in the 15th and 16th Centuries. Newton believed that light is composed of a stream of particles, which he called corpuscles. Though this idea was not originally Newton's, he developed it and showed that it was able to correctly model the laws of reflection and refraction, along with other optical phenomena. The big problem with Newton's corpuscular theory, though, was that it failed when one attempted to apply it to the diffraction and inter-ference of light. For this reason, over time, it was gradually real-ized that an older, rather more conceptually difficult theory was bet-ter suited to describing the nature of light. This was the wave-front theory put forward by Christian Huygens 26 years before Newton's model, which models light as a wave propagating through space. That these two greatly respected scientists could, within so short a time span, propose such vastly different models of the same phenomena, perfectly summarizes one of the big questions in the history of sci-ence. This is the question addressed by those philosophers we have glossed over, of whether matter, at its fundamental level, is contin-uous or composed of indivisible discrete units. Waves and particles, fields and forces, atoms and infinitely divisible matter; all fall under this continuous/discrete dichotomy.

The first great steps toward addressing this dichotomy came not from physics but from chemistry. In the early 1800s, John Dalton formulated his law of multiple proportions. This stated that, if two different chemical elements are capable of producing more than one compound, the ratio of masses of the two elements needed to form one compound is a simple multiple of the similar ratio for the other compound. Dalton's explanation for this law was that each element came in discrete amounts: atoms. Each compound was formed from some combination of the atoms of each element, and the exact com-pound depended on the particular arrangement of atoms. The theory was also able to explain why elements cannot be decomposed as com-pounds can. Each element has its own type of atom, and an atom of one type cannot be transformed into another. What Dalton's theory did not answer was the question of why there were different elements at all. There was simply a group of known elements with particular properties and no obvious unifying principle to explain those prop-erties. As we will see again later in this history, in the absence of an explanation for a large group of related phenomena, a good place to start is to catalog or categorize them to look for patterns. Even if those patterns cannot be explained, at least the underlying structure can begin to be glimpsed. Although many attempted such a categorization of the elements, the first person to construct a comprehensive *structured* catalog was Dmitri Mendeleev with the Periodic Table of Elements. He found that there were regular periodic patterns in the properties of the elements if they were arranged according to their relative mass. This pattern would later be explained with the discovery of the nucleus, composed of protons and neutrons, since the number of protons uniquely determines the position of an element on the table.

Mendeleev's periodic table allowed for predictions of elements' properties, but without an understanding of *why* the elements were arranged in this way; the table itself made no assumptions about the nature of the matter making up the elements. This was then the trend for many years: models were proposed that made use of atoms as a computational tool, and some of these made accurate predictions, but none was entirely convincing as an argument for the physical reality of atoms. This changed in 1905, when Albert Einstein wrote a paper explaining the origin of Brownian motion. Brownian motion had been observed many years earlier by botanist Robert Brown, while he was studying pollen grains. His sample was suspended in water and he was observing it through a microscope when he noticed that tiny particles (smaller than the pollen grains) exhibited what appeared to be random motion. The particles would drift for a short time and then suddenly change direction before drifting again. The reason for this motion, Einstein showed, was that the particles were sufficiently small as to have their trajectory altered by random collisions with water molecules. In fact, Einstein was not the first to suggest this as a mechanism for Brownian motion, but it was he who derived a statistical model of the motion of large numbers of molecules, demonstrating that the predicted motion of the small particles exactly matched observation.

Today, advances in microscopy and nanotechnology have moved us into an astonishing world in which physicists and materials scientists can not only see but even manipulate individual atoms. It is sometimes easy to forget that the majority of particle physics has taken place in relatively recent history. Indeed, the very existence of atoms was placed beyond reasonable doubt only after Einstein's 1905 paper. Our understanding of the nature of matter has made remarkable progress in a little over 100 years. So by 1905, you may be forgiven for thinking that the continuous/discrete dichotomy had been resolved. Matter was composed of atoms, and light was composed of waves. Not so fast, though. In the same year that he conclusively demonstrated the discrete nature of matter, Einstein also cast doubt on the wave nature of light. We will return to this point shortly.

1.2   ATOMIC STRUCTURE

Even before they had been definitively shown to exist, atoms were believed to have structure. A look at the periodic table shows that atoms appear to be built out of the smallest of their kind—the hydrogen atom—since many mass numbers are approximately integer multiples of the mass of hydrogen. This suggested that hydrogen was the fundamental unit of matter, and that heavier elements were somehow built out of hydrogen. Those masses with non-integer multiples of the hydrogen mass would later be explained with the discovery of the neutron. Since each element generally comes in several isotopes (forms with equal numbers of protons but differing numbers of neutrons), but a sample typically does not distinguish these isotopes, the measured atomic mass of an element is the mean of the different isotopes weighted by their relative abundance.

The next big milestone in particle physics was the discovery of the electron in 1896 by J. J. Thomson. The rays produced by a high-voltage cathode in a near vacuum were believed by some to be a stream of negatively charged molecules, while others believed they were some different kind of particle. In fact, the name "electron" had already been given to these hypothetical particles before Thomson's demonstration of their existence. By passing cathode rays through electric and magnetic fields and carefully varying the strength of these fields, Thomson was able to perform accurate measurements of the charge-to-mass ratio of the particles in the ray. Assuming the charge to be the same as in previous charge-to-mass ratio measurements of ions gives a mass for the electron of 0.0005 atomic mass units

(0.51 MeV), over a thousand times lighter than the hydrogen atom. This lent support to the idea that even the hydrogen atom was not fundamental, but was instead constructed from smaller constituents. In particular, Thomson proposed a model of the atom in which the negatively charged electrons were free to move around in a diffuse cloud of positive charge. This model came to be known as the plum-pudding model and is familiar to most students mainly because of the way in which it was overturned.

In the same year that Thomson made his discovery, an unrelated investigation into cathode rays led to the accidental discovery of radioactivity. The radiation given off by radioactive samples is characterized by its ability to ionize any material through which it passes. The phenomenon was of great interest to physicists, and its study was responsible for many discoveries, but it remained poorly understood until the discovery of nuclear decay. We now know radioactive decay to be related to the transmutation of the atomic nucleus. Due to instability, an atom of one element changes its identity to become an atom of a different element: a process previously thought impossible. With our modern understanding of the random and probabilistic nature of quantum phenomena, we also know that an individual such decay cannot be predicted, but that there is merely a finite and fixed probability of its occurence in any given unit time. This probability is the decay constant for the process, Γ. However, given the vast numbers of atoms in a typical sample, the law of large numbers implies that the sample as a whole will behave predictably. In particular, if the number of atoms at time t is N, then the expected value of the change in that number in a short time interval dt is given by

$$dN(t) = -\Gamma N(t). \tag{1.1}$$

Solving this first-order differential equation gives the characteristic exponential decay function for a radioactive sample:

$$N(t) = N(0)e^{-\Gamma t}. \tag{1.2}$$

We can also characterize this type of decay by its half-life, $t_{1/2}$. This is the time that it takes for half of a radioactive sample to decay, and is related to the decay constant by $t_{1/2} = \ln(2)/\Gamma$.

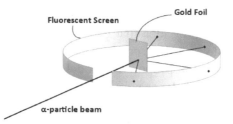

FIGURE 1.1 A schematic representation of the Rutherford scattering experiment.

In 1909, Ernest Rutherford, along with his collaborators, Geiger and Marsden, dispelled the plum-pudding myth with their famous gold-foil experiment. Rutherford's previous work on radioactivity had identified three distinct types of radiation: α, β, and γ. In this latest experiment, α-particles, which Rutherford himself had correctly identified as doubly-ionized helium atoms, were directed on to a gold foil target. The deflection of the alpha particles was then measured by observing their subsequent interaction with a fluorescent screen that would emit a flash at the point of impact (Figure 1.1). In this way, Rutherford demonstrated that the majority of α particles passed straight through the gold foil with minimum deflection, while a small number of particles received a large deflection. A simple calculation using classical mechanics, as in Exercise 1, will show that such large deflections can only occur if the α particle strikes an object with a mass much greater than itself, whereas striking a lighter object will have minimal impact on the α particle's trajectory. In the plum-pudding model, the only objects from which the α particles could be scattered were the electrons, known to have a mass several thousand times too small for large deflections. This suggested the existence of much heavier objects in the gold atoms. Furthermore, the rarity of such large deflections suggested that these massive subatomic components were also very small. This, then, is how the nuclear model of the atom was born. Rather than being spread out in a diffuse cloud, the positive charge of an atom is locked away in a massive and dense nucleus, around which the electrons orbit through electromagnetic attraction. In an effort to verify the point-like nature of the nucleus, Rutherford derived the scattering formula that bears his name, to describe the scattering of particles from a heavy point-like object as a function of energy and scattering angle. In the case of α particle scattering, this formula is given by

$$\frac{d\sigma}{d\Omega} = \frac{k^2 Z^2 e^4}{4E^2 \sin^4(\theta/2)}, \tag{1.3}$$

where E is the energy of the α particle, Z is the atomic number of the scattering center (nucleus), e is the fundamental charge, θ is the angle through which the α particle is scattered, and k is the Coulomb constant ($1/(4\pi\varepsilon_0)$ in SI units). The notation $d\sigma/d\Omega$ will be explained fully in Section 5.3.1, when we discuss cross-sections in detail, but we may think of it as a measure of how many scattering events occur for a given solid angle Ω. We will also see in Section 5.3.1 how Rutherford was able to arrive at this formula. Further experiments by Geiger and Marsden appeared to fit the Rutherford formula precisely, providing evidence that the model was at least on the right track. However, while Rutherford's model is certainly closer to the truth than the plum-pudding model, it immediately runs into problems. Specifically, if the electrons orbit the nucleus, then they should, according to classical electromagnetic theory, radiate away their kinetic energy as light, since they are accelerating in an electric field. Through this radiation, they should quickly lose all kinetic energy and fall into the nucleus. Clearly, there was still something missing in the model. This something was, of course, quantum mechanics.

Quantum mechanics had been developed around the same time, originally by Max Planck to explain the energy distribution of blackbody radiation, but soon also applied by Einstein to an explanation of the photoelectric effect. In order to explain how light could liberate electrons from a metal surface, he assumed that light comes in discrete packets. Einstein called these quanta, but today we call them photons. The photon is a massless particle that carries momentum p and energy E,[1] both related to the frequency f of the corresponding

---

[1] As we will see in Section 2.2, the mass (sometimes also called the rest mass or invariant mass) of a particle is a fundamental property that does not vary for a given particle type. For a photon, this mass is zero. However, the equivalence of mass and energy as demonstrated by special relativity shows that there is also an *effective* or *relativistic* mass of E/c^2. The momentum of a particle is the product of this effective mass with velocity, allowing a massless particle to carry a non-zero momentum.

wave description by

$$E = hf \quad \text{and} \quad p = \frac{hf}{c}, \tag{1.4}$$

where h is Planck's constant, and c the speed of light. A single photon is absorbed by an electron in the photoelectric effect and its energy is used partly to liberate the electron from the metal. Any leftover energy is carried away by the electron as kinetic energy through Einstein's equation,

$$E_{\text{kin}} = hf - \phi, \tag{1.5}$$

where ϕ is the energy required to liberate the electron, known as the material's work function. In this way, the quantum description of light was able to explain why the photoelectric effect will only occur above some material-dependent threshold frequency. If the frequency is too low, the energy of the photon is insufficient to liberate an electron. In contrast, the wave theory of light predicts that light of any frequency should be capable of producing this effect with the kinetic energy of electrons depending instead on the intensity of the incident light. The concept of single photons being emitted and absorbed by charged particles will turn out to be absolutely central to our modern understanding of electromagnetic interactions.

In 1911, Danish physicist Niels Bohr combined Rutherford's idea of an atomic nucleus with Planck's concept of discrete energy levels in quantum mechanics to arrive at a model of the atom in which electrons could only exist in certain orbits around the nucleus. This model was successful in that it was able to explain the structure of atomic emission spectra, each emission line corresponding to some "jump" between discrete energy levels. However, as understanding of quantum mechanics developed, ultimately it was realized that the Bohr model could not be the full story. Later developments showed that the strange dual nature of light, in which it behaves sometimes as a wave and sometimes as a stream of particles, also applies to matter. In order to account for this, quantum mechanics describes any system as a wave function that evolves according to the Schrödinger equation. When a measurement is taken of the system, the allowed values that can result are determined by expanding the wave function in terms of a set of basis states. In particular, the measured position of a particle can, in principle, take any value, but the probability of finding the particle at a particular point is proportional to the square of the wave function's value at that point. Another remarkable consequence of quantum mechanics is Heisenberg's uncertainty principle. This is a direct consequence of the wave-like nature of matter and states that any particle cannot have a simultaneously well-defined momentum and position. Certainty in one implies inherent uncertainty in the other, since a wave that is localized in space must necessarily be a sum of a large range of Fourier modes, corresponding to different momenta. Energy and time follow a similar reciprocal relationship: the duration of an event and the energy involved cannot be simultaneously known. The Bohr model is incompatible with this principle, since it models the electron as a particle with a definite position, albeit one that orbits the nucleus. The more modern model, then, relies on solving the Schrödinger equation for the bound system that is an atom. The orbits are replaced with a set of "orbitals" that are the probability distributions associated with distinct quantum states of the electron. With all this in mind, we do now have a definitive answer to the question that started this chapter: is matter continuous or discrete? The answer is a rather surprising "both!"

Rutherford continued his experiments with α particles and in 1917 found that directing them at nitrogen gas caused the emission of other charged particles. These he was able to identify as hydrogen ions and show that they were originating from the nitrogen atoms. He had shown that the heavier elements really are built out of hydrogen. Unlike the original formulation of this idea, however, it was the heavier elements' *nuclei* that were constructed from hydrogen nuclei. In this way, the proton was discovered to be the unit of positive charge in the nucleus, and to have a mass of around 938 MeV. The atom was now fairly well understood, and radioactivity was understood to be a nuclear process, with α particles identical to helium nuclei and β particles determined to be electrons. One suggestion for the process of β-decay, then, was that some electrons reside in the nucleus and are emitted during decay. A simple calculation using the uncertainty principle, however, shows that the electrons contained in the nucleus would have much higher energy than that measured in β decay. This puzzle was solved in the 1930s with the discovery of the neutron, completing the lineup of subatomic particles in ordinary matter. By firing α particles at a Beryllium target, James Chadwick was able to demonstrate that the nucleus also contains a neutral particle, the neutron, of around the same mass as the proton. It is this particle that decays during β decay, producing a proton and electron in the process. The electron is then emitted as a β particle. Further evidence for the proton and neutron as subnuclear particles came much later in the 1960s, when experimental energies were finally capable of producing a deviation from Rutherford's scattering formula (Equation 1.3) when scattering off the nucleus. These experiments directly demonstrated that the nucleus was not a point-like object but was built out of smaller components. For a better understanding of nuclear decay, physicists now turned to studying the forces that governed the behavior of the nucleus.

1.3 FORCES AND INTERACTIONS

In the classical theory of electromagnetic interactions, charges and currents produce electric and magnetic fields, respectively, and charges moving through these fields experience a force. If photons are to provide the quantum description of electromagnetism, we need a mechanism by which they can reproduce these same phenomena. This is achieved through the concept of virtual particle exchange, and the theory built around this idea is quantum electrodynamics (QED). If a particle is observed or detected, we say that it is a real particle. Its energy, momentum, and mass will necessarily obey the appropriate energy-momentum relation, in which case, we also say that the particle is "on the mass shell," or "on-shell." A virtual particle, on the other hand, is one that is emitted by one particle and absorbed by another. The electron, for example, is capable of emitting a photon, which may then be absorbed by a second electron or other charged particle. Since the photon has been absorbed, we can never detect it. In this way, the existence of the photon can only be inferred by its effects, and we say that it is virtual. Typically, a virtual photon will last only a short time and travel only a short distance. This means that the uncertainty in the photon's energy and momentum can be large, so much so that the photon need not obey the energy-momentum relation and may behave as though it has a non-zero mass. In this case, the virtual photon is said to be off-shell. Since the photon carries an energy and a momentum, and we know both of these quantities to be conserved, we can see how exchange of a virtual photon would lead to a deviation in the trajectories of the electrons. It is in this way that the photon mediates the electromagnetic force. Since this force applies only to charged particles, it must be the case that the emission and absorption of a photon is only possible in particles with a non-zero charge. We say that the photon "couples" to a charge but does not couple to particles without a charge. This image of photons being thrown between electrons is very intuitive for explaining the repulsion of like charges. Where its interpretation becomes less clear is when considering that the electromagnetic force can also be attractive. How, for instance, are we to describe the attraction between a proton and an electron in terms of photon exchange? The first part of the answer to this question is to realize that the uncertainty in the photon's momentum means that it can carry a negative momentum! Momentum conservation then guarantees that the change of momentum of the proton and electron at the point of emission and absorption is in such a way as to produce an attraction. The second part of the answer is that the way in which a photon couples to a particle depends on that particle's charge, and the relative sign of two charges will determine whether the resulting force is attractive or repulsive. There is no easy way to see why this should be the case, but the mathematics really does work out this way, as we will see later when considering QED amplitudes (see Section 9.3). In this way, then, forces are explained in particle physics via the exchange of virtual particles. Such interactions can be depicted using Feynman diagrams, which we can think of as stylized representations of the trajectories of the particles involved. Each external line (with one end not terminating on a vertex with other lines) represents a real particle that takes part in the interaction, while each internal line (with both ends terminating at vertices) represents a virtual particle. Fermions (matter-like particles) such as electrons are shown as solid lines, while photons and other exchange bosons (force-like particles) are shown as wiggly lines. With this in mind, then, we can show elastic electron scattering, $e^- + e^- \to e^- + e^-$, as

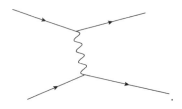

Each time a particle couples to the electromagnetic field through photon emission or absorption, it does so with a particular strength. This is a numerical value, which essentially measures the inherent probability of electromagnetic interactions for that particle at any given moment. This value is different depending on the particle's charge and so is separated into two parts. One is a dimensionless charge factor q, taken to be the ratio of the particle's charge to the fundamental unit charge. That is, $q = -1$ for the electron and $q = +1$ for the proton. The second part is a universal constant known as the electromagnetic coupling constant, e. In natural units, this coupling strength is related to the fine-structure constant, α, by $\alpha = \frac{e^2}{4\pi} \approx \frac{1}{137}$. Since electromagnetic scattering processes involve photon exchange, the leading-order contributions to scattering amplitudes are generally on the order of α, and higher-order corrections can be written as power series in α.

After the discovery of the proton and neutron, there were a number of unresolved problems regarding atomic structure. One such puzzle was the question of what held the nucleus together. What could overcome the electrostatic repulsion of such a small, dense concentration of positive charge? Clearly, there must be some additional attractive force to compensate. The behavior of this nuclear force was qualitatively different from the behavior of electromagnetism though. Most notably, the range of the force must be small, to account for the fact that its effects are not observed outside of the nucleus. In contrast, the range of the electromagnetic force is essentially infinite, though of course it decreases in *magnitude* over large distances. The nature of this force was probed further with experiments in which protons and neutrons were fired at each other. Since the neutron has no charge, the interaction of these two particles in such experiments must be mediated by the nuclear force. Given that the two are of similar mass, one would expect a glancing blow that results in both particles continuing with similar momenta to those before the collision. However, by measuring the interaction cross-section (the likelihood of an interaction) as a function of scattering angle, what was found was that back-scattering was almost as likely as forward scattering. In fact, the cross-section plot was almost symmetrical about a scattering angle of 90°.

In 1935, Hideki Yukawa proposed a model of the nuclear force that could account for both the short range and this cross-section problem, based on the exchange of a new kind of particle. In fact, there were to be three such particles. One was neutral to allow proton-proton and neutron-neutron interactions. The others were charged to allow the interconversion of protons and neutrons through their exchange. This could account for the back-scattering, since these events were then explained as those in which the proton and neutron had exchanged identities, as demonstrated in Figure 1.2. Their momenta were not really changed very much at all, just their clothing! There was a big difference, however, with the electromagnetic theory: whereas the photon is massless, Yukawa determined that his "mesons" must have a mass of around 100 MeV/c^2. To see how he arrived at this figure, consider the exchange of a particle of mass m. Assuming that this particle travels only a short distance at a speed approximately equal to c, the uncertainty in the particle's position is given by

$$\Delta x = c\Delta t = \frac{c\hbar}{\Delta E} = \frac{c\hbar}{\Delta mc^2}. \tag{1.6}$$

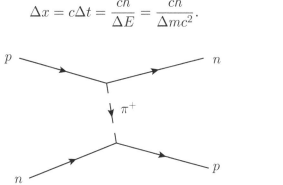

FIGURE 1.2 The exchange of one of Yukawa's charged mesons allows the proton and neutron to inter convert, solving the problem of back-scattering seen in nucleon interactions.

Turning this argument on its head, the mass of an exchange particle associated with a force of range Δx is

$$m = \frac{\hbar}{c\Delta x}, \tag{1.7}$$

which, for nuclear interactions with a range on the order of 1 fm, gives a mass of around 100 MeV. This force would also require its own coupling constant, g_N, and in order for the nucleus to overcome electrostatic repulsion, this coupling strength would have to be considerably greater than its electromagnetic equivalent.

The second big puzzle of nuclear interactions to consider is just how the neutron is capable of producing a proton and electron during β decay. The real problem lay in the fact that β radiation appeared to have a very broad momentum spectrum. If the relative frequency of emissions with a particular momentum is plotted against a range of momenta for α-decay, the result is a sharp peak at one momentum value. This is as we would expect for a single object undergoing a two-body decay, since the energy and momentum are fixed by conservation laws. However, when a similar plot is made of β-particle momenta, the peak is much broader and flatter. To Wolfgang Pauli, this suggested a third particle was being produced in the decay of the neutron. The particle would have to have no charge to ensure charge conservation, which would also explain why it had not been detected, since only charged particles make tracks in tracking chambers, as we will see in Section 5.1.2. In order to match the observed decays, the particle would also have to be very lightweight. In fact, to this day, Pauli's "neutrino" (whose existence has since been confirmed experimentally)[2] has been found to have such a small mass that its value cannot yet be directly measured. This may change in the near future, however, thanks to the recently completed *Karlsruhe Tritium Neutrino Experiment* (KATRIN), which is expected to measure the neu-

---

[2] The name "neutrino" is not Pauli's. In fact, Pauli called his particle the neutron, as what we would now call a neutron had not been discovered at that time. The name was later changed to neutrino to distinguish the two particles, and translates as "little neutral one." Another change in terminology means that we now call Pauli's particle an *anti*-neutrino for consistency with later developments.

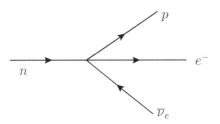

FIGURE 1.3 Feynman diagram for the Fermi "four-point" formulation of β decay.

trino mass to a precision of around 200 meV/c^2, through precision measurements of electron energies in the β decay of tritium nuclei. The current constraint on the neutrino mass is complicated by the fact that it is really an upper limit on the combined mass of three distinct *flavors* of neutrino (for reasons that will become clear when we discuss neutrino oscillations in Section 13.1). This current best guess comes from matching cosmological models to observations of, among other things, the cosmic microwave background, and shows that the combined neutrino masses can be no more than 1.2 eV/c^2. To put this in context, the electron has a mass of just over *half a million* electron volts. With the idea of the neutrino in place, Enrico Fermi suggested a mechanism for neutron decay, in which all four particles involved interacted at a single point, as in Figure 1.3. In this way, the neutron was thought to decay *directly* to a proton, electron, and (anti-)neutrino. This interaction had its own coupling strength, G_F, much smaller than the electromagnetic coupling.

Ignoring gravity, then, as particle physicists are somewhat prone to do, there are three known interactions that determine the behavior of particles. One of these is the familiar electromagnetic interaction, while the other two are short-range forces which only have significant effects within the nucleus. Based on their relative strengths, as encoded in the relevant couplings, the two nuclear forces are referred to simply as "weak" and "strong." The strong force is responsible for holding the nucleus together, while the weak interaction is responsible for nuclear decays.

1.4   STRANGE AND UNEXPECTED DEVELOPMENTS

Returning for a moment to an earlier time, another development was the discovery of spin angular momentum. The first suggestion that there was some as-yet unknown property of electrons, that would later be known as spin, came from Wolfgang Pauli in 1924. By postulating some two-fold degeneracy in electrons, he was able to explain the arrangement of electrons into their atomic orbitals. His discovery was the exclusion principle, which states that no two electrons (more generally, no two fermions) may occupy the same state. The two-fold degeneracy was required to reconcile this principle with the fact that electrons appeared to occupy states in pairs. The Stern-Gerlach experiment later showed, by splitting beams of particles with a magnetic field, that this two-fold degree of freedom in electrons is related to their angular momentum. Specifically, it was discovered that electrons have an intrinsic angular momentum, unrelated to their motion, and of a fixed *magnitude*. The quantization of angular momentum allows for the orientation of this "spin," as it became known, to take two distinct values in the presence of a magnetic field. The *magnitude* of a particle's spin, on the other hand, is an inherent property of that particle. An electron has a spin of $\frac{1}{2}$, while a photon has a spin of 1, and nothing can change that. One of the most important results in particle physics is the spin-statistics theorem. This states that particles can be grouped into two categories according to their spin, and that these two groups behave in profoundly different ways. The fermions have a spin that is half of an odd integer ($\frac{1}{2}, \frac{3}{2}, \ldots$) and obey the Pauli Exclusion Principle. The bosons have integer spin and do not obey the exclusion principle. That is, bosons can form systems in which more than one particle is in the same state, while fermions cannot.[3]

---

[3] More generally, the difference is that fermions obey the statistical mechanics of indistinguishable particles as derived by Fermi and Dirac, while bosons obey the Bose-Einstein statistics, hence the names. Bosons' ability to cluster into the lowest energy state or ground state is also the mechanism behind the non-classical state of matter known as a Bose-Einstein condensate.

In 1928, Paul Dirac, in an attempt to unify quantum mechanics and relativity, wrote down the equation that bears his name. Using this equation, Dirac was able to motivate the spin of the electron from a purely theoretical standpoint. The success of his equation came at a price, however, since it appeared to predict particle states of negative energy. Dirac's resolution of this problem led to one of several unexpected developments that characterized the next period in the history of particle physics. The solution that Dirac found was to postulate the existence of a new particle, somehow the equal and opposite of the electron. The particle, later dubbed the positron, should have the same mass as the electron but the opposite electric charge. What was more, every particle species should have a similar twin.[4] Dirac had predicted the existence of antimatter, and his prediction was validated four years later with the discovery of the positron in cosmic rays. Dirac's own understanding of antiparticles was quite different from the conventional viewpoint today. In Dirac's picture, the negative-energy states that are allowed by his equation are all populated by a sea of negative-energy electrons. Since this is the norm, we do not observe these sea electrons under normal circumstances. However, should one of these negative-energy states be given a boost in energy, it will raise the electron up to a positive energy state. This will also leave a noticeable hole in the sea, however, which is capable of moving around as other negative-energy electrons move into the hole. Relative to the negative sea, this hole will also appear to have a positive charge. Therefore, it is this hole that we see as a positron. Should a positive-energy electron chance upon the positron hole, it can fall into it, releasing its energy as it does so. In this way, we can see that a particle-antiparticle pair can be created apparently from nothing if given energy, and that they may subsequently annihilate one another in a burst of energy. Although Dirac's sea picture is no longer considered correct, it does provide an intuitive understanding of the difficult concept that is antimatter, since processes such as those described above really do occur:

$$e^+ + e^- \to \gamma + \gamma. \tag{1.8}$$

---

[4] To be clear, some particles are their own antiparticle twin. These include the photon and the neutral pion.

Incidentally, while positrons are denoted e^+ to contrast with electrons (e^-), the majority of antiparticles are shown by placing a bar over the particle name.

By 1935, things were looking fairly complete in the world of particle physics. There was a model to describe electromagnetism, and models to describe nuclear stability and decay, all in terms of particle exchange. All that remained to be seen was experimental confirmation of the existence of Yukawa's strong-force meson. This appeared to have arrived in 1936, when a charged particle of roughly the right mass was discovered in cosmic rays. Instead, this turned out to be something wholly unexpected: a cousin of the electron with similar properties but a much greater mass. The terminology became somewhat confused during this period, as particles named after their properties were later reclassified, and names were co-opted for other use. This can make reading published work from this period very confusing to the modern reader. The end result is that while "meson" originally referred to a particle of intermediate mass between the electron and proton, it now refers to a specific set of composite particles, which we will explore in Section 1.5. This left the electron's chunky cousin in need of name, and it later became the muon, denoted μ. Yukawa's meson was eventually found in 1947 and has also undergone a name change: we now know it as the pion.

The discovery of the muon was a prelude for what was to come. After the discovery of the pion, more unexpected particles were discovered. The first of these was the kaon (though not named as such at the time), which was again discovered in cosmic rays, but many more were to follow, including the Σ and Λ baryons, the Ξ baryons, the ρ mesons, and the η mesons. The existence of so many new particles came as a surprise to many, and the era gave rise to what became known as the "particle zoo." With no understanding of *why* these particles existed, there was no choice but to list them along with their properties. Things were getting messy, and it was beginning to look rather like the list of elements before the discovery of the periodic table. What was needed was a "periodic table of particles": a similar means of classifying particles and understanding their underlying trends. A first step in the right direction was to categorize the particles by their spin. In this way, a distinction was made between the leptons, the mesons, and the baryons. Originally, these names had referred to the relative mass of the particles: *lepto-*, meaning "light," *meso-*, "medium," and *bary-*, "heavy." However, over time, these distinctions began to blur, and the names were used in a different way. Lepton came to mean "electron-like," and so included the muon and electron. The majority of new particles, though, were categorized as mesons if they had integer spin, and baryons if they had half-integer spin. Collectively, these two categories form the hadrons. It was soon realized that there appeared to be a conservation law regarding baryons. The total number of baryons in a system appeared to remain unaltered before and after any interaction, as long as antibaryons were taken to count negatively against this sum. That is, if baryons are assigned a baryon number of 1, and antibaryons a baryon number of −1, then this number is conserved during physical processes. It was also realized that there is no such conservation law for mesons: a process may produce one or more mesons without violating any conservation law.

Similar conservation laws were found for the leptons, through careful observation of interactions involving muons, electrons, and neutrinos. In fact, rather than an overall "conservation of lepton number," which would be analogous to baryon conservation, it was found that there are independently conserved electron and muon numbers. The subtlety in this statement is that each of these leptons has associated with it a neutral particle. In the electron's case, this is the same neutrino involved in β decay. In the muon's case, however, it is an entirely new "mu neutrino," distinguished from the electron neutrino only by its lepton numbers. To clarify this statement, we assign an electron number of $L_e = +1$ to the electron and its neutrino, and $L_e = -1$ to their antiparticles. Similarly, we assign a muon number of $L_\mu = +1$ to the muon and μ neutrino, and $L_\mu = -1$ to their antiparticles. Both of these lepton numbers are then independently conserved.

A full understanding of the origin of the hadrons was still needed, and this is where Murray Gell-Mann came in. In 1961, Gell-Mann arranged these new particles according to two properties: isospin and strangeness. The isospin formalism requires an understanding of symmetry groups and so discussion of this will be deferred until

Chapter 6, where we will see exactly how Gell-Mann arranged the hadrons into groups. For now, we will simply assert that the hadrons have a well-defined property known as isospin, I_3, and we will categorize them according to this property. We will, however, discuss the concept of strangeness.

Strangeness

The strength of a force tells us a great deal about it. As well as the typical binding energies involved, it tells us the probability of an interaction occurring as well as the characteristic length-scale and time-scale of those interactions. This led to a problem with the new baryons that were being discovered: it was noticed that some of these baryons were produced at a high rate, but that their decays were typically rather slow processes. Since this was fairly strange behavior, Gell-Mann suggested the existence of a quantum number which he called "strangeness." This strangeness could be quantified: if a particle decayed via one of these strange processes to ordinary particles, it was given a strangeness of *magnitude* 1. Actually, in much the same way that the electron has the "wrong" charge of -1, history has dictated that strange baryons have a negative strangeness, with positive strangeness reserved for strange antibaryons. If a particle decays via a strange process to something that is itself strange, then it is said to have strangeness ± 2, and so on. Gell-Mann offered an explanation for the strange behavior of such hadrons based on this new quantum number. His idea was to suggest that there were two distinct types of interaction responsible for the production and decay of strange particles. The interaction responsible for their production had a short characteristic time-scale and large coupling, and also conserved strangeness. On the other hand, the process responsible for decay was weaker with a longer time-scale, and crucially did not obey the law of conservation of strangeness. We now identify these processes as the weak nuclear and strong nuclear interactions. In this way, the strong interaction could produce a pair of particles with opposite strangeness, which would then propagate away from their point of origin. Once the particle is separated from its twin, however, the strangeness-conserving process responsible for its production is no longer an option for decay. This leaves only the strangeness-violating weak process for their decay, with its characteristically longer time-scale.

With these concepts in place, we can now list some of the hadrons along with their properties (Table 1.1). Note that the values given are the currently measured values as opposed to the values as measured back in the 1960s. Indeed, some of the particles listed here had not even been discovered at that time. At the same time, this list is far from exhaustive, containing only a sample of the hadrons that have been observed.

1.5   QUARKS AND SYMMETRIES

By categorizing the known hadrons according to their baryon number, spin, isospin, and strangeness, Gell-Mann found a series of regular patterns, as in Figures 1.4–1.7. Since the baryons' antiparticles have the opposite baryon number, they form a separate plot. In the case of the mesons, however, particles and their own antiparticles are present in the same plot, since all have a baryon number of 0. In particular, each meson's antiparticle is in the position diametrically opposed to it, and those mesons in the center are their own antiparticles.

These patterns will be explored in more detail in Chapter 6, but the result was that Gell-Mann found he could explain the patterns in terms of a symmetry group, $SU(3)$. Just as the periodic table displays the patterns in the elements, Gell-Mann had drawn out the underlying symmetries hidden in the hadrons. Notice in the case of elements, though, that a true understanding of the periodic table only came with the discovery of the *structure* of the atom, since this did more than simply categorize the elements: it *explained* them. In a similar way, by 1961, Gell-Mann had a "periodic table" of the hadrons, but still lacked a mechanism that could explain it. This came two years later with the quark model.

Hadron	Mass (MeV)	Spin	Charge (q)	Isospin (I)	I_3	Strangeness (S)	Baryon number (B)
π^0	134.977	0	0	1	0	0	0
$\pi^\pm$	139.57	0	± 1	1	± 1	0	0
η	547.862	0	0	0	0	0	0
η'	957.78	0	0	0	0	0	0
$K^\pm$	493.667	0	± 1	$1/2$	$\pm 1/2$	± 1	0
ρ^0	775.5	1	0	1	0	0	0
$\rho^\pm$	775.4	1	± 1	1	± 1	0	0
n	939.565	$1/2$	0	$1/2$	$-1/2$	0	$+1$
p	938.272	$1/2$	$+1$	$1/2$	$+1/2$	0	$+1$
Σ^-	1197.449	$1/2$	-1	1	-1	-1	$+1$
Σ^0	1192.642	$1/2$	0	1	0	-1	$+1$
Σ^+	1189.37	$1/2$	$+1$	1	$+1$	-1	$+1$
Ξ^-	1321.71	$1/2$	-1	$1/2$	$-1/2$	-2	$+1$
Ξ^0	1314.86	$1/2$	0	$1/2$	$+1/2$	-2	$+1$
Δ^-	1232	$3/2$	-1	$3/2$	$-3/2$	0	$+1$
Δ^0	1232	$3/2$	0	$3/2$	$-1/2$	0	$+1$
Δ^+	1232	$3/2$	$+1$	$3/2$	$+1/2$	0	$+1$
Δ^{++}	1232	$3/2$	$+2$	$3/2$	$+3/2$	0	$+1$
Σ^{*-}	1387.2	$3/2$	-1	1	-1	-1	$+1$
Σ^{*0}	1383.7	$3/2$	0	1	0	-1	$+1$
Σ^{*+}	1382.8	$3/2$	$+1$	1	$+1$	-1	$+1$
Ξ^{*-}	1530	$3/2$	-1	$1/2$	$-1/2$	-2	$+1$
Ξ^{*0}	1530	$3/2$	0	$1/2$	$+1/2$	-2	$+1$
Ω^-	1672.45	$3/2$	-1	0	0	-3	$+1$

TABLE 1.1 Selected properties of a selection of hadrons. The number of significant figures given for the mass varies based on the precision with which the value is known.

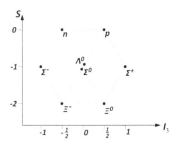

FIGURE 1.4 The spin-$\frac{1}{2}$ baryon octet.

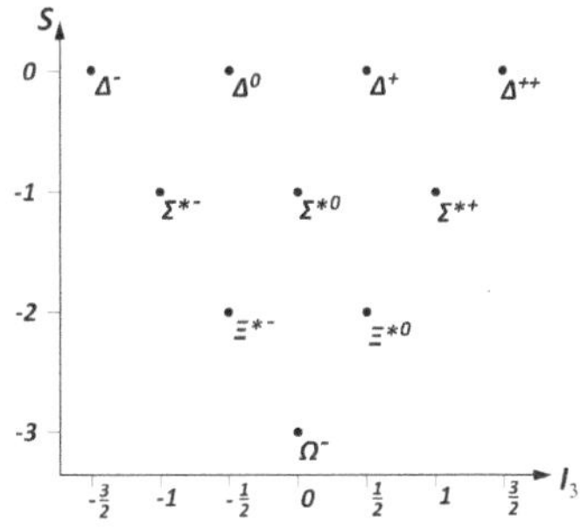

FIGURE 1.5 The spin-$\frac{3}{2}$ baryon decuplet.

What Gell-Mann realized was that he could explain the patterns of hadrons with the introduction of a set of three new particles, which he called quarks.[5] By assuming the quarks to have the properties

[5] The name "quark" is often said to have been taken from the line "Three quarks for Muster Mark" in the book *Finnegans Wake*, because of its fitting the requirement of three particles. However, Gell-Mann himself has stated that this is not quite the full story. In fact, he had chosen the pronunciation of the name he intended to use for these particles based on nothing more than a whim, and which he intended to rhyme with "cork." The spelling only came later when he saw the above passage from *Finnegans Wake*, and in fact the structure of the passage suggested that the author's pronunciation of that word did not match his own, but should instead rhyme with Mark. He neglected this fact and used the spelling "quark" to fit his chosen pronunciation. As a result of this, there really is a *right* way to pronounce quark (*cork*) and a *wrong* way (*Mark*). It is also worth mentioning that the pronunciation issue could have been avoided altogether if Zweig had beaten Gell-Mann to publication, since he independently made a similar discovery and preferred the name "aces."

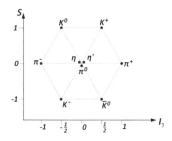

FIGURE 1.6 The spin-0 scalar meson nonet.

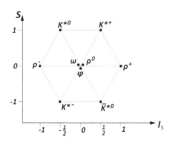

FIGURE 1.7 The spin-1 vector meson nonet.

Quark	Spin	Charge (q)	I_3	S	B
u	$1/2$	$+2/3$	$+1/2$	0	$+1/3$
d	$1/2$	$-1/3$	$-1/2$	0	$+1/3$
s	$1/2$	$-1/3$	0	-1	$+1/3$
$\bar{u}$	$1/2$	$-2/3$	$-1/2$	0	$-1/3$
$\bar{d}$	$1/2$	$+1/3$	$+1/2$	0	$-1/3$
$\bar{s}$	$1/2$	$+1/3$	0	$+1$	$-1/3$

TABLE 1.2 Quark Properties

listed in Table 1.2, all of the hadron properties listed in Table 1.1 (with the exception of mass) can be explained by taking baryons to be collections of three quarks, and mesons to be a combination of a quark and an antiquark. Each hadron property is then found by summing the corresponding properties of the individual quarks. The mass is an exception to this rule and will be considered in detail in Chapter 6. The quarks were named after the role they played in the model. The "up" and "down" quarks have isospin values of up ($+\frac{1}{2}$) and down ($-\frac{1}{2}$) respectively, while the "strange" quark has a non-zero strangeness. Notice that I have not listed the masses of the quarks, since this is a slightly complicated issue, as we will see in Chapter 6. With these assignments, we can see that the quark content of some sample hadrons is given by

Hadron	Quarks
p	uud
n	udd
Σ^+	uus
Σ^-	dds
π^+	$u\bar{d}$
π^-	$d\bar{u}$

While these ideas were widely regarded as a great step in understanding the hadrons, less popular was the suggestion that the quarks are actual physical particles. Even Gell-Mann saw the quarks more as mathematical constructs that fit the data. There is now plenty of evidence, however, that quarks are physical. We still cannot strictly claim to have seen an individual quark, but at least we now have a reason not to have seen one. Our current model of the strong force was developed around 1965, and grew out of the observation that quarks require an additional degree of freedom so as not to violate the Pauli Exclusion Principle. It was determined that each quark "flavor," u, d, and s, should come in three "colors," red, blue, and green. These names have nothing at all to do with color in the literal sense but are convenient names for the three varieties of each quark. This color charge was the basis for a color force: an interaction between quarks based on their color as electromagnetism is based on charge. Since it is based on color, and was modeled on quantum electrodynamics, it became known as quantum *chromo*dynamics. The mathematics of this new force conspired to keep quarks locked away inside hadrons in a process called confinement, explaining why no isolated quark has been observed. It also explained why quarks formed collections of three in baryons and quark-antiquark pairs in mesons: when a set of quarks contains one of each color, from outside the group the result is color neutral, just as a pair of opposing charges appears electrically neutral from sufficient distance. Similarly, a color and its anticolor also appear colorless. A lone quark, if there were such a thing, would attract two other quarks of other colors to itself to form a color-neutral baryon, or would attract its anticolor to form a meson. Once either of these objects is formed, the result is color neutral and no further interaction with external quarks occurs. Those interactions do continue inside the hadron, however, with the strong force mediators, the gluons, constantly carrying color information between the quarks. The only evidence of all this activity from outside the hadron is the residual strong force that "leaks" out and binds the hadron to other nearby hadrons through the exchange of pions and other mesons. In this way, the original formulation of the strong nuclear force is found to be merely the low-energy effect of a more fundamental interaction.

As evidence for physical quarks, an argument for why we cannot see them may seem fairly tenuous. However, there is additional evidence in the form of Rutherford scattering. Just as the nucleus was found to show structure when bombarded with high-energy particles, when the energy is increased far enough, the hadrons themselves similarly deviate from Rutherford's scattering formula. This demonstrates that the hadrons have structure in the form of smaller constituent parts. Once it is accepted that quarks are physical objects, we see that d and s are similar, in the same way that the muon is similar to the electron. For this reason, we arrange the fermions into generations of increasing mass.

Fermi's original four-fermion formulation of the weak interactions was found to be adequate to explain the majority of weak phenomena. However, later developments would show the four-point interaction to be inconsistent at high energies. Ultimately, these interactions were also explained in terms of a mediating particle. As with Yukawa's meson, the short range of these particles' influence was attributed to a large mass. The picture that emerged was of an interaction something like that shown in Figure 1.8, in which a quark emits a weak force carrier, changing flavor in the process. The carrier, known as a W^- (antiparticle W^+), which must be charged in order for the interactions to conserve charge, then decays to two leptons. To account for the decays of strange hadrons, it was realized that these weak interactions must occur within their own generation: $d \rightarrow u + W^-$, but also across generations: $s \rightarrow u + W^-$.

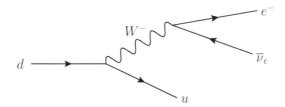

FIGURE 1.8 The weak interaction as a boson-mediated interaction.

Leptons, on the other hand, were found not to take part in these cross-generational interactions. It had also been noticed that the apparent couplings of quarks through the weak interaction were reduced when compared with the lepton couplings. To account for this difference, Nicola Cabibbo proposed a mixing of quarks, such that the states that take part in weak interactions are a linear combination of the physical quark states. We will see in Chapter 12 why such mixing is in fact a very natural consequence of the theory of weak interactions. This mixing was best explained if there was a fourth quark flavor, with charge $+\frac{2}{3}$ to match the up quark; this possibility was first put forward by Glashow and Bjorken on fairly shaky theoretical grounds, but yet completing the second generation of fermions. This hypothetical quark was later incorporated into a fuller understanding of weak interactions through the Glashow-Iliopoulos-Maiani mechanism, which displayed such elegance that they named the hypothetical particle "charm." A meson formed from a charm-anticharm pair, the J/Ψ,[6] was discovered in 1974, confirming the prediction, and showing the charm quark to be considerably heavier than its cousins. It was around this time that the quark model gained a near-universal following. Since this, two more leptons—the τ and its associated neutrino—and two more quarks—the top and bottom—have been found, bringing the number of fermion generations to three. These too have their own quantum numbers, so we now have quark flavor quantum numbers strangeness (S), charm (C), bottom ($\tilde{B}$), and top (T), where the tilde is placed on the symbol for bottom to distinguish it from the baryon number, B. In addition, we have a third

---

[6] The slightly unusual name, J/Ψ, is due simply to the fact that this particle was discovered independently by two groups and given two distinct names.

conserved lepton number, L_τ. The properties of all of these particles may be found in Appendix A.

1.6 THE STANDARD MODEL OF PARTICLE PHYSICS

During the 1960s, Glashow, Weinberg, and Salam developed a remarkable theory that unified the electromagnetic and weak interactions into one "electroweak" theory, which will be explored in detail in Chapter 11. The first prediction of this unified model to be experimentally verified was the existence of neutral weak currents. These are weak interactions in which no flavor change occurs: instead a fermion emits an electrically neutral weak force carrier called the Z^0 which interacts with another fermion, again with no change in flavor. In this way, Z^0 exchange is very similar to photon exchange but can occur between all particles, whereas photon exchange occurs only between charged particles. Such interactions allow for neutrinos to influence the motion of electrons, and it is in this way that the process was confirmed experimentally. Many years later in the 1980s, the $W^\pm$ and Z^0 bosons were directly detected by the Gargamelle bubble chamber after their production in the Large Electron-Positron (LEP) collider at CERN. There was, however, one piece of the model that remained elusive. In order to give masses to three of the mediators of the electroweak theory, while retaining a massless photon, the model made use of a specific spontaneous symmetry breaking process known as the Higgs mechanism, originally proposed by, among others, Brout, Englert, Higgs, and Kibble. This requires the introduction of a set of four spin-0 particles, three of which are "eaten" by the weak force carriers, thereby providing them with their mass. The fourth particle should remain physical and should, by that token, be detectable. This is, of course, the Higgs boson.

The combination of the electroweak theory with quantum chromodynamics led to the modern *Standard Model* of particle physics. With the exception of the discovery of neutrino oscillations, which imply a small but non-zero mass for neutrinos, and refined

Generation

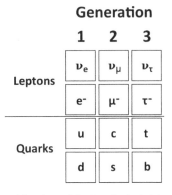

FIGURE 1.9 The fundamental fermions of the Standard Model.

measurements of the model's parameters, the model remains largely unchanged. It consists of the fermions listed in Figure 1.9, along with the photon, the $W^\pm$ and Z^0, eight gluons, and the Higgs. As well as the particle content, the Standard Model consists of a set of rules that govern the interactions of these particles. In particular, we find a set of conservation laws, whereby q, B, L_e, L_μ, and L_τ are conserved in all interactions. In addition, the isospin and flavor quantum numbers, I_3, S, C, $\tilde{B}$, and T, are also conserved in electromagnetic and strong interactions, but not in weak. These conservation laws can be understood in terms of "allowed" and "forbidden" vertices in Feynman diagrams. In particular, the vertices shown in Figure 1.10 are all allowed, where a solid line denotes a fermion, a dashed line denotes the Higgs, a wiggly line denotes either a photon (γ) or a weak boson ($W^\pm$, Z^0), and a springy line denotes a gluon. Any vertex not in this allowed list is forbidden in the Standard Model. If the initial and final states of a process can be connected to each other using only allowed vertices, then the process can occur, but if there is no set of allowed vertices that connects the states, then the process is forbidden and not observed. The fermion-fermion-W interaction is complicated by quark mixing (see Chapter 12), and the allowable weak vertices fall into two camps. First, there are those in which one fermion is a charged lepton and the other is the neutrino with the *same* lepton number, such as e^- and ν_e, or μ^- and ν_μ. The second type consists of one "up-type" and one "down-type" quark. That is, one fermion is u, c, or t, while the other is d, s, or b.

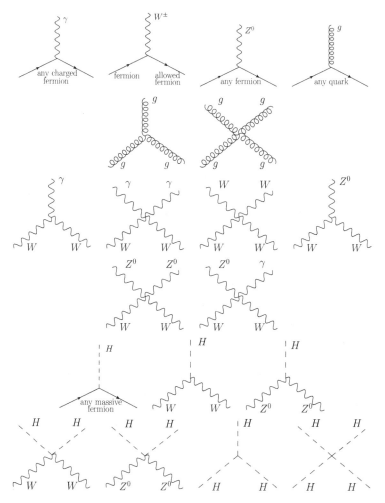

FIGURE 1.10 The allowed vertices of Standard Model interactions. See the text for an explanation of the "allowed fermion" label in the weak interaction.

The Standard Model has proven remarkably successful, but is by no means considered to be the complete theory of particle physics. For one thing, it makes no attempt to incorporate gravity. In fact, there is a very good reason for this, since even the mathematical framework that underpins the Standard Model runs into problems when it is applied to gravity. There are other issues with the Standard Model that will be partly addressed in Chapter 12.

1.7   THE CURRENT STATE OF THE FIELD

In the years since the Standard Model was formulated, there have been additional discoveries that the theory does not explain. One of these was the discovery that galactic rotation rates cannot be explained by the amount of visible matter within each galaxy. The most common view of this problem is that there must be a large amount of matter that is not visible, known as dark matter. The suggestion is that this should be some kind of as-yet undiscovered matter that does not participate in the Standard Model interactions, or possibly takes part only in the weak interaction. Although this idea has actually been around for a long time—the galactic rotation problem was discovered as far back as the 1920s—the idea has only really gained widespread acceptance, and the scale of the problem fully appreciated, since measurements were improved in the 1980s. While the Standard Model does not contain any suitable candidates for this dark matter, there are extensions of the model that are able to account for the presence of this mysterious phenomenon.

The biggest shock discovery, however, came in 1998 from observations of distant supernovae. Until this survey, it had been widely assumed that the mutual gravitational attraction of galaxies should act to slow down the expansion of the universe, and the aim of the survey was to measure this deceleration. What the team found was wholly unexpected: the deceleration parameter was negative. That is, the expansion of the universe is accelerating. This requires the presence of a new type of energy in the universe, dubbed dark energy in analogy with the dark matter that preceded it. In fact, though there are some alternative theories to explain the perceived universal acceleration, it is now widely believed that over 68% of the total energy in the universe is accounted for by dark energy. We will see in Chapter 13 that dark energy may also have an explanation intimately tied to particle physics.

The final outstanding piece of the Standard Model was discovered experimentally in 2012. CERN's Large Hadron Collider (LHC) was built as a general-purpose collider, but one of the problems it set out to tackle was direct observation of the Higgs boson, which has now been achieved. This was (and still is!) not the LHC's only goal, however. Given the theoretical problems with the Standard Model, it is widely expected that there should exist additional physics "Beyond the Standard Model." Some of the theoretical extensions of the Standard Model that have been proposed include supersymmetry, grand unified theories, axions, and other dark matter candidates. All of these ideas, along with others, generally predict the existence of new as-yet unobserved particle species. At the time of writing, none of these proposals has been confirmed experimentally. In a sense, this leaves particle physics today in somewhat of a limbo: it is expected that new physics lies just beyond the reach of our most powerful experiments, but there is so far no hint as to what that new physics may look like. To be clear, though, this not is a disappointing end to this history. On the contrary, not knowing quite what to expect from the current and next generation of particle physics experiments makes this a very exciting time to be active in this field.

EXERCISES

1. Suppose the α particle in Rutherford's gold-foil experiment has initial velocity $\mathbf{v}$ and collides with a stationary target of mass m_t. The α particle then moves off with velocity $\mathbf{v}_\alpha$ and the target has velocity $\mathbf{v}_t$.

(a) Using conservation of momentum and conservation of energy, show that

$$|\mathbf{v}_t|^2 \left(1 - \frac{m_t}{m_\alpha}\right) = 2\,|\mathbf{v}_\alpha| \cdot |\mathbf{v}_t| \cos\theta,$$

where θ is the angle between $\mathbf{v}_t$ and $\mathbf{v}_\alpha$.

(b) Hence show that if $m_\alpha \gg m_t$, then θ is acute and there is little momentum transfer.

(c) Similarly, show that if $m_\alpha \ll m_t$, then θ is obtuse.

(d) How do these scenarios relate to models of the atom?

2. Radium-226 has a half-life of 1600 years. What is the decay constant for this material in decays s^{-1}? How long would it take a sample to decay to 1% of its original size?

3. A free neutron decays via β-decay with a half-life of around 15 minutes. Why is this not necessarily true of the neutrons in atomic nuclei?

4. The nucleus is on the order of a few fm across. The β decay of a nucleus causes the emission of an electron, typically with an energy of around a few MeV. By considering the uncertainty in the position and momentum of an electron confined to the nucleus, show that the emitted electron cannot reside in the nucleus before emission, but must be produced at the moment of decay.

5. **(a)** By considering the conservation of charge, strangeness, baryon number, and lepton numbers, determine whether the following processes can occur. If the process cannot occur, state which conservation law forbids it.

$$p + n \rightarrow \pi^+ + \pi^0$$
$$e^+ + e^- \rightarrow \mu^+ + \bar{\nu}_\mu$$
$$K^+ \rightarrow \pi^+ + \pi^0$$

$$e^+ + e^- \rightarrow p + \bar{p}$$
$$K^+ \rightarrow \pi^+ + \pi^+ + \pi^-$$
$$p + \bar{p} \rightarrow \nu_\mu + \bar{\nu}_\tau$$
$$\mu^+ - \bar{\nu}_e \rightarrow \nu_\mu + e^+$$
$$\mu^+ + \bar{\nu}_e \rightarrow \nu_\mu + e^+$$

6. Draw possible tree-level Feynman diagrams (no closed loops) for the following interactions.

$$\tau^+ \rightarrow e^+ + \nu_e + \bar{\nu}_\tau$$
$$p + \bar{p} \rightarrow n + \bar{n}$$
$$e^+ + e^- \rightarrow e^+ + e^-$$

In the last example, there are two diagrams that contribute: can you find both?

CHAPTER 2

SPECIAL RELATIVITY

This is the first of two chapters of general background that will be useful for the remainder of the book. These chapters are not intended as introductions to these subjects but more as a recap of material that the reader should be reasonably familiar with. They will not be an in-depth look at the background material but will cover only those aspects of the material that will prove most useful for later chapters. This chapter revisits the ideas of special relativity, introducing the concept of four-vectors and tensors. Most importantly for our purposes, it will introduce the index notation that will be used throughout much of the rest of this book. For a more in-depth introduction to this rich subject, the reader is encouraged to take a look at the text *The Special Theory of Relativity* in this series.

2.1   LORENTZ TRANSFORMATIONS

2.1.1   Scalars, Vectors, and Reference Frames

The reader is almost certainly familiar with the concepts of a "vector" and a "scalar." The way in which these concepts are typically introduced is to describe a scalar as "just a number" and a vector as a "number with a direction." In practice, one likely thinks of a scalar as a single number and a vector as a collection of components. However, these two descriptions don't quite match. In reality, there are objects we can write down that consist of a collection of components that do not correspond to the idea of a number with direction. The reason for this is that the real essence of a vector is captured by its behavior under coordinate transformations. It is not enough to list a series of components: we must also check that these components behave the right way if we look at the object from a different perspective. To illustrate this idea, consider a two-dimensional vector **v**. When this vector is viewed from a particular reference frame, S, it may have a set of components v^1, v^2. However, if viewed from a different reference frame, $\overline{S}$, it has a different set of components $\overline{v}^1, \overline{v}^2$. (We use superscripts to identify the components for reasons that will become clear shortly. Also, we are using bars to denote the second frame rather than the more common "primed" notation simply to avoid clashes with superscipt indices.) If the reference frames are related by a rotation through an angle θ, then we would expect the relationship between components in each frame to be of the form $\overline{v}^1 = v^1 \cos \theta + v^2 \sin \theta$ and $\overline{v}^2 = v^2 \cos \theta - v^1 \sin \theta$. We can say then that an object is only a vector *if it transforms in this way*. The transformation can be written in a concise form using a rotation matrix $\underline{\underline{\mathbf{R}}}$ where

$$\overline{\mathbf{v}} = \underline{\underline{\mathbf{R}}} \cdot \mathbf{v}$$

$$\text{or} \quad \begin{pmatrix} \overline{v}^1 \\ \overline{v}^2 \end{pmatrix} = \begin{pmatrix} \cos \theta & \sin \theta \\ -\sin \theta & \cos \theta \end{pmatrix} \begin{pmatrix} v^1 \\ v^2 \end{pmatrix}. \quad (2.1)$$

What is this matrix notation really saying though? It says that the matrix $\underline{\underline{\mathbf{R}}}$ consists of four numbers which we can label by their row and column: $R^1{}_1$, $R^1{}_2$, $R^2{}_1$, and $R^2{}_2$, and that some (but not all) of these numbers multiply other numbers v^1 and v^2. Specifically,

$$\overline{v}^1 = R^1{}_1 v^1 + R^1{}_2 v^2$$
$$\overline{v}^2 = R^2{}_1 v^1 + R^2{}_2 v^2, \quad (2.2)$$

which we can summarize as

$$\overline{v}^i = \sum_{j=1}^{2} R^i{}_j v^j \quad \text{for } i = 1, 2, \quad (2.3)$$

where the index i labels the new components and j labels the old components. This is shorthand for two equations: one with $i = 1$ and one with $i = 2$. Notice that the i index appears on both sides of the equation. The index j plays a very different role. There is no j on the left because j is summed over on the right, so there is no free j to relate the left to. In fact, we could change the name of j to anything we like without changing the meaning of the equation. For this reason, j is called a dummy index, while i is a free index. Equations like the one above are so common in this notation that, by convention, we leave out the $\sum$ sign. A repeated index automatically implies a sum over the dummy index. This is known as the Einstein summation convention. Notice that $R^i{}_j$ is just a measure of how the i-th new coordinate depends on the j-th old coordinate. So

$$R^i{}_j = \frac{\partial \bar{x}^i}{\partial x^j}. \tag{2.4}$$

Now although two observers in different frames will disagree on the components of a vector, there is one property of the vector on which they agree: namely, its length. This value is independent of any particular reference frame, and it is this invariance property that classifies it as a *scalar*. To put this another way, the transformation law for a scalar between different reference frames is simply $\bar{\ell} = \ell$. Notice that this is *not* true of a vector component such as v^1: this transforms instead as one component of a vector, not as a scalar.

For ease of generalization to later cases, let's now consider a vector with infinitesimal components dv^1, dv^2. The length of this vector is given by

$$\begin{aligned} d\ell &= \sqrt{dv^2} \\ &= \sqrt{(dv^1)^2 + (dv^2)^2} \\ &= \sqrt{dv^i \delta_{ij} dv^j} \end{aligned} \tag{2.5}$$

where δ_{ij} is the Kronecker delta, defined to be 1 if $i = j$ and 0 otherwise. More generally, the scalar product of any two vectors is given by

$$\mathbf{u} \cdot \mathbf{v} = u^i \delta_{ij} v^j. \tag{2.6}$$

The delta is playing the role of a metric, an object that tells us how to form scalars from vectors in a given coordinate system. For Cartesian coordinates in Euclidean space, this metric is very simple. However, for other systems, it takes different forms. For example, in polar coordinates r, θ, infinitesimal distances are given by

$$d\ell^2 = dv_r^2 + r^2 dv_\theta^2 = dv^i p_{ij} dv^j, \tag{2.7}$$

where p_{ij} is now the metric and is given by $p_{rr} = 1$, $p_{\theta\theta} = r^2$, $p_{r\theta} = p_{\theta r} = 0$.

In order to generalize these ideas to relativity, it is useful to consider here one more concept: the gradient. The gradient of a scalar quantity, ϕ, is the vector formed from the derivatives of the scalar with respect to the coordinates, $(\partial_1 \phi, \partial_2 \phi)$, denoted $\partial_i \phi$. By the chain rule, this transforms to a new reference frame according to

$$\begin{aligned} \overline{(\partial_i \phi)} &= \frac{\partial \phi}{\partial \overline{x}^i} \\ &= \frac{\partial x^j}{\partial \overline{x}^i} \frac{\partial \phi}{\partial x^j} \\ &= \frac{\partial x^j}{\partial \overline{x}^i} (\partial_j \phi), \end{aligned} \tag{2.8}$$

which is *not* the same as the transformation of our original vector v^i (Equation 2.4). We have two different types of vector that transform differently: the first is called contravariant and the second covariant. We don't think about this difference very often in Euclidean space since the two types of vector behave almost identically. We will see shortly that the same is not true in relativity.

Notice also that in order to form a scalar quantity ϕ from two vectors u^i, v^i, then one must be contravariant and the other must be covariant, since then

$$\overline{\phi} = \overline{u}^i \overline{v}_i$$

$$= \frac{\partial \overline{x}^i}{\partial x^k} u^k \frac{\partial x^\ell}{\partial \overline{x}^i} v_\ell$$

$$= \frac{\partial \overline{x}^i}{\partial x^k} \frac{\partial x^\ell}{\partial \overline{x}^i} u^k v_\ell$$

$$= \frac{\partial x^\ell}{\partial x^k} u^k v_\ell \qquad (2.9)$$

$$= \delta_k^\ell u^k v_\ell$$

$$= u^k v_k$$

$$= \phi.$$

From this, we can see that another way to view the metric is as an object that converts a contravariant vector to a covariant vector. We can similarly define an inverse metric δ^{ij} that converts covariant to contravariant.

If different observers are to agree on the laws of physics, then the two sides of an equation must have the same transformation rule. We must be able to take two equal quantities (F and ma, for instance) and transform to a different reference frame to find that they are still equal! To put it another way, dummy indices must always appear one-up-one-down, and free indices must match on opposite sides of an equation. Let's check a few:

$$\mathbf{F} = m\mathbf{a} \text{ becomes } F^i = ma^i,$$

$$\mathbf{E} = -\boldsymbol{\nabla}V - \frac{\partial \mathbf{A}}{\partial t} \text{ becomes } E^i = -\partial^i V - \frac{\partial A^i}{\partial t}, \text{ and} \qquad (2.10)$$

$$\mathbf{B} = \boldsymbol{\nabla} \times \mathbf{A} \text{ becomes } B^i = \varepsilon^{ijk} \partial_j A_k,$$

where ε^{ijk} is the Levi-Civita symbol, defined to be antisymmetric and 1 if i, j, k are a cyclic permutation of $1, 2, 3$. That is

$$\varepsilon^{123} = \varepsilon^{231} = \varepsilon^{312} = 1,$$

$$\varepsilon^{321} = \varepsilon^{213} = \varepsilon^{132} = -1, \qquad (2.11)$$

$$\varepsilon^{ijk} = 0 \text{ if any of } i, j, k \text{ are equal.}$$

We can take these ideas further and define objects with more than one index, that transform according to (e.g.)

$$\overline{T}^{ij} = \frac{\partial \overline{x}^i}{\partial x^k} \frac{\partial \overline{x}^j}{\partial x^\ell} T^{k\ell}. \tag{2.12}$$

Such an object is known as a rank-2 tensor, and we can similarly define rank-3 tensors and so on. Note that "vector" is just another name for a rank-1 tensor and "scalar" means rank-0 tensor.

2.1.2 Special Relativity

Einstein's special theory of relativity begins from the assumption that the speed of light is a constant value, regardless of the reference frame of the observer. This assumption leads directly to the concept of a Lorentz transformation. The form of a Lorentz transformation is most easily written down if we make certain assumptions about the two reference frames we wish to transform between. Specifically, let's assume that one observer, $\overline{S}$, is moving with speed v relative to the other, S, along a direction that they both call the x-direction, such that the two observers coincide at a time they both call 0. Then the transformation between frames is given by

$$\overline{x} = \frac{x - vt}{\sqrt{1 - v^2/c^2}}, \quad \overline{t} = \frac{t - vx/c^2}{\sqrt{1 - v^2/c^2}}. \tag{2.13}$$

Any transformation between frames not perfectly aligned in this way can be found by combining appropriate rotations with the above standard transformation. Since this transformation mixes up the time and spatial coordinates, it is no longer appropriate to treat time and space as separate entities. Instead they become blended together into a four-dimensional space-time. For this to make sense, the units for space and time must match: it would make little sense in three-dimensional Euclidean space to measure one direction in feet and another in meters!

One way to correct the mismatched units for time and distance is to use a scaling factor that everyone can agree on. Since special relativity is *founded* on the principle that all observers agree on the speed of light, we can re-scale time coordinates by c, giving us a four-dimensional vector (or four-vector) (ct, x, y, z). However, another approach is to recognize that nature is trying to tell us something: maybe we were using the wrong units to begin with. If distances and times are really two sides of the same coin, we should really use *exactly* the same unit for each. In SI units, one meter is defined to be the distance that light travels in one 299,792,458th of a second, but this number is completely arbitrary, chosen simply so that the meter is a typical everyday length for most humans and the second is a typical everyday time. If we had chosen the fraction to be simply 1, we could instead measure both times and distances in the same unit. Whether we choose to call that unit seconds, meters, or something else is up to us. In fact, we will use a similar argument in Chapter 3 to show that an appropriate unit for length and time is MeV^{-1}. With these "natural units," the standard Lorentz transformation looks a little more symmetrical:

$$\overline{x} = \frac{x - vt}{\sqrt{1 - v^2}}, \quad \overline{t} = \frac{t - vx}{\sqrt{1 - v^2}}, \tag{2.14}$$

and the space-time four-vector is simplified to (t, x, y, z).

2.1.3 Minkowski Space

The space that these four-vectors inhabit is not a simple generalization of Euclidean space to four dimensions. Instead, it is a space with a fundamentally different metric, known as Minkowski space. By considering the Lorentz transformation, one finds that the quantity

$$ds^2 = dt^2 + dx^2 + dy^2 + dz^2 = dx^i \delta_{ij} dx^j \tag{2.15}$$

is *not* invariant. That is, it is not a scalar. The correct form of the scalar quantity associated with four-vectors is, in fact,

$$ds^2 = dt^2 - dx^2 - dy^2 - dz^2 = dx^\mu g_{\mu\nu} dx^\nu. \tag{2.16}$$

Notice that we have changed the indices on the four-vector to Greek letters: this is common practice to distinguish Minkowski vectors from Euclidean vectors.

The metric, then, is not the Kronecker delta as for Euclidean space but instead takes the form:

$$g_{\mu\nu} = \begin{cases} 1 & \mu = \nu = 0 \\ -1 & \mu = \nu \neq 0 \\ 0 & \mu \neq \nu \end{cases}, \tag{2.17}$$

or, in matrix form

$$g_{\mu\nu} = \begin{pmatrix} 1 & 0 & 0 & 0 \\ 0 & -1 & 0 & 0 \\ 0 & 0 & -1 & 0 \\ 0 & 0 & 0 & -1 \end{pmatrix}. \tag{2.18}$$

As in the Euclidean case, we can also define an inverse metric, $g^{\mu\nu}$, by $g_{\mu\rho}g^{\rho\nu} = \delta^\nu_\mu$, which converts covariant vectors to contravariant. The Lorentz transformation, just like the rotation in the two-dimensional case, can be written as

$$\overline{x}^\mu = \Lambda^\mu{}_\nu x^\nu, \tag{2.19}$$

where

$$\Lambda^\mu{}_\nu = \frac{\partial \overline{x}^\mu}{\partial x^\nu}. \tag{2.20}$$

We have already seen that covariant vectors transform according to the inverse of this derivative, so we can say that

$$\overline{x}_\mu = \left(\Lambda^{-1}\right)^\nu{}_\mu x_\nu. \tag{2.21}$$

Also, notice that a Lorentz transformation followed by its inverse necessarily brings us back to where we started, so we have $\left(\Lambda^{-1}\right)^\mu{}_\rho \Lambda^\rho{}_\nu = \delta^\mu_\nu$.

It is worth noting at this point that, while we have referred to "the" Lorentz transformation, the term is really much broader than the way in which we have used it. A Lorentz transformation is any transformation of the above form that preserves the scalar product of two vectors. To preserve this product, the transformation must clearly preserve the metric. Therefore, we require:

$$\Lambda^{\mu}{}_{\nu}\Lambda^{\rho}{}_{\sigma}g_{\mu\rho} = g_{\nu\sigma}. \tag{2.22}$$

Such transformations include the boosts (one reference frame moving relative to another) but also includes ordinary three-dimensional rotations. To see that this is the case, we need only consider the transformation:

$$\Lambda^{\mu}{}_{\nu} = \begin{pmatrix} 1 & 0 & 0 & 0 \\ 0 & \cos\theta & \sin\theta & 0 \\ 0 & -\sin\theta & \cos\theta & 0 \\ 0 & 0 & 0 & 1 \end{pmatrix}, \tag{2.23}$$

which gives a rotation in the x, y-plane through an angle θ, and is easily shown to obey Equation 2.22. Angles are of course a very natural way to describe rotations, since to find the combined transformation due to two successive rotations about the same axis, we need only add the two angles. It turns out that there is a similarly neat way to combine successive boosts if we define the *rapidity*, ξ, of a boost by $\xi = \tanh^{-1} v$, where v is the relative speed of the two reference frames. With this definition, boosts may be written in the form

$$\Lambda^{\mu}{}_{\nu} = \begin{pmatrix} \cosh\xi & -\sinh\xi & 0 & 0 \\ -\sinh\xi & \cosh\xi & 0 & 0 \\ 0 & 0 & 1 & 0 \\ 0 & 0 & 0 & 1 \end{pmatrix}. \tag{2.24}$$

The typical route taken by a text on relativity at this point is to show that various other constructs transform as four-vectors. Here, we will simply state that the following objects (among others) all obey the correct transformation laws:

$$\begin{aligned} p^{\mu} &= (E, p_x, p_y, p_z) = (E, \mathbf{p}), \\ A^{\mu} &= (V, A_x, A_y, A_z) = (V, \mathbf{A}), \\ j^{\mu} &= (\rho, j_x, j_y, j_z) = (\rho, \mathbf{j}), \end{aligned} \tag{2.25}$$

where E is a particle's energy, $\mathbf{p}$ is momentum, V is the electric potential, $\mathbf{A}$ is the electromagnetic vector potential, ρ is charge density, and $\mathbf{j}$ is charge current density.

However, two important quantities that do not generalize to four-vectors in the relativistic case are the electric and magnetic fields, $\mathbf{E}$ and $\mathbf{B}$. In fact, these are found to transform as the components of an antisymmetric rank-2 tensor:

$$F^{\mu\nu} = \begin{pmatrix} 0 & -E_x & -E_y & -E_z \\ E_x & 0 & -B_z & B_y \\ E_y & B_z & 0 & -B_x \\ E_z & -B_y & B_x & 0 \end{pmatrix}, \qquad (2.26)$$

known as the Maxwell field strength tensor. With this in mind, it is reasonably straightforward to show that Maxwell's equations can be written:

$$\partial_\mu F^{\mu\nu} = j^\nu, \qquad F^{\mu\nu} = \partial^\mu A^\nu - \partial^\nu A^\mu. \qquad (2.27)$$

It should be noted that, in the above equations, we have extended the idea of natural units to set the permeability of free space μ_0 also to a value of 1.

Since the metric is not quite as simple as in the Euclidean case, it is now especially important to distinguish between contra- and covariant four-vectors. This is because if we use the metric to lower an index on, for example, the momentum four-vector, we get:

$$p_\mu = g_{\mu\nu}p^\nu$$

$$= \begin{pmatrix} 1 & 0 & 0 & 0 \\ 0 & -1 & 0 & 0 \\ 0 & 0 & -1 & 0 \\ 0 & 0 & 0 & -1 \end{pmatrix} \begin{pmatrix} E \\ p_x \\ p_y \\ p_z \end{pmatrix}$$

$$= \begin{pmatrix} E \\ -p_x \\ -p_y \\ -p_z \end{pmatrix}. \qquad (2.28)$$

However, as long as we are careful to make such distinctions, we are free to move indices up and down with metrics as we please.

A final point on notation is that, having made a careful distinction in Section 2.1.1 between the contravariant and covariant vectors in Euclidean space, we now lower all Euclidean indices, since we know that there is actually no such distinction to be made in Euclidean space. This allows us such useful notations as the following:

$$
\begin{aligned}
A^\mu &= (A_0, A_1, A_2, A_3) = (A_0, \mathbf{A}), \\
A_\mu &= (A_0, -A_1, -A_2, -A_3) = (A_0, -\mathbf{A}), \quad \text{and} \\
A^2 &= A_\mu A^\mu = A_0^2 - A_1^2 - A_2^2 - A_3^2 \\
&= A_0^2 - \mathbf{A} \cdot \mathbf{A}.
\end{aligned}
\tag{2.29}
$$

2.2 ENERGY AND MOMENTUM IN MINKOWSKI SPACE

We have seen that the energy and momentum of a particle together form a four-vector: the four-momentum. Since we know that energy and momentum are both conserved quantities, we can summarize both of these conservation laws in one equation:

$$
\sum_i p_i^\mu = \sum_f p_f^\mu,
\tag{2.30}
$$

where i indexes a set of initial four-momenta, and f indexes a similar set of final momenta. An important consequence of the four-vector nature of the four-momentum is that its square is an invariant quantity. That is, for a given particle, there is a constant m such that

$$
p^2 = p^\mu p_\mu = m^2.
\tag{2.31}
$$

We know this invariant quantity as the particle's mass. In terms of the components of four-momentum, the above equation relates energy, momentum, and mass according to:

$$
E^2 = p_i p_i + m^2 = |\mathbf{p}|^2 + m^2.
\tag{2.32}
$$

In particular, in the case of a particle at rest, the particle still has energy simply due to its mass, namely $E = m$. Therefore, we can attribute any additional energy coming from the energy-momentum relation to the particle's motion. That is, the kinetic energy of a particle is given by $E_k = E - m$.

Example Calculation

Consider the decay of a heavy particle A into two lighter particles B and C. Assuming that A was at rest when it decayed, what is the energy of each of the decay products? This is a classic problem whose solution is a great example of the use of four-vectors. First, we can use conservation of four-momentum to write:

$$p_A^\mu = p_B^\mu + p_C^\mu. \tag{2.33}$$

We then rearrange and square, remembering that the square of a particle's four-momentum is equal to the square of its mass:

$$\begin{aligned}
p_A^\mu - p_B^\mu &= p_C^\mu \\
(p_A - p_B)^2 &= p_C^2 \\
p_A^2 + p_B^2 - 2p_A \cdot p_B &= p_C^2 \\
m_A^2 + m_B^2 - 2p_A \cdot p_B &= m_C^2.
\end{aligned} \tag{2.34}$$

Since we assumed that A was at rest, we know that its three-momentum is zero: $\mathbf{p}_A = 0$, and therefore its four-momentum is given by $(E_A, 0, 0, 0)$. Squaring this, we find

$$p_A^2 = E_A^2 - \mathbf{p}_A \cdot \mathbf{p}_A = E_A^2 = m_A^2, \tag{2.35}$$

so we can further simplify p_A to $(m_A, 0, 0, 0)$. This simple form for p_A reduces Equation 2.34 to

$$m_A^2 + m_B^2 - 2m_A E_B = m_C^2, \tag{2.36}$$

so we find we can express E_B as

$$E_B = \frac{m_A^2 + m_B^2 - m_C^2}{2m_A}. \tag{2.37}$$

Since this expression involves only invariant quantities, we can now determine the energy of particle B (as long as we know the masses of the particles involved). By symmetry, a similar argument holds for particle C.

2.2.1 Invariant Mass

We can extend these methods by introducing a quantity known as the invariant mass, W. For any collection of particles, W is defined by

$$W^2 = \sum_k p_k^\mu p_{k\mu} = \sum_k E_k^2 - \mathbf{p}_k^2. \tag{2.38}$$

To be clear, this quantity is only actually a mass in the special case that the collection consists of only one particle. In larger collections, the physical significance of W is a little more subtle. It is the mass equivalent to the total energy of the system as measured in the center-of-momentum frame, if that energy were to be concentrated in a single particle. However, the fact that it is an invariant quantity makes it useful for calculations.

As an example of the use of the invariant mass, consider a short-lived particle X that typically decays via

$$X \rightarrow A + B, \tag{2.39}$$

with a lifetime that is too short for direct detection. How can we infer the existence of such a particle? Well, regardless of the four-momentum of the particle X, the invariant mass for the A and B particles gives

$$W^2 = (E_A + E_B) - (\mathbf{p}_A + \mathbf{p}_B)^2$$
$$\left(p_A^\mu + p_B^\mu\right)\left(p_{A\mu} + p_{B\mu}\right) = p_X^\mu p_{X\mu} = m_X^2, \tag{2.40}$$

by conservation of four-momentum. If the A and B particles are measured as having originated at the same point with a combined invariant mass of m_X^2, it is evidence that they could have been produced by the decay of a particle with mass m_X. Of course, one such signal on its own may well be due to an A and a B particle coincidentally cross-ing paths. The power of the invariant mass is only harnessed when looking at many thousands of signals or more. If the same invariant mass appears in an A, B pair many times, it begins to provide evidence of an undetected particle species X. These ideas will be built upon when we consider resonances in Chapter 6.

Another use of the invariant mass is in determining the mass of a particle that has escaped detection, for example by traveling through an area not covered by a detector. Particles escaping an experiment without being directly detected may still be inferred through conservation of momentum. When the momentum in a collision appears to be unconserved, there must be some additional particle or particles to account for the discrepancy. The invariant mass of this collection of undetected particles is known as the missing mass, and is a particularly useful concept when one particle is responsible for the majority of the missing momentum. For example, in a collision $A + B \rightarrow C + D + X$ where X is undetected, we construct the invariant mass:

$$W^2 = (E_A + E_B - E_C - E_D)^2 - (\mathbf{p}_A + \mathbf{p}_B - \mathbf{p}_C - \mathbf{p}_D)^2 . \quad (2.41)$$

In this case, W is the mass of the missing particle, $W^2 = m_X^2$.

EXERCISES

1. Show that $\varepsilon_{ijk}\varepsilon_{k\ell m} = \delta_{i\ell}\delta_{jm} - \delta_{im}\delta_{k\ell}$.

2. Show that Equation 2.24 is equivalent to the standard Lorentz transformation given in Equation 2.14.

3. **(a)** Using the transformation laws for the contravariant four-vector A^μ and the covariant four-vector B_ν, derive the transformation law for the rank-2 tensor $X^\mu{}_\nu = A^\mu B_\nu$.
   (b) Hence show that the object X^μ_μ is a scalar.

4. Evaluate $g_{\mu\nu}g^{\mu\nu}$.

5. (a) For a symmetric constant rank-2 tensor $a_{\mu\nu}$, show that

$$\partial_\rho \left(a_{\mu\nu} A^\mu A^\nu \right) \equiv 2a_{\mu\nu} A^\mu \partial_\rho A^\nu.$$

(b) For an antisymmetric rank-2 tensor $b^{\mu\nu}$ show that $a_{\mu\nu} b^{\mu\nu} = 0$.

6. Calculate $\partial_\rho \exp \left(g_{\mu\nu} x^\mu x^\nu \right)$ where $g_{\mu\nu}$ is the metric and x^μ is the space-time coordinate four-vector.

7. (a) By considering the case when $\nu = 0$, show that $\partial_\mu F^{\mu\nu} = j^\nu$ gives one of the inhomogeneous Maxwell equations.

(b) By considering the case when $\nu \neq 1$, show that it also gives the other inhomogenous equation.

8. The antisymmetric rank-2 tensor $L^{\mu\nu}$ is defined by

$$L^{\mu\nu} = x^\mu p^\nu - x^\nu p^\mu,$$

where x^μ is a position vector and p^μ is the four-momentum.
(a) How many independent components does $L_{\mu\nu}$ have?
(b) What is the physical significance of the components L_{ij} and L_{0i}?

9. A Higgs boson with initial three-momentum 200 MeV decays to two photons. Assuming the photons have equal energy, find the energy of each photon and the angle between their trajectories.

10. In Experiment A, a particle of mass m is accelerated toward its antiparticle, which is stationary. In Experiment B, the same type of particle and its antiparticle are accelerated toward each other with equal and opposite momentum.
(a) Write down the four-momentum of each particle in both cases.
(b) Use conservation of four-momentum to find the invariant mass, W, for each system, in terms of the particle energies, momenta, and masses.

(c) Hence determine the minimum total energy required for each experiment to produce a particle of type X with mass m_X.

(d) Write the required energy in experiment A in terms of that in experiment B and show that

$$E_A = \frac{3}{4}E_B^2 - m^2.$$

(e) Which type of experiment is more energy-efficient?

11. The position of an object is described by a four-vector x^μ. We can define a four-velocity $U^\mu = dx^\mu/d\tau$ where τ is the proper time for the object (the time as experienced by an observer moving with the object). We can also define a four-force $dp^\mu/d\tau$.

(a) What are the components of U^μ?

(b) By considering the energy and momentum of a particle of charge Q in an electromagnetic field, show that the covariant form of the Lorentz force is

$$\frac{dp^\mu}{d\tau} = QU_\nu F^{\mu\nu},$$

where $F^{\mu\nu}$ is the electric field strength tensor.

CHAPTER 3

QUANTUM MECHANICS

While special relativity gave our understanding of space and time a gentle prod in the right direction, quantum mechanics beat our ideas of the universe into submission. We will not delve into the history or origins of quantum mechanics, as other texts do a better job of this than a single chapter can manage. As in the last chapter, the aim here is to provide a reminder of some ideas that the reader should be familiar with, focusing in particular on those points that are most relevant to later chapters. For a comprehensive introduction to the subject, I direct the reader to the quantum mechanics text in this series: *Quantum Mechanics*.

3.1   STATES AND OPERATORS

A good place to start in understanding quantum mechanics is wave-particle duality: particles exhibit wave-like behavior and waves can sometimes act like particles. This is very counterintuitive, since waves and particles are very different beasts on the surface. Waves are characterized by smooth continuous properties, whereas particles are discrete objects with definite properties. The way to make (at least some) sense of this duality is to understand that everything acts in a manner that is not quite like either of these concepts, but as something that has properties of both. There is a correspondence between the particle-like and the wave-like properties of a quantum object, which can be summed up in two equations. First, the energy of a particle is related to the angular frequency of its corresponding wave by $E = \hbar\omega$, and second, the momentum of a particle is related to the wave-vector by $\mathbf{p} = \hbar\mathbf{k}$. As with relativity, we take the hint that nature is giving us, and redefine our units in such a way as to get rid of the unnecessary constants. In particular, we choose $\hbar = 1$, reducing these equations to $E = \omega$ and $\mathbf{p} = \mathbf{k}$. In order to reconcile the wave-like and particle-like aspects of a system, quantum mechanics introduces a new concept, which is able to reproduce both types of behavior in different situations. The new concept is that of the quantum state.

A system has a Hilbert space $\mathcal{H}$ of possible states that it can be in, essentially equivalent to the phase-space of classical mechanics.[1] We denote individual states in this space using the "ket" notation: $|\psi\rangle$. In fact, the physically distinct quantum states are given only by the *direction* in the Hilbert space. In other words, the state of a system corresponds to a ray of vectors in the Hilbert space, rather than a particular vector. For this reason, it is necessary to choose a consistent normalization for the physical states of a system. The simplest choice is for all physical state vectors to be normalized according to $\langle\psi\,|\,\psi\rangle = 1$. The Hilbert space is a complex vector space, so we must also have a way to write the Hermitian conjugate of a state. For this we use the "bra" $\langle\psi| = |\psi\rangle^{\dagger}$. The inner product of two state vectors is then written simply as $\langle\psi_1\,|\,\psi_2\rangle$. We also have a set of operators $\widehat{O}$ that act as a linear map from $\mathcal{H}$ to $\mathcal{H}$, mapping state vectors to other state vectors. These are linear in the sense that $\widehat{O}\,(a\,|\psi_1\rangle + b\,|\psi_2\rangle) = a\widehat{O}\,|\psi_1\rangle + b\widehat{O}\,|\psi_2\rangle$. It should be noted that this ket notation is specifically designed to be independent of any particular representation we may have in mind for the quantum state, so if we wish to formulate quantum mechanics in terms of wavefunctions and differential operators, then we are free to do so: the state vector

---

[1] As far as a mathematician is concerned, a Hilbert space is any vector space equipped with an inner product, along with one or two other important properties. While Hilbert spaces can exist with any finite number of dimensions, physicists generally reserve the term for the infinite-dimensional vector space of quantum mechanical state vectors.

can simply be replaced with the wavefunction. Similarly, some situations are simpler to consider in terms of vectors and matrices, and we are equally justified in making this replacement instead. The two representations are equivalent, inasmuch as they contain the same information and make the same predictions.

The behavior of the state vector is determined by some linear differential equation, until any act of measurement is performed, at which point it is said to "collapse" into a particular state. It is these two distinct types of behavior that give rise to the wave-like and particle-like properties of quantum systems. The particle-like properties of the system are reproduced by identifying observable quantities, A, of the system with Hermitian operators, $\widehat{A}$, where the possible measured values are the eigenvalues, λ, of the operator: that is, there exists a state $|\psi\rangle$ such that $\widehat{A}|\psi\rangle = \lambda|\psi\rangle$. Similarly, $|\psi\rangle$ is then termed an eigenstate of $\widehat{A}$. For a particular observable, the Hermiticity and linearity of the corresponding operator guarantees that it has a set of eigenstates that form a complete orthonormal set. That is, they form a suitable basis for the Hilbert space. As such, any quantum state can be written as a linear combination of eigenstates of $\widehat{A}$, $|\psi\rangle = a_1|\psi_1\rangle + a_2|\psi_2\rangle + \ldots$. When the measurement is taken, the measured value will be one of the eigenvalues, say λ_i, with eigenstate $|\psi_i\rangle$. The probability of this particular outcome is given by a_i^2 where a_i is the coefficient of $|\psi_i\rangle$ in the linear decomposition of the original state $|\psi\rangle$.[2] At the same time, the state "collapses" into the corresponding eigenstate $|\psi_i\rangle$. As a result, if the same measurement is performed again immediately (before the system has time to evolve), the same result will be obtained, since the linear combination for the collapsed state is simply $|\psi\rangle = 1|\psi_i\rangle$. Simple probability theory then tells us that the expectation value for this measurement (the average value if the same measurement could be taken multiple times on the same state) is given by $\left\langle \widehat{A} \right\rangle = \left\langle \psi \left| \widehat{A} \right| \psi \right\rangle$.

Suppose we wish to know more than one piece of information about a quantum system. This will mean taking measurements of more than one quantity, and therefore applying more than one

---

[2] This is true so long as state vectors are normalized such that $\langle \psi | \psi \rangle = 1$.

operator to the state. In particular, suppose we wish to measure the observables represented by the operators $\widehat{A}$ and $\widehat{B}$. If we measure $\widehat{A}$ first, we will collapse the system into some eigenstate of $\widehat{A}$, say $|\psi_1\rangle$. If we then measure $\widehat{B}$, we collapse the system to a $\widehat{B}$ eigenstate, say $|\psi_2\rangle$. If $|\psi_2\rangle$ consists of a linear combination of more than one eigenstate of $\widehat{A}$, a second measurement of $\widehat{A}$ may well collapse the system to a *different* eigenstate from $|\psi_1\rangle$. Measuring $\widehat{B}$ has altered the value of $\widehat{A}$. The only way we can guarantee that the measurement of $\widehat{B}$ does not affect the value of $\widehat{A}$ is if the two operators share the same eigenstates. If this is the case then it can be shown that the operators must commute. In fact, the converse is also true, so we find that the eigenstates of two operators are compatible *if and only if* the operators commute. This is a key concept in quantum mechanics, as it says that the values of two incompatible observables are not only simultaneously unknowable, but in fact cannot even be simultaneously defined. A state with a well-defined value for one observable can never be a state with a well-defined value for a second incompatible observable. In turn, this implies that the state of a quantum system is uniquely defined only by some subset of the observable quantities of the system.

One final point about the observable operators is that they obey the same relations with each other as their corresponding classical quantities. This is important in the next section.

3.2 THE SCHRÖDINGER EQUATION

In order to produce the wave-like properties of a system, quantum mechanics postulates that the time-evolution of a state vector between measurements is determined by a wave equation. In particular, one set of states that should provide solutions to the wave equation are the plane-waves:

$$\psi = \psi_0 e^{-i(\omega t - \mathbf{k} \cdot \mathbf{x})} = \psi_0 e^{-i(Et - \mathbf{p} \cdot \mathbf{x})} = \psi_0 e^{-ip \cdot x} \qquad (3.1)$$

(in the wave-formulation of quantum mechanics). Often, we will want to work specifically in the wavefunction representation, in which case we will denote the wavefunction as ψ rather than $|\psi\rangle$.

By acting with the differential operator $i\frac{\partial}{\partial t}$, we find that

$$i\frac{\partial}{\partial t}\psi_0 e^{-i(Et-\mathbf{p}\cdot\mathbf{x})} = E\psi_0 e^{-i(Et-\mathbf{p}\cdot\mathbf{x})}$$

$$\implies i\frac{\partial}{\partial t}|\psi\rangle = E|\psi\rangle. \tag{3.2}$$

That is, the state (wavefunction) is an eigenstate (eigenfunction) of the operator $i\frac{\partial}{\partial t}$ with an eigenvalue equal to the energy of the state. We identify this operator, therefore, as the energy operator

$$\widehat{E} = i\frac{\partial}{\partial t}. \tag{3.3}$$

The plane wave is similarly an eigenfunction of the operator $-i\boldsymbol{\nabla}$ with (vector-valued) eigenvalue $\mathbf{p}$, so we identify this as the momentum operator

$$\widehat{\mathbf{p}} = -i\boldsymbol{\nabla}, \tag{3.4}$$

or in index notation

$$\widehat{p_i} = -i\partial_i. \tag{3.5}$$

We now construct the wave equation by taking the non-relativistic energy-momentum relation:

$$E = \frac{1}{2m}\mathbf{p}^2 + V, \tag{3.6}$$

where V is the potential energy, then promoting each term to the relevant operator,

$$\widehat{E} = \frac{1}{2m}\widehat{\mathbf{p}}^2 + \widehat{V}, \tag{3.7}$$

and giving the operators something to act on:

$$\widehat{E}\,|\psi\rangle = \frac{1}{2m}\widehat{\mathbf{p}}^2\,|\psi\rangle + \widehat{V}\,|\psi\rangle$$

$$i\frac{\partial}{\partial t}\,|\psi\rangle = -\frac{1}{2m}\boldsymbol{\nabla}^2\,|\psi\rangle + V\,|\psi\rangle \qquad (3.8)$$

$$= -\frac{1}{2m}\partial_i\partial_i\,|\psi\rangle + V\,|\psi\rangle .$$

We have arrived at the time-dependent Schrödinger equation. Notice that the operator $\widehat{V}$ simply has the effect of multiplying by the potential energy V.

All of this should be familiar, and it is reviewed here simply to ensure that the reader can appreciate the logic behind the equation. This will be useful when we wish to derive similar equations in a relativistic setting.

3.3   PROBABILITY CURRENT

Possibly slightly less familiar is the concept of probability density current. Recall that the probability of finding the particle described by a wavefunction in a given region R is given by

$$P(R) = \int_R \langle\psi\,|\,\psi\rangle\,\mathrm{d}^3x \qquad (3.9)$$

where the wavefunction is normalized such that

$$\int_{\substack{\text{all}\\\text{space}}} \langle\psi\,|\,\psi\rangle\,\mathrm{d}^3x = 1. \qquad (3.10)$$

So we can interpret $\langle\psi\,|\,\psi\rangle = |\psi|^2 = \psi^*\psi$ as a probability density function. As a system evolves in time, the probability of finding the particle at a given point will change. The probability density will fall in some places and rise in others, but all the while, the total amount of probability density must remain the same since the probability of finding the particle *somewhere* must always be unity. We can imagine, then, a flow of probability density around the system, with a probability density current carrying from place to place the likely position of the particle. We can find an expression for this probability density current by recognizing that the conservation of overall probability implies a continuity equation.

Continuity equations take the form

$$\frac{\partial}{\partial t}\rho + \boldsymbol{\nabla} \cdot \mathbf{j} = 0, \tag{3.11}$$

where ρ is the density of some conserved quantity and $\mathbf{j}$ is the associated density current. To see why this implies continuity, we integrate over some region V,

$$\int_V \mathrm{d}^3 x\, \frac{\partial}{\partial t}\rho = -\int_V \mathrm{d}^3 x\, \boldsymbol{\nabla} \cdot \mathbf{j}$$
$$= -\int_{\text{boundary}} \mathrm{d}\mathbf{S} \cdot \mathbf{j}, \tag{3.12}$$

by the divergence theorem. Therefore,

$$\frac{\partial}{\partial t}Q = -\int_{\text{boundary}} \mathrm{d}\mathbf{S} \cdot \mathbf{j}, \tag{3.13}$$

where $Q = \int_V \mathrm{d}^3 x\, \rho$. So the value of Q within the region V can only change if there is a net non-zero value of the quantity $\mathbf{j}$ at the boundary of V. We can see that Q is the conserved quantity and $\mathbf{j}$ is the current. In particular, if V is taken to be all space, then the right side must vanish and the total value of Q for the universe is constant. Put another way, Q is conserved locally and also therefore globally.

So we wish to find a suitable $\mathbf{j}$ when $\rho = |\psi|^2$. To do so, consider the Schrödinger equation and its complex conjugate in their wave-mechanical representation:

$$i\frac{\partial}{\partial t}\psi = -\frac{1}{2m}\partial_i\partial_i\psi + V\psi \quad \text{and} \quad -i\frac{\partial}{\partial t}\psi^* = -\frac{1}{2m}\partial_i\partial_i\psi^* + V\psi^*. \tag{3.14}$$

Now we multiply the first of these by ψ^* and the second by ψ:

$$i\psi^*\frac{\partial}{\partial t}\psi = -\frac{1}{2m}\psi^*\partial_i\partial_i\psi + V\psi^*\psi \quad \text{and}$$
$$-i\psi\frac{\partial}{\partial t}\psi^* = -\frac{1}{2m}\psi\partial_i\partial_i\psi^* + V\psi^*\psi. \tag{3.15}$$

Subtracting one from the other, we find

$$i\left(\psi^*\frac{\partial}{\partial t}\psi + \psi\frac{\partial}{\partial t}\psi^*\right) = -\frac{1}{2m}\left(\psi^*\partial_i\partial_i\psi - \psi\partial_i\partial_i\psi^*\right)$$
$$i\frac{\partial}{\partial t}\left(\psi^*\psi\right) = -\frac{1}{2m}\partial_i\left(\psi^*\partial_i\psi - \psi\partial_i\psi^*\right) \tag{3.16}$$
$$\implies \frac{\partial}{\partial t}\rho = -\partial_i j_i,$$

where $\rho = \psi^*\psi$ and we have identified the current as

$$j_i = -\frac{1}{2mi}\left(\psi^*\partial_i\psi - \psi\partial_i\psi^*\right), \tag{3.17}$$

or

$$\mathbf{j} = -\frac{1}{2mi}\left(\psi^*\boldsymbol{\nabla}\psi - \psi\boldsymbol{\nabla}\psi^*\right). \tag{3.18}$$

This, then, is the probability density current that quantifies how probability moves around in the system. We will see later that the correct interpretation of a similar conserved current in the relativistic case is an important part of understanding particle physics.

3.4 ANGULAR MOMENTUM AND SPIN

A particularly important, and relevant, example that demonstrates some of the ideas of quantum mechanics is that of angular momentum. The (classical) orbital angular momentum of an object is given by

$$\mathbf{L} = \mathbf{r} \times \mathbf{p}, \quad \text{or} \quad L_i = \varepsilon_{ijk}r_j p_k, \tag{3.19}$$

where **r** is the object's position vector and **p** is the object's momentum. Since operators bear the same relations to each other as their classical counterparts, we can define the orbital angular momentum operator as

$$\widehat{L}_i = \varepsilon_{ijk} r_j \widehat{p}_k = -i\varepsilon_{ijk} r_j \partial_k. \qquad (3.20)$$

That is,

$$\begin{aligned} L_x &= -iy\partial_z + iz\partial_y \\ L_y &= -iz\partial_x + ix\partial_z \\ L_z &= -ix\partial_y + iy\partial_x. \end{aligned} \qquad (3.21)$$

Looking at the commutation relations for the individual components of angular momentum, we find

$$\begin{aligned} [L_x, L_y] &= iL_z \\ [L_y, L_z] &= iL_x \\ [L_z, L_x] &= iL_y. \end{aligned} \qquad (3.22)$$

Since no two of these components commute, it is impossible to measure two distinct components of angular momentum simultaneously. In fact, it is meaningless even to consider the value of one component if another component is known: it is simply not defined. However, it *is* possible to measure simultaneously one component and the overall magnitude of the angular momentum $\mathbf{L}^2$, since $\mathbf{L}^2$ commutes with all three of the components. We will see later that this is an example of a *Casimir*, and will also see how this idea generalizes to other systems. So while it makes no sense to ask what all individual components of angular momentum are, we can ask what one component is (conventionally the z component) as well as the overall magnitude. This means that angular momentum eigenstates have two eigenvalues associated with them: one for each simultaneously measurable quantity. It can be shown that the possible values of $\mathbf{L}^2$ are constrained to be of the form $\ell(\ell + 1)$ where ℓ is an integer, while the possible values of L_z are constrained to be integers m such that $-\ell \leq m \leq \ell$. These constraints arise from consistency of the eigenstate solutions to the Schrödinger equation under rotation through 2π. Specifically, it may be derived from the spherical harmonics (see Section 6.4.3).

Now consider a particle with an intrinsic angular momentum, not due to the particle orbiting any point, but just an inherent property of the particle itself. This is found to be the case for most particles—indeed, it is a part of the identity of most fundamental particles—and the typical picture that springs to mind is of a particle spinning on its axis. For this reason, the property is referred to as spin, even though the picture is somewhat misleading. The particle cannot be literally spinning, because fundamental particles are either point particles or are so small that their spatial extent cannot be detected. If the latter is true, then the particle would be spinning so fast to account for the intrinsic angular momentum that the surface of the particle would be traveling considerably faster than the speed of light. On the other hand, if the particles truly are point-like, then it is not clear what it would even *mean* for them to spin. So particles do not literally spin, but they do have an intrinsic, measurable, angular momentum *called* spin.

In the case of spin, the constraints on the possible eigenvalues that apply to orbital angular momentum do not necessarily apply, and we must use a different approach to find the eigenvalues. First, we introduce a set of operators to represent the spin. If spin is a type of angular momentum, then the commutation relations for angular momentum operators must apply and we find

$$[S_x, S_y] = iS_z$$
$$[S_y, S_z] = iS_x \qquad (3.23)$$
$$[S_z, S_x] = iS_y.$$

Also, we construct the operator that describes the magnitude of the spin $\mathbf{S}^2 = S_x^2 + S_y^2 + S_z^2$ and find

$$\left[\mathbf{S}^2, S_x\right] = \left[\mathbf{S}^2, S_y\right] = \left[\mathbf{S}^2, S_z\right] = 0, \qquad (3.24)$$

as expected.

Consider now an eigenstate $|\psi\rangle$ of both S_z and $\mathbf{S}^2$, with eigenvalues m_s and S^2 respectively. We construct a pair of "ladder operators"

$$S_\pm = S_x \pm iS_y \qquad (3.25)$$

and find the following commutation relations:

$$[S_z, S_\pm] = \pm S_\pm \quad \text{and} \quad [\mathbf{S}^2, S_\pm] = 0. \tag{3.26}$$

Acting on the eigenstate $|\psi\rangle$ with S_+ produces a new state which, through use of the commutation relations, we can easily verify is also an eigenstate of both S_z and $\mathbf{S}^2$ with eigenvalues $m_s + 1$ and S^2 respectively. That is:

$$S_z S_+ |\psi\rangle = (m_s + 1) S_+ |\psi\rangle \quad \text{and} \quad \mathbf{S}^2 S_+ |\psi\rangle = S^2 S_+ |\psi\rangle. \tag{3.27}$$

This is why S_+ is known as a ladder operator (specifically a raising operator); it has the effect of incrementing the z component of spin. Similarly, S_- is a lowering operator, which decreases the z component of spin by one unit. Both operators, however, leave the magnitude of the spin unaltered.

So for a given value of S^2, there exists a series of eigenstates with this magnitude of spin and differing values of m_s—that is, differently oriented spins. For this value of S^2, there must exist a maximum m_s, since if m_s increases without bound, the z component of spin will eventually outgrow the spin's magnitude. Let's call that maximum eigenvalue $m_{\max}$ and the corresponding eigenstate $|\psi_{\max}\rangle$. Since the incremental relation, Equation 3.27, holds even in this case, we have

$$S_z S_+ |\psi_{\max}\rangle = (m_{\max} + 1) S_+ |\psi_{\max}\rangle, \tag{3.28}$$

but this implies the existence of a state $S_+ |\psi_{\max}\rangle$ with an eigenvalue $(m_{\max} + 1)$, which is greater than our assumed maximum value $m_{\max}$. The only way to resolve this apparent contradiction is for $S_+ |\psi_{\max}\rangle$ to vanish. We can use this fact to relate $m_{\max}$ to S^2. First, inverting Equation 3.25, we can write the operators for the x and y components of spin as

$$S_x = \frac{S_+ + S_-}{2} \quad \text{and} \quad S_y = \frac{S_+ - S_-}{2i}, \tag{3.29}$$

which in turn allows us to write $\mathbf{S}^2$ as

$$\mathbf{S}^2 = S_- S_+ + S_z + S_z^2 \tag{3.30}$$

or equivalently as

$$\mathbf{S}^2 = S_+ S_- - S_z + S_z^2. \tag{3.31}$$

Acting on $|\psi_{\text{max}}\rangle$ with the first form of $\mathbf{S}^2$ above, we find

$$\begin{aligned}
\mathbf{S}^2 |\psi_{\text{max}}\rangle &= (S_- S_+ + S_z + S_z^2) |\psi_{\text{max}}\rangle \\
&= (m_{\text{max}} + m_{\text{max}}^2) |\psi_{\text{max}}\rangle \\
&= m_{\text{max}}(m_{\text{max}} + 1) |\psi_{\text{max}}\rangle,
\end{aligned} \tag{3.32}$$

since $S_+ |\psi_{\text{max}}\rangle = 0$. So we can identify $S^2 = m_{\text{max}}(m_{\text{max}} + 1)$. We can perform this analysis again, using the eigenstate with the minimum allowed z component of spin m_{min}. This leads, in a similar fashion, to $S^2 = m_{\text{min}}(m_{\text{min}} - 1)$. Equating these expressions we find

$$m_{\text{max}}(m_{\text{max}} + 1) = m_{\text{min}}(m_{\text{min}} - 1). \tag{3.33}$$

It should be fairly obvious that, for a given m_{max}, the only values of m_{min} that satisfy this relation are $m_{\text{min}} = m_{\text{max}} + 1$, which we can immediately rule out as nonsensical, and $m_{\text{min}} = -m_{\text{max}}$.

Since m_{min} and m_{max} are necessarily separated by an integer value, this restricts the possible values for m_{max} to half of an integer value. Putting this all together, we find that, without the additional restrictions imposed on orbital angular momentum, the allowed values of spin (and, indeed, of angular momentum in general) are that $S^2 = s(s + 1)$ for some half-integer value of s, and m_s takes all values in integer steps from s to $-s$.

This is an important result since, as we saw in Chapter 1, the spin quantum number s is a fundamental property of elementary particles. In fact, due to the spin-statistics theorem and the Pauli exclusion principle, it is the quantum number that determines the behavior of the particle arguably more than any other.

3.5 SPIN $\frac{1}{2}$ PARTICLES AND THE PAULI MATRICES

The particles of the standard model all have spin quantum numbers of 0, $\frac{1}{2}$, or 1, and of these, the only fermions are the spin-$\frac{1}{2}$

particles. Since fermions make up what we typically think of as matter, it is important to have a good understanding of the quantum mechanical behavior of spin-$\frac{1}{2}$ particles. In particular, in many physical systems, it is electrons that underpin the mechanics of the system, so we must be able to model the behavior of electrically charged spin-$\frac{1}{2}$ particles. We will see in Section 3.8 how to include electromagnetic interactions, but for now, let's find a way to model the spin of an electron.

Since spin is an angular momentum, the spin operators must obey the correct angular momentum commutation relations (Equation 3.23). In addition, each operator must have only two eigenstates with eigenvalues of $\pm\frac{1}{2}$. In the language of Section 4.3.1, we wish to find a two-dimensional representation of the algebra defined by Equation 3.23. This is achieved by the matrices $\Sigma_i = \frac{1}{2}\sigma_i$, where σ_i are the Pauli matrices:

$$\sigma_1 = \begin{pmatrix} 0 & 1 \\ 1 & 0 \end{pmatrix}, \quad \sigma_2 = \begin{pmatrix} 0 & -i \\ i & 0 \end{pmatrix}, \quad \sigma_3 = \begin{pmatrix} 1 & 0 \\ 0 & -1 \end{pmatrix}.$$
(3.34)

Notice that σ_3 is chosen to be diagonal since it is conventionally the z axis along which the spin is measured. The reader is invited to check that the above matrices Σ_i do indeed satisfy Equation 3.23. Furthermore, the eigenvalues for each operator are easily shown to be $\pm\frac{1}{2}$ as required.

Since we are representing the spin operators with 2×2-matrices, we must similarly use a two-component column matrix for the state vector. Since this state vector is introduced specifically to describe spin, it is known as a *spinor*. This is a term whose meaning has expanded greatly since its original introduction, and we will be seeing much more of the concept in Chapter 8. In this representation, we can show that the eigenstates of Σ_3 are given by $\begin{pmatrix} 1 \\ 0 \end{pmatrix}$ (with eigenvalue $+\frac{1}{2}$), and $\begin{pmatrix} 0 \\ 1 \end{pmatrix}$ (with eigenvalue $-\frac{1}{2}$). Additionally, the total spin operator is given by

$$\Sigma^2 = \frac{3}{4}\begin{pmatrix} 1 & 0 \\ 0 & 1 \end{pmatrix}.$$
(3.35)

That this operator is proportional to the identity simply demonstrates that *any* state in the relevant Hilbert space is an eigenstate of this operator. The proportionality factor of $\frac{3}{4}$ is equal to $s(s+1)$, with $s = \frac{1}{2}$ as we would expect.

3.6 THE HAMILTONIAN

One way to express Schrödinger's equation is in the form:

$$i\frac{\partial}{\partial t}|\psi\rangle = \widehat{H}|\psi\rangle, \tag{3.36}$$

where $\widehat{H}$ is the Hamiltonian operator, representing the sum of kinetic and potential energy in the system. An unfortunate side effect of the rise of quantum mechanics is that many students today are introduced to the Hamiltonian via quantum mechanics, only later (if at all) meeting the Hamiltonian in its original classical setting. Hamiltonian mechanics is an alternative formalism to Newtonian mechanics; whereas Newtonian mechanics makes use of forces and velocities, Hamiltonian mechanics works directly with energies and momenta. The typical procedure is as follows:

- For a system with N degrees of freedom, make a list of N coordinates x_k that parametrize the system.

- Let p_k be the momentum associated with the k-th degree of freedom.

- Construct the Hamiltonian or total energy of the system, H, in terms of coordinates x_k and momenta p_k.

- Solve Hamilton's equations:

$$\dot{x}_k = \frac{\partial H}{\partial p_k}, \quad \dot{p}_k = -\frac{\partial H}{\partial x_k}. \tag{3.37}$$

A little thought will reveal that these equations are nothing more than the definition of velocity and Newton's 2nd law respectively. As an example, the Hamiltonian for a single particle with potential energy V is given by:

$$H = \frac{\mathbf{p}^2}{2m} + V = \frac{p_i p_i}{2m} + V, \tag{3.38}$$

which, if substituted into Hamilton's equations, gives precisely the classical behavior we would expect. We also see that this Hamiltonian immediately gives back the Schrödinger equation when the dynamic variables are promoted to operators, as discussed in Section 3.2.

The Schrödinger equation given in Equations 3.8 and 3.36 is time-dependent and describes the evolution of the wavefunction if left to its own devices. When the energy of the system is measured, however, the relevant equation is the time-independent Schrödinger equation, given by

$$E\psi = -\frac{1}{2m}\partial_i \partial_i |\psi\rangle + V |\psi\rangle = \widehat{H} |\psi\rangle, \tag{3.39}$$

where E is the measured energy. In other words, as with any measurement, the possible outcomes of measuring the energy of a system are the eigenvalues of a quantum mechanical operator, and in this case the relevant operator is the Hamiltonian.

A further important property of the Hamiltonian is related to its commutation properties. The expected value of an observable quantity is conserved if and only if the corresponding quantum operator commutes with the Hamiltonian. Though not a rigorous proof, the following gives an idea of why this is the case:

$$\begin{aligned}
\frac{\partial}{\partial t}\left\langle \psi \left| \widehat{A} \right| \psi \right\rangle &= \frac{\partial \langle \psi|}{\partial t} \widehat{A} |\psi\rangle + \langle \psi| \widehat{A} \frac{\partial |\psi\rangle}{\partial t} \\
&= i\left\langle \psi \left| \widehat{H}\widehat{A} \right| \psi \right\rangle - i\left\langle \psi \left| \widehat{A}\widehat{H} \right| \psi \right\rangle \\
&= i\left\langle \psi \left| \left[\widehat{H}, \widehat{A}\right] \right| \psi \right\rangle.
\end{aligned} \tag{3.40}$$

Thus, if $\left[\widehat{H}, \widehat{A}\right] = 0$, there is no change in the expectation value of $\widehat{A}$.

3.6.1 The Lagrangian

A final point to make here is that there is an alternative description of classical mechanics that relies on a different, but related, concept. Rather than constructing the Hamiltonian as the sum of kinetic and potential energy, we can construct another quantity, known as the Lagrangian, as the *difference* between kinetic and potential energy: $L = E_{\text{kin}} - V$. Rather than using a set of coupled first-order equations to find the behavior of the system with respect to one of its degrees of freedom, we now use a second-order equation, known as the Euler-Lagrange equation. This equation is derived from the principle of least action, which states that the path followed by a system is that which has the smallest value of the action, S, defined by $S[\text{path}] = \int_{\text{path}} L \mathrm{d}t$. The equation takes the form

$$\frac{\mathrm{d}}{\mathrm{d}t}\left(\frac{\partial L}{\partial \dot{x}_k}\right) = \frac{\partial L}{\partial x_k}, \tag{3.41}$$

where L is considered to be a function of x_k and $\dot{x}_k$ for all k, as well as time. This means that x_k and $\dot{x}_k$ are treated as *independent* variables. We can also write the momentum in the k-th direction in terms of the Lagrangian as $p_k = \frac{\mathrm{d}L}{\mathrm{d}\dot{x}_k}$, and we can move between Hamiltonian and Lagrangian formalisms by means of the Legendre transformation: $H = p\dot{x} - L$. The reader is encouraged to explore these ideas more in Exercise 7.

An advantage of the Lagrangian formalism is that the Lagrangian transforms as a scalar. This is in contrast to the Hamiltonian, which, being the total energy of the system, transforms in the same way as the time coordinate. In the non-relativistic case, this is essentially the same as a scalar. However, the power of the Lagrangian formalism really becomes apparent in relativistic mechanics, when the Lagrangian is invariant under Lorentz transformations while the Hamiltonian behaves as just one component of a four-vector. For this reason, particle physics, at the junction between relativity and quantum mechanics, is most naturally described in the Lagrangian formalism. However, since relativity necessarily treats space and time on an equal footing, the form of the Euler-Lagrange equation given above clearly cannot apply to relativistic systems, since the time coordinate receives special treatment. The correct relativistic generalization of the equation utilizes a Lagrangian *density*, $\mathcal{L}$, related to the Lagrangian by $L = \int_V d^3x\, \mathcal{L}$. The Euler-Lagrange equation is now given by

$$\partial_\mu \left(\frac{\partial \mathcal{L}(\phi_i(x))}{\partial(\partial_\mu \phi_i(x))} \right) = \frac{\partial \mathcal{L}(\phi_i)}{\partial \phi_i} \quad \forall i, \tag{3.42}$$

where the ϕ_i are a set of fields that depend on all four space-time coordinates x. For the sake of any readers unfamiliar with this formalism, where practical I will provide alternatives to the Lagrangian approach. This will also serve to keep the interpretation of equations clear. However, as we progress, later chapters will rely more heavily on the Lagrangian formalism. Fortunately, all that we really require is the Euler-Lagrange equation (Equation 3.42).

3.7 QUANTUM MECHANICS AND ELECTROMAGNETISM: THE SCHRÖDINGER APPROACH

A physical system with electromagnetism can be made quantum mechanical in the same way that any system is quantized: we impose the relevant commutation relations on the classical Hamiltonian. We must, however, first find a suitable classical Hamiltonian for a system incorporating electromagnetism. To include a conservative force such as an electrostatic field in a Hamiltonian is straightforward, since we know that a charge qe in a potential V has a potential energy qeV. This is simply included as a potential term in the Hamiltonian. The magnetic part of the electromagnetic interaction is not so obvious, however, since this is not a conservative force. The magnetic force, of course, depends on the velocity of a charged particle through the magnetic field. This can be accounted for with a Hamiltonian of the form

$$\begin{aligned} H &= \frac{1}{2m}\left(\mathbf{p} - qe\mathbf{A}\right)^2 + qeV \\ &= \frac{1}{2m}\left(p_i - qeA_i\right)\left(p_i - qeA_i\right) + qeV, \end{aligned} \tag{3.43}$$

where m is the mass of the charged particle, qe and $\mathbf{p}$ are its charge and momentum, and V and $\mathbf{A}$ are the scalar and vector potentials respectively.

There is no intuitive reason for using this Hamiltonian, but it is easily shown via Hamilton's equations (Exercise 7) to lead to the correct behavior of the particle in an electromagnetic field. Notice that we have replaced the physical momentum of the particle with $\mathbf{p} - qe\mathbf{A}$. Since the left-hand side of the equation is just the total energy in the system, another way to view what we have done is that we have made an additional substitution $E \mapsto E - qeV$. In four-vector notation (although it should be stressed that this is still a non-relativistic Hamiltonian), we have switched p^μ for $p^\mu - qeA^\mu$ where p^μ and A^μ are the four-momentum and electromagnetic four-potential. This substitution (known as minimal substitution or minimal coupling) is found to be sufficient for the inclusion of electromagnetism in more general systems as well.

Promoting this Hamiltonian to an operator leads to a time-independent Schrödinger equation for spinless particles in an electromagnetic field:

$$
\begin{aligned}
E\psi &= -\frac{1}{2m}\partial_i\partial_i\psi + \frac{iqe}{2m}\left(\partial_i(A_i\psi) + A_i\partial_i\psi\right) + \frac{q^2e^2 A_i A_i}{2m}\psi + qeV\psi \\
&= -\frac{1}{2m}\partial_i\partial_i\psi + \frac{iqe}{m}A_i\partial_i\psi + \frac{iqe}{2m}\left(\partial_i A_i\right)\psi \\
&\quad + \frac{(qe)^2}{2m}A_i A_i\psi + qeV\psi.
\end{aligned} \tag{3.44}
$$

One of the successes of this equation is that it correctly predicts the normal Zeeman effect. The energy levels of a charged particle with orbital angular momentum undergo Zeeman splitting in the presence of a magnetic field. This is the mechanism behind the fine structure of spectral lines, and the above Schrödinger equation is adequate to predict this behavior. To see this, consider a magnetic vector potential given by

$$
A_k = -\frac{1}{2}\varepsilon_{k\ell m}x_\ell n_m, \tag{3.45}
$$

where $\mathbf{x}$ is the position vector, and $\mathbf{n}$ is a constant vector.

By applying the curl operator to this potential, the magnetic field is found to be

$$B_i = \varepsilon_{ijk}\partial_j A_k$$
$$= -\frac{1}{2}\varepsilon_{ijk}\varepsilon_{k\ell m}\partial_j x_\ell n_m = n_i, \tag{3.46}$$

so a potential of this form gives us a constant magnetic field. It is also easily shown that the divergence of this potential is zero: $\partial_i A_i = 0$. If we assume that there is also no electric field, the third and final terms in Equation 3.44 thus vanish, while the construct $A_i\partial_i\psi$ in the second term becomes

$$A_i\partial_i\psi = \frac{1}{2}B_k\varepsilon_{kji}x_j\partial_i\psi$$
$$= \frac{i}{2}B_k\widehat{L}_k\psi, \tag{3.47}$$

by the definition of the angular momentum operator in Equation 3.20.

Substituting into the Schrödinger equation, then, gives

$$E\psi = -\frac{1}{2m}\partial_i\partial_i\psi - \frac{qe}{2m}B_i\widehat{L}_i\psi + \frac{(qe)^2}{2m}A_iA_i\psi. \tag{3.48}$$

The second term in this equation is the term responsible for Zeeman splitting, since a non-zero angular momentum clearly affects the energy eigenvalues of the system. When considering electrons, the Zeeman splitting term is more commonly written as

$$\mu_i B_i, \tag{3.49}$$

where μ is the orbital magnetic moment of the electron, given by

$$\mu_i = \mu_B L_i. \tag{3.50}$$

Here, $\mu_B = e/(2m_e)$ is the Bohr magneton. This can be thought of as the natural unit for describing the magnetic moment,[3] or as the constant of proportionality when converting angular momenta to magnetic moments.

---

[3] Specifically, the magnetic moment of the electron, since the mass appearing in the expression is the electron mass. Similar expressions can be used for other particles.

Unfortunately, the above electromagnetic Schrödinger equation is limited. Where it falls down is in describing the anomalous Zeeman effect. This is an additional splitting of energy levels, which was termed anomalous since, at the time of its discovery, it had no explanation. This is because the effect is due to the particle's intrinsic angular momentum, which had not yet been discovered. A suitable term may be introduced to the electromagnetic Schrödinger equation to account for this behavior, of the form

$$\mu_{(S)i}B_i. \tag{3.51}$$

Here, $\mu_{(S)}$ is the spin magnetic moment, given by

$$\mu_{(S)i} = g_s\mu_B S_i, \tag{3.52}$$

where $\mathbf{S}$ is the spin angular momentum. Notice the introduction of the new factor g_s, however. This is the "spin g-factor," found experimentally to be $g_s \approx 2$.[4] This means that spin is somehow twice as effective as orbital angular momentum at producing a magnetic moment. This is the reason that the electromagnetic Schrödinger equation cannot be considered entirely adequate: the introduction of this term is not only *ad hoc*, but includes a factor whose value must be deduced from experiment, with no theoretical justification.

3.8 QUANTUM MECHANICS AND ELECTROMAGNETISM: THE PAULI EQUATION

For a more satisfying account of the behavior of electrons in a magnetic field, then, we need a generalization of the Schrödinger equation that correctly incorporates spin. We have already seen that the way to describe the spin of an electron is via the Pauli matrices, requiring the use of a two-component wavefunction. To derive our

---

[4] The deviation of g_s from 2 is known as the anomalous magnetic moment, and is a subject to which we will return in Chapter 9.

new equation, note that, in the absence of an electromagnetic field, it should reduce to the free Schrödinger equation, only with a two-component wavefunction (and an implicit identity matrix)

$$i\frac{\partial}{\partial t}\psi = \frac{1}{2m}\hat{p}_i\hat{p}_i\psi. \tag{3.53}$$

A useful identity involving the Pauli matrices now comes into play. For any two vectors **A** and **B**, we have

$$(\sigma_i A_i)(\sigma_j B_j) = A_i A_i + i\varepsilon_{ijk}\sigma_i A_j B_k. \tag{3.54}$$

Noting that $\varepsilon_{ijk}\sigma_i p_j p_k$ vanishes due to the symmetry of $p_j p_k$, this allows us to write Equation 3.53 in the form

$$i\frac{\partial}{\partial t}\psi = \frac{1}{2m}\left(\sigma_i \hat{p}_i\right)^2 \psi. \tag{3.55}$$

Now reintroducing electromagnetic interactions via minimal substitution gives the Pauli equation:

$$i\frac{\partial}{\partial t}\psi = -\frac{1}{2m}\left(\sigma_i\left(\partial_i + iqeA_i\right)\right)\left(\sigma_j\left(\partial_j + iqeA_j\right)\right)\psi + qeV\psi. \tag{3.56}$$

We will now demonstrate that this equation correctly predicts the spin magnetic moment. First, we again use the identity (3.54), but this time notice that the second term will not vanish:

$$\left(\sigma_i\left(\partial_i + iqeA_i\right)\right)\left(\sigma_j\left(\partial_j + iqeA_j\right)\right)$$
$$= \left(\partial_i + iqeA_i\right)\left(\partial_i + iqeA_i\right) + i\varepsilon_{ijk}\sigma_i\left(\partial_j + iqeA_j\right)\left(\partial_k + iqeA_k\right). \tag{3.57}$$

The symmetry of the bracketed factors, together with the antisymmetric Levi-Civita symbol, would appear to cause the second term to vanish. While this would be true for ordinary vectors, the object $(\partial_i + iqeA_i)$ is a differential operator, and this argument fails to apply. This may seem strange, but it is equivalent to the statement that, while the cross-product of a vector with itself must necessarily vanish, the cross-product of a vector with its own derivative need not. So we have

$$i\varepsilon_{ijk}\sigma_i \left(\partial_j + iqeA_j\right)\left(\partial_k + iqeA_k\right)$$
$$= i\varepsilon_{ijk}\sigma_i \left(\partial_j\partial_k + iqeA_j\partial_k + iqe\partial_j A_k - (qe)^2 A_j A_k\right) \quad (3.58)$$
$$= i\varepsilon_{ijk}\sigma_i \left(iqeA_j\partial_k + iqe\partial_j A_k\right),$$

where the first and last terms *have* vanished by symmetry. Recalling that the derivative acts on everything to the right, applying the above to the wavefunction gives

$$i\varepsilon_{ijk}\sigma_i \left(\partial_j + iqeA_j\right)\left(\partial_k + iqeA_k\right)\psi$$
$$= -qe\varepsilon_{ijk}\sigma_i A_j\partial_k\psi - qe\varepsilon_{ijk}\sigma_i A_k\partial_j\psi - qe\varepsilon_{ijk}\sigma_i \left(\partial_j A_k\right)\psi \quad (3.59)$$
$$= -qe\varepsilon_{ijk}\sigma_i \left(\partial_j A_k\right)\psi,$$

where all but the final term cancel due to the antisymmetry of ε_{ijk}.

We are now in a position to put everything together. The energy levels predicted by the Pauli equation are given by its time-independent counterpart. For an electron in the presence of a constant magnetic field and no electric field, we can let $q = -1$, $A_i = -\frac{1}{2}\varepsilon_{ijk}x_j B_k$ and $V = 0$. This leads to

$$E\psi = -\frac{1}{2m}\left(\partial_i - ieA_i\right)^2\psi - \frac{e}{2m}\varepsilon_{ijk}\sigma_i\left(\partial_j A_k\right)\psi$$
$$= -\frac{1}{2m}\partial_i\partial_i\psi + \frac{e}{2m}A_i\partial_i\psi B_i\widehat{L}_i\psi + \frac{e^2}{2m}A_i A_i\psi - \frac{e}{2m}\sigma_i B_i\psi.$$
$$(3.60)$$

Using the definition of the Bohr magneton and the fact that $S_i = \frac{1}{2}\sigma_i$, we can write this as

$$E\psi = -\frac{1}{2m}\partial_i\partial_i\psi + \frac{e}{2m}B_i\widehat{L}_i\psi + \frac{e^2}{2m}A_i A_i\psi - \frac{2e}{2m}S_i B_i\psi$$
$$= -\frac{1}{2m}\partial_i\partial_i\psi + \mu_B B_i\widehat{L}_i\psi + \frac{e^2}{2m}A_i A_i\psi - 2\mu_B B_i\widehat{S}_i\psi,$$
$$(3.61)$$

and we see that we have reproduced the electromagnetic Schrödinger equation with the spin magnetic moment included and the correct spin g-factor. While the Pauli equation is thus successful at describing the behavior of spin-$\frac{1}{2}$ particles, it is also not the full story, since it fails to incorporate relativity. We will see in Chapter 8, however, how the Pauli equation can be derived as the low-energy limit of a more general relativistic equation (Exercise 13). It is also worth mentioning here that the magnetic moments of the proton and neutron cannot be derived in this way, since this method applies only to fundamental particles, whereas the proton and neutron are composite. Instead the proton and neutron magnetic moments are generally expressed as

$$\mu_p = g_p \mu_N \mathbf{S} \quad \text{and} \quad \mu_n = g_n \mu_N \mathbf{S}, \tag{3.62}$$

where μ_N is the nuclear magneton, given by

$$\mu_N = \frac{e}{2m_p}. \tag{3.63}$$

The same magneton is used for both, since there is no obvious analogy of the Bohr magneton specifically for the neutron. The g-factors are determined experimentally to be $g_p = 5.586$ and $g_n = -3.826$. Notice that a non-zero moment for the neutron is strong evidence of internal structure, since a neutral fundamental particle should have no magnetic moment.

EXERCISES

1. Verify that the Pauli matrices satisfy the appropriate commutation relations to be used as angular momentum operators.

2. Let $|\uparrow\rangle$ and $|\downarrow\rangle$ be the spin-up and spin-down eigenstates of Σ_z with eigenvalues $+1/2$ and $-1/2$ respectively.
   (a) Find $|\uparrow\rangle$ and $|\downarrow\rangle$ in the form of two-component spinors.
   (b) Hence show that

$$\Sigma_x |\uparrow\rangle = \frac{1}{2} |\downarrow\rangle \quad \Sigma_x |\downarrow\rangle = \frac{1}{2} |\uparrow\rangle$$
$$\Sigma_y |\uparrow\rangle = \frac{i}{2} |\downarrow\rangle \quad \Sigma_y |\downarrow\rangle = -\frac{i}{2} |\uparrow\rangle.$$

3. If the overall spin operator is Σ, where $\Sigma^2 = \Sigma_x^2 + \Sigma_y^2 + \Sigma_z^2$,
show that $\Sigma^2 = \frac{3}{4} \begin{pmatrix} 1 & 0 \\ 0 & 1 \end{pmatrix}$.

4. When two particles are combined, we must consider the spin of the combined state.
- **(a)** Show that $\Sigma_z^{(\text{total})} |\uparrow\uparrow\rangle = 1 |\uparrow\uparrow\rangle$ where $\Sigma_z^{(\text{total})} = \Sigma_z^{(A)} + \Sigma_z^{(B)}$.
- **(b)** Show that the overall spin of the combined state is 1.
- **(c)** Find the z-component of spin for the mixed spin states $|\uparrow\downarrow\rangle$, $|\downarrow\uparrow\rangle$, and show that these states are not eigenstates of $\left| \Sigma^{(\text{total})} \right|^2$.
- **(d)** Show that both the symmetric spin state $\frac{1}{\sqrt{2}} (|\uparrow\downarrow\rangle + |\downarrow\uparrow\rangle)$ and the antisymmetric spin state $\frac{1}{\sqrt{2}} (|\uparrow\downarrow\rangle - |\downarrow\uparrow\rangle)$ *are* $\left| \Sigma^{(\text{total})} \right|^2$ eigenstates and find their eigenvalues.

5. In the previous questions, we have chosen the representation of $S_i \equiv \Sigma_i$ arbitrarily to represent our spin operators: we could have chosen any other set of matrices that obey the correct commutation relations. Show that the relations in Exercise 2(b) hold regardless of representation by deriving them straight from the commutation relations for angular momentum.

6. Show that $(\sigma_i A_i)(\sigma_j B_j) = A_i A_i + i\varepsilon_{ijk}\sigma_i A_j B_k$.

7. (a) Show that the Hamiltonian given in Equation 3.43 leads via Hamilton's equations (Equation 3.37) to the Lorentz force law for a charge in an electromagnetic field.
- **(b)** Write down an appropriate Lagrangian for the same system and show that this also leads to the Lorentz force via the Euler-Lagrange equation.

CHAPTER 4

SYMMETRIES AND GROUPS

4.1 THE IMPORTANCE OF SYMMETRY IN PHYSICS

Symmetry plays an important role in various areas of physics—none more so than particle physics. In fact, it could be argued that particle physics is essentially the study of which symmetry groups exist and how they are represented. While this statement may seem strange now, hopefully its meaning will become more transparent by the end of this chapter and even clearer by the end of the book.

Part of the importance of symmetries in physics stems from a key result in Lagrangian mechanics known as Noether's theorem, after its discoverer, mathematician Emmy Noether. This states that, for each continuous symmetry in the Lagrangian for a theory, there is a corresponding conserved quantity. In fact, Noether's theorem also gives us a way to compute the conserved quantity directly from the relevant symmetry. Neither the formal proof of the theorem, nor the exact nature of the correspondence between symmetries and conservation laws will be given here.[1] However, as a rule of thumb, any two quantities related by the Heisenberg uncertainty principle, such

---

[1] For a formal proof, the reader is directed to any introductory text on Lagrangian mechanics.

as time and energy, or position and momentum, give a clue as to a symmetry/conserved quantity pair. As particular examples:

symmetry transformation	conserved quantity
spatial translation	momentum
time-translation	energy
rotation	angular momentum
boost	COM motion
QM phase transformation	charge

The final example in this list may not be familiar but is arguably the most important for our purposes. We will return to this in later chapters when we discuss conserved currents and gauge theories.

4.2 DISCRETE SYMMETRIES

Discrete symmetries are those symmetries under which we can only make discrete transformations of the system without affecting its appearance. As a simple example, the rotational symmetries of a square are discrete: we can rotate the square through 90°, 180°, 270°, or 360° without altering its appearance, although not through any intermediate angle. There are a number of important discrete symmetries that play a role in particle physics, but before we explore them, let's first take some time to understand the mathematical structure of discrete symmetries.

4.2.1 Mathematical Structure of Discrete Symmetries

The study of symmetry relies on the mathematical concept of a group. While we will not go into the general theory of groups, we will consider the group structure of symmetry transformations in particular. Any two symmetry transformations, or elements of the group, can be combined to form a third symmetry transformation. This is known as the closure property of groups. For example, in the case above of rotations of a square, a rotation through 90° leaves the square invariant, as does a rotation of 180°. But we can also combine these transformations, by performing one after the other, to produce a rotation of 270°, which is also a symmetry transformation. This is part of what we mean when we say that the transformations form a group, but it is not the full story. We also require an identity element in the group, which when combined with any other transformation gives back that same transformation. That is, the identity transformation combined with, say, a 90° rotation should produce a 90° rotation, and so forth. It should be fairly obvious that the identity element in our case is a rotation of 0° (or 360°) or the transformation "do nothing." In fact the "do nothing" transformation is the identity element in all symmetry groups. The next thing we need for a group structure is a concept of "inverse" elements. We need to be able to undo any transformations that we perform to get back to the identity. In our example, a rotation through 270° undoes a rotation of 90° and vice versa, while a rotation of 180° is its own inverse. The elements of a discrete group are often best summarized in a table. The following table shows the result of combining each possible pair of elements of the rotation group of a square.

	0°	90°	180°	270°
0°	0°	90°	180°	270°
90°	90°	180°	270°	0°
180°	180°	270°	0°	90°
270°	270°	0°	90°	180°

The final property required of a group is associativity. This specifies that the bracketing of combinations of transformations has no effect on the result:

$$(R_1 \cdot R_2) \cdot R_3 = R_1 \cdot (R_2 \cdot R_3). \tag{4.1}$$

Importantly, however, a group does not necessarily have the property of commutativity. While in our example the order in which transformations are applied does not affect the result, in more general groups, the order *does* matter. In the former case, we say that the group is Abelian, while in the latter it is non-Abelian. That is, in a non-Abelian group, we have $R_1 \cdot R_2 \neq R_2 \cdot R_1$.

4.2.2 Discrete Symmetries in Particle Physics

There are several discrete symmetries that play a particularly important role in particle physics, for a couple of different reasons.

First, the eigenvalues of elementary particles under these discrete symmetries (where they exist) play an important role in characterizing and classifying particles. Second, these discrete symmetries are not exact. That is to say that, while they appear to be genuine symmetries on the surface, a closer investigation shows them to be violated in some way. As an analogy, the average human face is close to symmetrical, to the extent that, when shown a picture of a face and its mirror image, most people will not immediately see any distinction between them. It is only on closer inspection, when observers gradually begin to notice the odd freckle here and crooked tooth there, that they realize there is a difference. Particle physics is very good at pointing out these differences.

It should be noted that discrete symmetries are exempt from Noether's theorem, since Noether's theorem applies *only* to continuous symmetries. However, we can show that some discrete symmetries also lead to conserved quantities. To see this, we must consider what we actually mean for a system to have a particular symmetry. Since a symmetry transformation leaves the system invariant, the transformed system and the original system should evolve identically in time. In practice, this means that what we are really interested in are those transformations that do not affect the Hamiltonian. Suppose we have a system, $|\psi\rangle$, and a transformation that changes this to a new system $|\phi\rangle$. Then we can define an operator $\hat{O}$ that effects the transformation, such that $|\phi\rangle = \hat{O}|\psi\rangle$. If this is a symmetry transformation, we do not expect the Hamiltonian to be affected. We can deduce, then, that the Hamiltonian and the transformation operator commute, and we already know that commuting operators share the same eigenstates. The states that have a well-defined time evolution under the Hamiltonian thus also have a well-defined constant eigenvalue under the transformation operator $\hat{O}$. If this operator is additionally Hermitian, then it is an observable, and the eigenvalue becomes a measurable conserved quantity.

Parity

The first discrete symmetry that we consider is, to many, the most obvious. When considering symmetry, what almost certainly comes to mind before anything else is a mirror image. A parity transformation is essentially the same as taking the mirror image of a system. More precisely, it is the simultaneous inversion of all three spatial coordinates, which is equivalent to a reflection followed by a rotation. Parity-invariance is the property of a system that ensures it retains its original appearance after a parity transformation. In the case of a system of particles described by a state vector $|\psi(t, \mathbf{x})\rangle$, a parity transformation is effected by the parity operator, $\mathbb{P}$. If we assume that parity is a genuine symmetry, then the state $|\psi(t, \mathbf{x})\rangle$ must be an eigenstate under $\mathbb{P}$:

$$\mathbb{P} |\psi(t, \mathbf{x})\rangle = P |\psi(t, -\mathbf{x})\rangle . \qquad (4.2)$$

Since it is obvious that two reflections bring us back to where we started, we must also have

$$\mathbb{P}\mathbb{P} |\psi(t, \mathbf{x})\rangle = |\psi(t, \mathbf{x})\rangle = P^2 |\psi(t, \mathbf{x})\rangle , \qquad (4.3)$$

which restricts the possible values of P to ± 1. All systems, then, must take one of these two values under parity transformations. In addition, whichever one a system takes, the value is then conserved, so parity provides a good quantum number for characterizing states. In the case of individual particles, the intrinsic parity is a particle's eigenvalue under the parity transformation. That is, each particle species has a quantum number associated with it, which is related to its transformation under parity and which is itself also called "parity." Notice, incidentally, that the defining invariance of a scalar under continuous transformations does not require invariance under parity. We make a distinction, therefore, between true scalars with even parity ($P = +1$) and pseudo-scalars with odd parity ($P = -1$). Similarly, the natural behavior of a vector is to change sign under a parity transformation, as we can see if we consider an arrow and its mirror image. So we define a vector to have odd parity and a pseudo-vector or axial vector to have even parity.

It can be shown that the parity of the photon must be -1. This follows essentially from the fact that the photon is the quantum mechanical description of the electric and magnetic fields, which are vector-valued and pseudo-vector-valued respectively. That is, while both fields have three components and transform the same way under rotations, the electric field changes sign under parity, while the magnetic field does not. Since $\mathbf{E}$ and $\mathbf{B}$ are given by $\mathbf{E} = -\frac{\partial}{\partial t}\mathbf{A} - \nabla V$ and $\mathbf{B} = \nabla \times \mathbf{A}$, we see that the potential must transform as a four-vector under parity. Since, as we will see in Section 7.2, it is this potential that is promoted to the role of wavefunction for the photon, this means that the photon must also transform as a four-vector under parity. Hence, the photon has odd intrinsic parity. The parity of fermions, on the other hand, cannot be so easily pinned down. Since fermions are necessarily produced in fermion-antifermion pairs, only the parity of the pair can be determined. In fact, a particle-antiparticle pair is found to have opposing parities, such that the product is -1. Which has the even parity and which has the odd is, however, undetermined and must be set by convention. The standard is to assign a parity of $+1$ to quarks and leptons, and a parity of -1 to antiquarks and antileptons.

While parity invariance appears to be an exact symmetry as far as the electromagnetic and strong interactions are concerned, it has been known since 1957 that the weak interaction violates parity. Such a possibility was suggested a year earlier by Lee and Yang when they noted that the existence of parity violation in the weak interactions would solve one of the then-outstanding problems of particle physics: the "τ-θ problem." It should be stressed that the τ particle in this puzzle is an old name: it is *not* the lepton that we now call the τ. The τ-θ problem arose simply from the assumption of parity conservation. Two strange particles (τ^+ and θ^+) were known to decay via different routes: $\theta^+ \to \pi^+ + \pi^0$ and $\tau^+ \to \pi^+ + \pi^+ + \pi^-$. Since the parities of these final states were known to be $+1$ and -1 respectively, parity conservation ruled out the possibility that the τ^+ and θ^+ were the same particle, despite having apparently identical properties otherwise. In particular, increasingly precise measurements of the particles' masses continued to suggest that they were one and the same. Lee and Yang noted that, while experiments confirmed conservation of parity in both the electromagnetic and strong interactions, there was a lack of experimental evidence regarding its conservation or otherwise in the weak interactions. They proposed an experimental test of parity conservation in the weak sector, which was performed by Wu *et al* through careful measurement of the radioactive decay of cobalt nuclei. If parity were respected by weak interactions, the decays of spin-aligned nuclei should emit radiation in the forward (aligned with spin) and backward (anti-aligned) directions with equal likelihood. In fact, the experiment found that the majority of radiation was emitted in one direction, indeed violating parity. With the conclusive demonstration of parity violation in the weak interactions, the solution to the τ-θ puzzle was trivial: they are the same particle, which we would now call a kaon.

All of this means, then, that the quantum number associated with parity is not really conserved in nature. However, since the weak force is so feeble compared with the electromagnetic and strong force, and interactions mediated by it so infrequent in comparison with these other forces, the amount of violation is sufficiently small that the quantum number is still approximately conserved. For this reason, it is still used to characterize particles.

Charge Conjugation

Charge conjugation is not quite as obvious as parity, but is another symmetry that was formerly widely believed to be exact. The $\mathbb{C}$ transformation consists of replacing all particles in a system with their antiparticles. Since the properties of antiparticles were thought to be equal and opposite to those of the corresponding particles, it would make intuitive sense that nature is invariant under such a transformation. However, around the time of the previous Wu experiment, physicists came to realize that the neutrinos emitted in β-decay were all "left-handed," while all anti-neutrinos were "right-handed." This concept of handedness will be made more precise in a later chapter, but for now, it is enough to say that this violates charge conjugation symmetry. However, as with parity, the violation only occurs in the weak sector. As such, it is a valid approximation to use the $\mathbb{C}$ eigenvalue as a quantum number, acknowledging the fact that it is not necessarily conserved during weak interactions. In this case, though, we find that there are only certain classes of particles for which such an eigenvalue exists at all. As an example, consider a particle state $|X\rangle$, where the particle X has a distinct antiparticle. In this case, the effect of the charge conjugation operator is simply

$$\mathbb{C} \, |X\rangle = |\overline{X}\rangle . \qquad (4.4)$$

Since X has a distinct antiparticle, $\overline{X}$, this is clearly not proportional to the original state. In other words, this is *not* an eigenstate. The only way a particle can be an eigenstate of the charge conjugation operator is if it is its own antiparticle: for example, the photon or the neutral pion. In this case, the argument proceeds exactly as for the parity operator, and we find that the eigenvalue must be ± 1. Again, this eigenvalue is conserved in electromagnetic and strong interactions, and so functions as a useful quantum number for classification. The difference, however, is that it is only defined for a small subset of all particle species.

It is worth mentioning here that, with the discovery of both $\mathbb{C}$ and $\mathbb{P}$ violation in the weak interaction, it was then suggested that the combined symmetry of $\mathbb{CP}$ may still be exact. This is also now known not to be the case in the weak interactions, and more will be said on this when we discuss $\mathbb{CP}$ violation in Chapter 12.

Time Reversal

Time reversal, $\mathbb{T}$, can be thought of in two different ways, one very abstract and one rather more practical. The abstract interpretation is as a literal reversal of the direction of the time coordinate. The more practical approach to $\mathbb{T}$ symmetry is to consider a reversal of particle momenta. For example, an interaction of the form

$$A(\mathbf{p}_1) + B(\mathbf{p}_2) \rightarrow C(\mathbf{p}_3) + D(\mathbf{p}_4), \qquad (4.5)$$

where $\mathbf{p}_k$ are the three-momenta of each particle, is related by time-reversal to the process

$$C(-\mathbf{p}_3) + D(-\mathbf{p}_4) \rightarrow A(-\mathbf{p}_1) + B(-\mathbf{p}_2). \qquad (4.6)$$

While it is obvious that there is a definite arrow of time in nature, a distinction should be drawn between time-irreversibility due to the violation of $\mathbb{T}$ in the equations of motion, and that due to the initial conditions of the system. The definite direction that time has on a macroscopic scale because of the 2nd law of thermodynamics is related to the boundary conditions of the universe, rather than an inherent asymmetry in the laws of mechanics. For this reason, it was not an unreasonable assumption that $\mathbb{T}$ symmetry is respected in nature, since it is not violated in classical mechanics. In fact, the principle of detailed balance states that the transition rates for the two processes given previously should be identical if $\mathbb{T}$ symmetry is exact.

However, with the discovery of $\mathbb{CP}$-violation, it is understood that $\mathbb{T}$ must also be violated in nature, due to the $\mathbb{CPT}$ theorem (see as follows). To date, this violation has yet to be measured directly, however, since it is only expected to occur in the weak interactions, and producing the initial conditions to test the principle of detailed balance in the weak sector has proven prohibitively difficult. Furthermore, unlike the other discrete symmetries discussed previously, the violation of $\mathbb{T}$ cannot be probed by looking for violation of a quantum number, because there *is no quantum number associated with* $\mathbb{T}$. The reason for this is simply that the operator $\mathbb{T}$ is not Hermitian (in fact, it is anti-Hermitian), and as such does not have an observable quantity associated with it, conserved or otherwise.

The $\mathbb{CPT}$ Theorem

While $\mathbb{C}$, $\mathbb{P}$, $\mathbb{CP}$, and $\mathbb{T}$ are all violated by the weak interaction, there is good reason to believe that the combined symmetry of $\mathbb{CPT}$ is exact. Of course, we must be careful to make such claims, since it was previously thought to be obvious that the individual discrete symmetries must be exact, and it came as something of a surprise that they are not. However, the difference is that there exists a theorem that demonstrates $\mathbb{CPT}$ must be exact, based only on very general principles, such as Lorentz-invariance. This is not to say that we should not test the symmetry, though, since experiment is the ultimate arbiter in physics. Indeed, there are currently experiments running to test the validity of the $\mathbb{CPT}$ theorem. One possible effect of a violation of $\mathbb{CPT}$ would be a difference in behavior of matter and antimatter, and such differences are being tested at Berkeley and at CERN's anti-proton decelerator facility.

A consequence of the CPT theorem is another useful symmetry property: crossing symmetry is the symmetry in the transition amplitude between related processes. In particular, the amplitude $\mathcal{M}$ for a particular interaction involving an incoming particle, X, with momentum p, is identical to the amplitude for a similar process involving an outgoing *anti*-particle, $\overline{X}$, with momentum $-p$. That is,

$$\mathcal{M}\left[X(p) + A_1(p_1) + \ldots + A_m(p_m) \rightarrow B_1(q_1) + \ldots B_n(q_n)\right]$$
$$= \mathcal{M}\left[A_1(p_1) + \ldots + A_m(p_m) \rightarrow B_1(q_1) + \ldots B_n(q_n) + \overline{X}(-p)\right].$$
$$(4.7)$$

Note that the quantity $-p$ necessarily implies a negative energy (assuming p has positive energy), so this result is best thought of as a calculational tool more than a physically meaningful insight. Note also that this applies only to the *amplitudes* for the two processes. It does not imply identical transition rates for the two interactions, since they will generally have different phase spaces. For these two reasons, it is important to remember that the transition rate for an interaction does not tell us anything useful about the rate for an interaction related by crossing symmetry. Indeed, it is trivial to give an example in which one of the interactions is forbidden. For example, Bhabha scattering ($e^+ + e^- \rightarrow e^+ + e^-$) is an allowed scattering process, while the process $e^- \rightarrow e^+ + e^- + e^-$ is kinematically forbidden for physical energies, since the available phase space is empty.

As we will see in Chapter 7, the amplitude for a particular process can be calculated directly from the appropriate Feynman diagram. With this in mind, crossing symmetry can be inferred simply by considering that such calculations depend only on the topology of the diagrams. It is then fairly obvious that diagrams such as

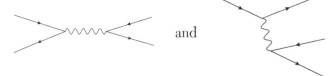

and

give essentially identical results.

4.3 CONTINUOUS SYMMETRIES

4.3.1 Mathematical Structure of Continuous Symmetries

Beginning again with the simple example of rotation, let's now replace the square with a circle. Symmetric rotations are now possible through any angle we care to choose, and so become continuously parametrized. As with discrete symmetries, we can still combine two rotations to produce a third rotation and as such, we can see that continuous symmetries also form a group structure. However, the group also now has the additional structure of a smooth manifold. Such groups are known as Lie groups in honour of Sophus Lie, who carried out much of the early work on such groups. As before, the identity element, $\mathbb{1}$, is the "do nothing" transformation of rotation through an angle of $0°$. Since the transformations are parametrized by the angle θ, we can write a general rotation as $R = R(\theta)$. We can choose this parametrization in such a way that a zero value gives the identity ($R(0) = \mathbb{1}$), which fits with the intuitive notion of using the rotation angle as the parameter. Since the set of rotations we are considering is now continuous, we can consider an element that is only infinitesimally removed from the identity, and since the group has a smooth structure, we know that this will correspond to an infinitesimal value of the parameter. That is, we can consider a rotation through an infinitesimal angle $\delta\theta$, effected by the rotation group element δR, which is only infinitesimally removed from $\mathbb{1}$. Hence, to first order in $\delta\theta$, we can say that $\delta R = \mathbb{1} + i\delta\theta\, T$, where $iT = \frac{\partial R}{\partial \theta}$ gives the non-trivial part of the transformation. We refer to T as the *generator* of the rotations. The factors of i are merely a convention—in fact, a convention unique to physics, with mathematicians generally choosing a different definition and absorbing the i into the generator.

We now imagine constructing a finite rotation through an angle θ by means of repeated application of this infinitesimal rotation. In particular, we let $\delta\theta = \lim_{N\to\infty} \theta/N$, and construct the rotation

$$
\begin{aligned}
R(\theta) &= R(\delta\theta)R(\delta\theta)R(\delta\theta)\ldots \\
&= (\mathbb{1} + i\delta\theta T)\,(\mathbb{1} + i\delta\theta T)\,(\mathbb{1} + i\delta\theta T)\ldots \\
&= \lim_{N\to\infty}\left(\mathbb{1} + i\frac{\theta}{N}T\right)^n.
\end{aligned}
\tag{4.8}
$$

This last expression is hopefully familiar to many readers as the exponential function. So we can identify

$$
R(\theta) = e^{i\theta T},
\tag{4.9}
$$

which is known as the exponential map.

What if we want to describe rotations in three dimensions? Now the problem becomes considerably more complicated as we must specify both an axis and an angle of rotation. A moment's thought will convince the reader that, although there are infinitely many axes around which we can rotate, it is possible to achieve any rotation by way of three independent rotations about fixed axes (an example of three such rotations would be the Euler angles). This means we have only three *independent* axes of rotation to consider, and therefore three parameters to specify. If we let T_1, T_2, and T_3 be the generators of rotations about the x, y, and z axes respectively, then we can use these to build up a general rotation about any axis. In particular, the axis of rotation may be specified by some linear combination of these generators $\delta\theta_1 T_1 + \delta\theta_2 T_2 + \delta\theta_3 T_3$. We can then construct a finite rotation in the same way as before:

$$
R(\theta_1, \theta_2, \theta_3) = \exp\left(\sum_{k=1}^{3} i\theta_k T_k\right).
\tag{4.10}
$$

There is an important difference between the two-dimensional and the three-dimensional rotations. Any two rotations in two dimensions must necessarily be about the same axis and, as such, it does not matter which order we apply two transformations. That is, the rotations in two dimensions commute, giving an Abelian group. In three dimensions, the rotations do not commute and we have a non-Abelian group structure. This is captured in the fact that the generators of the group also do not commute. In the case of three-dimensional rotations, the commutation relations for the generators are given by:

$$[T_1, T_2] = iT_3. \tag{4.11}$$

This is said to form the *Lie algebra* for the group. This algebra is enough to determine the local structure of the group. The only property of the group not uniquely captured in the previous relations is its overall topology.

Group Representations

So far, we have discussed groups in very abstract terms. In practice, we generally want to consider the effect of a transformation on a particular object such as a tensor, which we write in a specific basis. For example, suppose we wish to find the effect of a rotation on a vector. The obvious approach would be to write it in column vector form and act on it with a rotation matrix. This rotation matrix would then be a particular *representation* of the abstract rotation that we wish to perform. To be more concrete, a representation is a map from the group of (abstract) transformations, R, to a group of matrices, $M(R)$, such that the product of the representations of two transformations is the representation of the combination of the transformations:

$$M(R_1)M(R_2) = M(R_1 R_2). \tag{4.12}$$

Clearly, in a different coordinate system, the representation would be different. However, the idea generalizes much more than this; we can consider the effect of a rotation on other tensors, in which case even the dimension of the matrix representing the transformation can change. The only thing that must be the same for all representations of a group is that the representation of the generators must obey the appropriate Lie algebra.

As an example, the group of rotations in three dimensions has a three-dimensional representation:

$$T_1 = \begin{pmatrix} 0 & 0 & 0 \\ 0 & 0 & -i \\ 0 & i & 0 \end{pmatrix}, \quad T_2 = \begin{pmatrix} 0 & 0 & i \\ 0 & 0 & 0 \\ -i & 0 & 0 \end{pmatrix}, \quad T_3 = \begin{pmatrix} 0 & -i & 0 \\ i & 0 & 0 \\ 0 & 0 & 0 \end{pmatrix},$$

$$(4.13)$$

that acts on the three-dimensional vectors in the way that we would expect. A rather obvious six-dimensional representation is given by

$$T_1 = \begin{pmatrix} 0 & 0 & 0 & 0 & 0 & 0 \\ 0 & 0 & -i & 0 & 0 & 0 \\ 0 & i & 0 & 0 & 0 & 0 \\ 0 & 0 & 0 & 0 & 0 & 0 \\ 0 & 0 & 0 & 0 & 0 & -i \\ 0 & 0 & 0 & 0 & i & 0 \end{pmatrix},$$

$$T_2 = \begin{pmatrix} 0 & 0 & i & 0 & 0 & 0 \\ 0 & 0 & 0 & 0 & 0 & 0 \\ -i & 0 & 0 & 0 & 0 & 0 \\ 0 & 0 & 0 & 0 & 0 & i \\ 0 & 0 & 0 & 0 & 0 & 0 \\ 0 & 0 & 0 & -i & 0 & 0 \end{pmatrix},$$

$$T_3 = \begin{pmatrix} 0 & -i & 0 & 0 & 0 & 0 \\ i & 0 & 0 & 0 & 0 & 0 \\ 0 & 0 & 0 & 0 & 0 & 0 \\ 0 & 0 & 0 & 0 & -i & 0 \\ 0 & 0 & 0 & i & 0 & 0 \\ 0 & 0 & 0 & 0 & 0 & 0 \end{pmatrix}, \qquad (4.14)$$

but this clearly splits into two three-dimensional representations, since the matrices are block diagonal. That is, the upper three and lower three components of the six-component vector space acted on by this representation will only mix among themselves and never with each other. In this case, we say that the representation is reducible, whereas an irreducible representation is one that cannot be broken down in this way. We will see later that irreducible representations are a key concept in particle physics.

Far less obvious is that this group's associated algebra also has a *two*-dimensional representation, given by:

$$T_1 = \frac{1}{2} \begin{pmatrix} 0 & 1 \\ 1 & 0 \end{pmatrix}, \quad T_2 = \frac{1}{2} \begin{pmatrix} 0 & -i \\ i & 0 \end{pmatrix}, \quad T_3 = \frac{1}{2} \begin{pmatrix} 1 & 0 \\ 0 & -1 \end{pmatrix},$$

$$(4.15)$$

which the reader should recognize as the spin-$\frac{1}{2}$ spin operators. This is no coincidence and is a point to which we will return in Section 4.3.2.

Classification of Lie Groups

In Chapter 6, we will consider the various ways in which particles are classified and categorized, and it will become clear that a particle is really nothing more than one part of an irreducible representation of various Lie algebras. There are several different Lie algebras that we must consider in particle physics, including the special unitary and special orthogonal groups. The Lie algebras fall into more than just these two categories, but these are the two that are most important for our purposes, so we now explore them a little.

The group of transformations on an n-dimensional real vector space that preserves the inner product $\mathbf{u} \cdot \mathbf{v} = u_i v_i$ can be represented by the $n \times n$ orthogonal matrices—that is, those obeying $M^T M = \mathbb{1}_n$. For this reason, such a group is known as the orthogonal group on n elements, or $O(n)$, and corresponds to our usual concept of rotations. However, in addition to the proper rotations, these groups also contain reflections; if we wish to restrict our attention to proper rotations, we must consider only those matrices with the additional property that their determinant is equal to 1. These are known as the special orthogonal groups on n elements, or $SO(n)$.

A similar class of groups exists for the complex-valued vector spaces. However, the relevant property in this case is that the matrices be unitary ($M^\dagger M = \mathbb{1}_n$), so we have the groups $U(n)$ and

$SU(n)$, where again the S denotes that the latter is the *special* unitary group in which all elements have unit determinant.

The generators of $SU(n)$ must be Hermitian, as this guarantees the unitarity of the group elements, and traceless, as this gives the elements unit determinant. We can also determine how many generators are required. An $n \times n$ complex matrix has n^2 complex values, or equivalently $2n^2$ real parameters. For that matrix to be unitary, it must obey the property $M^\dagger M = \mathbb{1}_n$, which reduces the number of independent parameters to n^2. The additional property of having unit determinant further reduces this number to $n^2 - 1$. This, then, is the number of generators for the group $SU(n)$. A similar argument can be made for orthogonal groups, giving the number of generators of $SO(n)$ as $\frac{1}{2}n(n-1)$, which must be traceless and antisymmetric. As well as the determinant and Hermiticity/orthogonality properties of the generators, it is also necessary to choose an overall normalization. The standard convention is to choose the generators such that $\mathrm{tr}(T_i T_j) = \frac{1}{2}\delta_{ij}$.

We could go further than the previous two categories of Lie groups and ask if similar groups exist based on the quaternions, or even the octonions. In fact they do, but we will not delve into this here, as such groups are not directly relevant to us.

4.3.2 Continuous Symmetries in Particle Physics

A pertinent question might well be "what do these groups have to do with particles?" To answer this question, consider the infinitesimal rotation of a position vector about the z axis in a three-dimensional system, through an angle $\delta\theta$, as in Figure 4.1. This alters the values of x and y according to

$$x \mapsto x + \delta\theta y \quad \text{and} \quad y \mapsto y - \delta\theta x. \tag{4.16}$$

Now a quantity that depends on that position vector (e.g., a wavefunction) will vary, to first order, according to

$$\psi(\mathbf{x}) \mapsto \psi(\mathbf{x}) + i\delta\theta\left(-ix\frac{\partial\psi(\mathbf{x})}{\partial y} + iy\frac{\partial\psi(\mathbf{x})}{\partial x}\right). \tag{4.17}$$

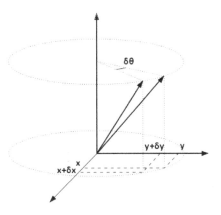

FIGURE 4.1 Rotation of a vector through an infinitesimal angle $\delta\theta$ about the z axis, demonstrating the transformation of the x and y components.

This is of the form that we would expect for an infinitesimal transformation, with a generator given by

$$T = -ix\frac{\partial}{\partial y} + iy\frac{\partial}{\partial x} = x\widehat{p}_y - y\widehat{p}_x. \tag{4.18}$$

So the generator of rotations about the z axis is precisely the quantum mechanical operator for the z component of angular momentum. As stated at the start of this chapter, a rule of thumb is that Noether's theorem connects those variables that are related by the uncertainty principle. This rule goes further, and we can say that the generator for a symmetry transformation is the quantum mechanical operator for the relevant conjugate variable.[2]

So $SO(3)$ symmetry is closely related to angular momentum, and therefore to spin. Reconsidering the results of Chapter 3 in group-theoretic terms, we can see that spin-1 particles form a three-dimensional representation of $SO(3)$ (Equation 4.13), while spin-$\frac{1}{2}$ particles form a two-dimensional representation of its associated algebra.

---

[2] Those readers unfamiliar with the concept of conjugate variables from Lagrangian mechanics can think of conjugates as again those variables related by the uncertainty principle.

Generators, Quantum Numbers, and Representations

In Section 4.3.1 and previously, it was stated that there is a two-dimensional representation of the algebra of $SO(3)$. Let us now clarify this statement a little. Strictly, the three-dimensional rotation group, $SO(3)$, does *not* have a two-dimensional representation. However, the group is locally isomorphic to the two-dimensional unitary group, $SU(2)$. This is equivalent to the statement that the algebras of these two groups are identical, and the groups are only distinguished by their global structure, or topology. The *algebra* has a two-dimensional representation, but this representation strictly applies only to the *group* $SU(2)$. So why must we consider this representation at all when discussing spin? The answer is that spin is a purely quantum mechanical concept and so is encoded in the wavefunction for a particle, which is complex-valued. As such, the types of symmetry transformations we may perform on a wavefunction are generally unitary transformations. For example, consider a two-component spinor-valued wavefunction $|\psi\rangle$, transformed according to $|\psi\rangle \mapsto M|\psi\rangle$, where M is an element of the special unitary group on two elements, $SU(2)$. Then the Hermitian conjugate spinor transforms as $\langle\psi| \mapsto \langle\psi| M^\dagger$ and the physically significant quantity $\langle\psi|\psi\rangle$ is invariant, since

$$\langle\psi|\psi\rangle \mapsto \left\langle \psi \left| M^\dagger M \right| \psi \right\rangle \tag{4.19}$$

and $M^\dagger M$ is the identity by definition. Hopefully, this argument will persuade the reader that $SU(2)$ is more suited to describing spin than $SO(3)$. A more technical reason is that $SU(2)$ is the "double cover" of $SO(3)$. In other words, the topology of $SU(2)$ in a sense "wraps around" $SO(3)$ twice, and this double wrapping is, in effect, canceled out when considering the physical quantity $\langle\psi|\psi\rangle$.

We have already seen a two-dimensional representation of $SU(2)$ in the spin-$\frac{1}{2}$ spin operators (Equation 3.34), so it should come as no surprise that these are the generators for $SU(2)$. However, recall also that in Section 3.4, we saw that it is more convenient to use just one of these generators, along with raising and lowering operators. Since only one of the observables corresponding to the three spin operators can be measured at any one time, it is useful to have a set of operators that return eigenstates of just one of the generators. It is also no coincidence that we generally choose S_z as the measured quantity, since it is diagonal in the standard representation.

We can represent all this in the form of a diagram. This may well seem unnecessarily abstract at first, but it generalizes nicely to larger groups. In what follows, we will deliberately gloss over much of the mathematics, choosing instead to give the reader an intuitive understanding of the specific concepts relevant to the remainder of the text. In particular, we should acknowledge that the diagrams in what follows are well-defined but lie in a different space (the dual space) from the one that actually contains the generators. This need not concern us, however, and it is simpler to think of the diagrams as representing the generators themselves. On a one-dimensional "graph" of the spin quantum number S_z, we represent the raising and lowering operators with arrows, demonstrating the effect that they have on the value of S_z. The generator S_z is shown as a dot at the origin, since it does not change the value of its observable.

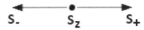

We can use the same axis to depict the different representations of $SU(2)$. A spin-$\frac{1}{2}$ particle is shown by its two allowed values of S_z:

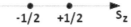

while a spin-$\frac{3}{2}$ particle is shown by its four allowed values:

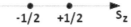

and so on.

Although the individual generators do not commute, we can find one more object that does commute with our chosen S_z generator.

This is the "quadratic Casimir," given by $\sum_i T_i^2$, and its eigenvalue serves to identify the particular representation of $SU(2)$. That is, the Casimir's eigenvalue tells us which of the above diagrams we need to look at. This is equivalent to the fact that the individual components of a particle's angular momentum cannot all be simultaneously measured, but we can measure one component and the overall magnitude. The Casimir gives that magnitude.

Let us now consider the larger symmetry group $SU(3)$, corresponding to a three-fold symmetry in which a particle has available to it three states and their linear combinations. Such a group has eight generators, $T_i = \frac{1}{2}\lambda_i$, where λ_i are the Gell-Mann matrices, given by

$$\lambda_1 = \begin{pmatrix} 0 & 1 & 0 \\ 1 & 0 & 0 \\ 0 & 0 & 0 \end{pmatrix} \quad \lambda_2 = \begin{pmatrix} 0 & -i & 0 \\ i & 0 & 0 \\ 0 & 0 & 0 \end{pmatrix} \quad \lambda_3 = \begin{pmatrix} 1 & 0 & 0 \\ 0 & -1 & 0 \\ 0 & 0 & 0 \end{pmatrix}$$

$$\lambda_4 = \begin{pmatrix} 0 & 0 & 1 \\ 0 & 0 & 0 \\ 1 & 0 & 0 \end{pmatrix} \quad \lambda_5 = \begin{pmatrix} 0 & 0 & -i \\ 0 & 0 & 0 \\ i & 0 & 0 \end{pmatrix} \quad \lambda_6 = \begin{pmatrix} 0 & 0 & 0 \\ 0 & 0 & 1 \\ 0 & 1 & 0 \end{pmatrix}$$

$$\lambda_7 = \begin{pmatrix} 0 & 0 & 0 \\ 0 & 0 & -i \\ 0 & i & 0 \end{pmatrix} \quad \lambda_8 = \frac{1}{\sqrt{3}} \begin{pmatrix} 1 & 0 & 0 \\ 0 & 1 & 0 \\ 0 & 0 & -2 \end{pmatrix}. \tag{4.20}$$

If a particle can be in one of three distinct states, then it really has only two degrees of freedom: if it is not in state 1 or state 2, then it must be in state 3. As such, a system with $SU(3)$ symmetry is categorized by two quantum numbers. It is no coincidence that, of the eight generators of $SU(3)$, the largest subgroup of these that commutes consists of just two generators. If we are to label a state by two quantum numbers, then those numbers must be simultaneously measurable, and so the quantum operators must commute. Notice that λ_3 and λ_8 are diagonal: this tells us that these correspond to mutually compatible quantum numbers of the system, according to which the states will be labeled. These quantum numbers will also be the appropriate axes for our generator graph. As in the case of $SU(2)$, the remaining operators are not in a particularly useful form for our current purposes, and we are better off forming linear combinations to act as ladder operators. Now, however, we find that these do not raise or lower just one of the quantum numbers, but a combination of the two. Specifically:

- $I_\pm$ alters the T_3 value by ± 1,

- $U_\pm$ alters the T_3 value by $\pm\frac{1}{2}$ *and* the T_8 value by $\pm\frac{\sqrt{3}}{2}$, and

- $V_\pm$ alters the T_3 value by $\pm\frac{1}{2}$ *and* the T_8 value by $\mp\frac{\sqrt{3}}{2}$,

where it is left to the reader to find expressions for these operators in Exercise 5.

Measuring either quantum number has no effect on its value, so we place two dots at the center of the diagram to denote the diagonal generators, and the effect of the ladder operators on the quantum numbers is again shown pictorially as in Figure 4.2.

The irreducible representations of this group must consist of sets of states whose quantum numbers are related to each other by the ladder operators. From this fact, we can see how to write out the representations of $SU(3)$ graphically. Some examples are given in Figures 4.3–4.5, but there are infinitely many more besides. Such diagrams are properly referred to as weight diagrams, the weight of a state being the set of its eigenvalues under the group's generators.

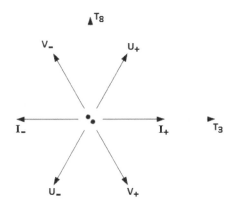

FIGURE 4.2 The ladder operators of $SU(3)$ in pictorial form.

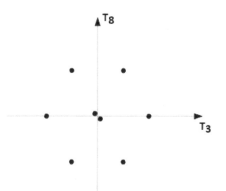

FIGURE 4.3 The eight-dimensional irreducible representation of $SU(3)$. Also known as the adjoint representation or **8**.

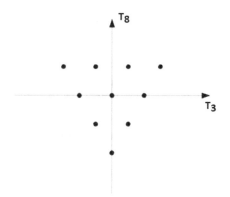

FIGURE 4.4 A 10-dimensional irreducible representation of $SU(3)$, **10**.

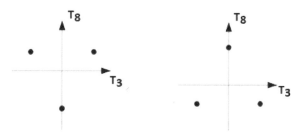

FIGURE 4.5 Two three-dimensional irreducible representations of $SU(3)$, known as the fundamental or **3** and anti-fundamental or $\overline{\mathbf{3}}$.

The irreducible representations may be combined to form larger representations, though these are typically reducible. The representations in Figure 4.5 are known as the fundamental and anti-fundamental representations, since they are the simplest, out of which the larger representations may be built. This happens in the same way that tensors are built from vectors. In fact, a vector is the equivalent of the fundamental representation of the symmetry group of space-time transformations, and tensors are higher representations. To see how this works, consider a three-component column vector ψ^i that transforms under the three-dimensional representation of $SU(3)$. That is, under $SU(3)$, ψ^i transforms according to

$$\psi^i \mapsto \psi^{i\prime} = M^i{}_j \psi^j, \tag{4.21}$$

where $M^i{}_j$ is some element of the three-dimensional representation of the group. Then the nine-component object $\psi^i \psi^j$ must transform according to

$$\begin{aligned} \psi^i \psi^j \mapsto \left(\psi^i \psi^j\right)' &= \left(M^i{}_k \psi^k\right)\left(M^j{}_\ell \psi^\ell\right) \\ &= M^i{}_k M^j{}_\ell \left(\psi^k \psi^\ell\right). \end{aligned} \tag{4.22}$$

Since we can write $M^i{}_k M^j{}_\ell$ as the exterior product of two 3×3 matrices, it is equivalent to a nine-dimensional matrix representation. However, if we were to do this, we would find that this representation is reducible. This is most easily seen by considering again the construct $\psi^i \psi^j$. Since this is constructed from two identical column vectors, it makes sense to talk about the symmetric and antisymmetric parts of the object. That is, we can construct the symmetric part of the product as $\left(\psi^i \psi^j\right)_s = \left(\psi^i \psi^j + \psi^j \psi^i\right)/2$, and the antisymmetric part as $\left(\psi^i \psi^j\right)_a = \left(\psi^i \psi^j - \psi^j \psi^i\right)/2$, with $\psi^i \psi^j = \left(\psi^i \psi^j\right)_s + \left(\psi^i \psi^j\right)_a$. Now consider the action of the transformation $M^i{}_k M^j{}_\ell$ on the symmetric part of this product:

$$\begin{aligned} M^i{}_k M^j{}_\ell \left(\psi^k \psi^\ell\right)_s &= M^i{}_k M^j{}_\ell \left(\psi^k \psi^\ell + \psi^\ell \psi^k\right)/2 \\ &= \left(\psi^{i\prime} \psi^{j\prime} + \psi^{j\prime} \psi^{i\prime}\right)/2 \\ &= \left(\left(\psi^i \psi^j\right)' + \left(\psi^j \psi^i\right)'\right)/2, \end{aligned} \tag{4.23}$$

which is just the symmetric part of the transformed tensor. This demonstrates that the symmetric and antisymmetric parts of the tensor transform entirely independently of each other. We can thus deduce that the nine-dimensional representation is reducible to a six-dimensional symmetric representation and a three-dimensional antisymmetric representation. Similar results may be found with higher-order constructs.

As another example, combining the three-dimensional fundamental with the three-dimensional anti-fundamental also gives a nine-dimensional representation. In this case, it makes no sense to talk about symmetric and antisymmetric parts, as we have combined vectors from different spaces. However, this time the trace of the representation transforms independently of the rest of the representation under $SU(3)$: in fact, it is invariant. As such, the representation is clearly reducible, so the trace is separated out as an $SU(3)$ singlet, while the remaining eight degrees of freedom form an $SU(3)$ octet. In group-theoretic notation, then, we say that these representations combine according to $3 \otimes \overline{3} = 8 \oplus 1$. Similarly, we find that $3 \otimes 3 \otimes 3 = 10 \oplus 8 \oplus 8 \oplus 1$, where 10 is symmetric, 1 is antisymmetric, and the two 8s have mixed symmetry properties. We will see in Chapter 6 that the hadrons fall into some of these irreducible representations, and it is this fact that led Gell-Mann to propose the quark model.

Another representation of particular importance is that shown in Figure 4.3. This is the representation that the generators themselves form, and is known as the adjoint. As we will see in Chapter 10, the exchange bosons for the three non-gravitational fundamental interactions form adjoint representations of their associated groups.

Lorentz Invariance, Mass, and Spin

Of course, we know that the particles we wish to describe do not exist just in three-dimensional space, but in four-dimensional spacetime. What are the relevant groups to consider when we wish to combine relativity with quantum mechanics? The four-dimensional rotation group $SO(4)$ might seem an obvious answer but, by definition, this preserves the inner product $u \cdot v = u_1 v_1 + u_2 v_2 + u_3 v_3 + u_4 v_4$, rather than the correct Lorentz-invariant product $u \cdot v = u_0 v_0 - u_1 v_1 - u_2 v_2 - u_3 v_3$. To make this distinction clear, we denote the group of proper Lorentz transformations $SO(1,3)$ to highlight the presence of one positive and three relative negative terms in the scalar product.[3] This group is known as the (proper) Lorentz group. If we include space-time translations, this is extended to the Poincaré group. An important result in particle physics (whose proof is beyond the scope of this text) is that the irreducible representations of the Poincaré group are labeled by their mass and their spin.

An outline of the result is as follows. The generators of space-time translations are the energy and momentum operators, p^μ. The irreducible representations of the translation group are labeled by their eigenvalues under the operator p^2, which of course are just given by m^2. The generators of Lorentz transformations are given by

$$\widehat{J}^{\mu\nu} = (x^\mu \widehat{p}^\nu - x^\nu \widehat{p}^\mu) = i(x^\mu \partial^\nu - x^\nu \partial^\mu). \qquad (4.24)$$

Notice that this object is antisymmetric and so has only six independent components. These correspond to the three rotations and three boosts. In fact, the generators for rotations in the i,j-plane are given by

$$\widehat{J}^{ij} = (x^i \widehat{p}^j - x^j \widehat{p}^i) = -i\left(x_i \frac{\partial}{\partial x_j} - x_i \frac{\partial}{\partial x_i}\right), \qquad (4.25)$$

while boosts in the i direction are generated by

$$\widehat{J}^{0i} = (t\widehat{p}^i - x^i \widehat{E}) = -i\left(t\frac{\partial}{\partial x_i} + x_i \frac{\partial}{\partial t}\right). \qquad (4.26)$$

These generators can be shown to obey the algebra

$$\left[\widehat{J}^{\mu\nu}, \widehat{J}^{\rho\sigma}\right] = -i\left(g^{\mu\rho}\widehat{J}^{\nu\sigma} - g^{\mu\sigma}\widehat{J}^{\nu\rho} - g^{\nu\rho}\widehat{J}^{\mu\sigma} + g^{\nu\sigma}\widehat{J}^{\mu\rho}\right), \qquad (4.27)$$

as must any other representations of this group.

---

[3] As with rotations, "proper" here refers to the fact that the group considers only those transformations that do not involve a reflection. In the case of the proper Lorentz group, this means no spatial inversion or time reversal.

More generally, a representation will typically have a non-zero spin component, which provides additional terms. A useful means of extracting these terms from the representation is provided by the Pauli-Lubański vector,[4]

$$W_\mu = \frac{1}{2}\varepsilon_{\mu\nu\rho\sigma}\widehat{J}^{\nu\rho}\widehat{p}^\sigma, \tag{4.28}$$

since it can be shown that

$$W^2 = -m^2\widehat{S}^2, \tag{4.29}$$

where $\widehat{S}^2$ is the total spin operator. As such, the eigenstates of W^2 are labeled by $-m^2s(s+1)$, where s is the spin quantum number. Notice that W^2 plays the role of Casimir for the Lorentz group.

It follows that the representations of the Poincaré group are labeled by their mass and spin. This result was first established by Eugene Wigner, who also showed that these labels determine the number of degrees of freedom of the representation. In particular, any massless representation has two degrees of freedom, regardless of spin, whereas representations with $m \neq 0$ have $2s + 1$ degrees of freedom.

EXERCISES

1. Draw a table that shows the effect of combining the symmetries of a square with each other if we allow both rotations and reflections.

2. **(a)** Given that finite elements of a Lie group may be written in the form of Equation 4.10, show that the generators of $SO(3)$ may be written as

$$T_k = -i\frac{\partial R}{\partial\theta_k}R^{-1}.$$

---

[4] Strictly a pseudo-vector, since it transforms as such under parity transformations.

(b) Hence show that the generator of rotations about the z axis is given by

$$T_3 = \begin{pmatrix} 0 & -i & 0 \\ i & 0 & 0 \\ 0 & 0 & 0 \end{pmatrix}.$$

3. **(a)** Given that infinitesimal transformations may be written as $\delta R = \mathbb{1} + i\delta\theta_i T_i$, expand the boost transformation in Equation 2.24 to first order in ξ and find the generator for boosts in the x direction.

   (b) By a similar method, show that an arbitrary infinitesimal Lorentz transformation may be written in the form $\Lambda^\mu{}_\nu = \delta^\mu_\nu + g^{\mu\rho}\,\delta\omega_{\rho\nu}$, where

$$\omega_{\rho\nu} = \begin{pmatrix} 0 & -\xi_1 & -\xi_2 & -\xi_3 \\ \xi_1 & 0 & \theta_3 & -\theta_2 \\ \xi_2 & -\theta_3 & 0 & \theta_1 \\ \xi_3 & \theta_2 & -\theta_1 & 0 \end{pmatrix}$$

   is an antisymmetric matrix that parametrizes the transformation in terms of the rotation angles, θ_i, and the rapidities, ξ_i.

4. The generators of an $SU(N)$ group must be a set of linearly independent, Hermitian, and traceless matrices. Show that the Pauli and Gell-Mann matrices form two such sets.

5. **(a)** Construct the ladder operators $I_\pm = T_1 \pm iT_2$, $U_\pm = T_4 \pm iT_5$ and $V_\pm = T_6 \pm iT_7$, where $T_i = \lambda_i/2$ are the $SU(3)$ generators.

   (b) Find the commutators of $U_\pm$ with T_3, T_8 and each other.

   (c) Hence show that U_+ raises the values of T_3 and T_8 by $1/2$ and $\sqrt{3}/2$ respectively.

6. We can write the anti-fundamental ($\overline{3}$) of $SU(3)$ in terms of the fundamental (3), as $\psi_i = \varepsilon_{ijk}\psi^j\psi^k$, where ε_{ijk} is the Levi-Civita symbol.

   (a) Write down an expression for an arbitrary element of $3 \otimes \overline{3}$ and hence show that it cannot be written as the sum of symmetric and antisymmetric parts.

   (b) Show that the trace of $3 \otimes \overline{3}$ ($\psi_i\psi^i$) transforms independently of the rest of the representation.

CHAPTER 5

EXPERIMENTAL PARTICLE PHYSICS

While the majority of this book will take a theoretical approach to particle physics, it is of course experiment that is the ultimate judge of a physical theory, and we would not have the understanding that we do of the nature of particles if it were not for their experimental detection. For this reason, it is worth taking some time to appreciate how particles are produced and detected in experiments, and how their properties are measured. This chapter will look at some of the ways in which individual particles can interact with bulk matter, and how these interactions may be exploited to build particle detectors. It will look at both historic and contemporary detector design, and why some of the early detectors have been superseded. It will then look at some aspects of the design of particle accelerators and how they are used to produce the particles for analysis. Since this is not the main focus of the rest of the book, the first part of this chapter will give less mathematical detail, and is intended only to give a flavor of what experimental particle physics entails. The final part looks at the experimentally measurable quantities of particle physics, and how we might hope to calculate these quantities theoretically.

5.1   DETECTORS

Everything we know about particle physics comes from our ability to observe the particles whose properties we wish to understand. To this end, there are a variety of detector types that exploit particles' interactions with bulk matter in order to capture a record of the particles' behavior. In order to understand the construction and implementation of these particle detectors, we must first look at the nature of these particle-matter interactions.

5.1.1   Interactions of Particles with Matter

The following is a limited list of some of the types of interactions that particles undergo when traveling through bulk matter. While it is not exhaustive, it should be sufficient for the reader to appreciate the principles behind some of the detection methods we will discuss. In what follows, since we are dealing with experimental physics, and particularly to highlight the difference between relativistic and non-relativistic cases, we will work temporarily in non-natural (SI) units.

Ionization of Matter by Charged Particles

When a charged particle, whether positive or negative, travels through a material, it exerts an electrostatic force on the electrons in the outer orbitals of the material's atoms. If this force is sufficiently strong, the electrons may be removed from their atoms leading to ionization of the material. Since the removal of an electron from an atom requires energy, and since there is only one place the energy could have come from in this situation, this ionization also results in a loss of kinetic energy of the particle. Consider a particle with charge qe moving at speed v past an atom with an impact parameter b. This latter quantity is simply a measure of the distance of closest approach if the particle's trajectory were a straight line. Put another way, the impact parameter is the orthogonal separation of the atom from the tangent to the particle's initial trajectory. We parametrize the situation as in Figure 5.1.

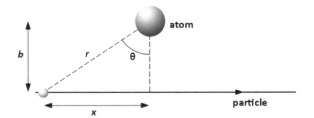

FIGURE 5.1 Energy loss of charged particle due to ionization.

Let x denote the displacement of the particle in its direction of motion from the point of closest approach. The momentum transferred to an electron in the atom in the x direction averages out to 0 as the particle travels by. The momentum transfer in the direction perpendicular to the particle's motion is given by

$$
\begin{aligned}
p_e &= \int F \cos\theta \mathrm{d}t \\
&= \int_{-\infty}^{\infty} \frac{qe^2}{4\pi\varepsilon_0\left(x^2+b^2\right)} \frac{b}{\left(x^2+b^2\right)^{1/2}} \frac{\mathrm{d}x}{v} \\
&= \frac{2qe^2}{4\pi\varepsilon_0 vb}.
\end{aligned}
\tag{5.1}
$$

The energy transferred to the electron, then, is

$$
E = \frac{p_e^2}{2m_e} = \frac{2}{m_e}\left(\frac{qe^2}{4\pi\varepsilon_0 v}\right)^2 \frac{1}{b^2}.
\tag{5.2}
$$

This is the energy lost by the charged particle to a single electron. Since there are many electrons in the material, consider a cylindrical shell of radius b and thickness db:

The number of electrons in this shell is

$$nZ \times 2\pi b \, db \, dx, \tag{5.3}$$

where n is the number density of atoms in the medium and Z is their atomic number, so the rate of energy loss per unit length traveled in the material is

$$
\begin{aligned}
-\frac{dE}{dx} &= \int_{b_{\min}}^{b_{\max}} nZ2\pi b \frac{2}{m_e} \left(\frac{qe^2}{4\pi\varepsilon_0 v^2} \right)^2 \frac{1}{b} db \\
&= \frac{4\pi nZ}{m_e v^2} \left(\frac{qe^2}{4\pi\varepsilon_0} \right)^2 \int_{b_{\min}}^{b_{\max}} \frac{1}{b} db \\
&= \frac{4\pi nZ}{m_e v^2} \left(\frac{qe^2}{4\pi\varepsilon_0} \right)^2 \ln \left(\frac{b_{\max}}{b_{\min}} \right),
\end{aligned}
\tag{5.4}
$$

where we must include a maximum and minimum b to account for quantum mechanical effects. In particular, the minimum impact parameter is determined by the electron's de Broglie wavelength relative to the particle, since the value of b is not well defined below this limit due to uncertainty. More precisely, since the maximum momentum that may be imparted to the electron during the collision is $p_{\max} = 2m_e v$ (see Exercise 1), the uncertainty principle tells us that the smallest physically meaningful b is given by $b_{\min} = \hbar/(2m_e v)$. For the maximum b, if the particle is too distant from the atom, it will not impart enough energy to ionize the atom, since energy levels are discrete. To a good approximation, the interaction between particle and electron occurs only over a short distance where $|x| < b$, so that the duration of the interaction is $\sim 2b/v$. Comparing this with the frequency of the electron's orbit, we find

$$
\begin{aligned}
\frac{2b_{\max}}{v} &= \frac{1}{f} \\
\implies b_{\max} &= \frac{v}{2f} = \frac{hv}{2I}
\end{aligned}
\tag{5.5}
$$

where I is the average ionization energy for all electrons in the atom. Since I is measured empirically, by convention we absorb the residual factor of π into its definition to arrive at the Bethe formula

$$-\frac{\mathrm{d}E}{\mathrm{d}x} = \frac{4\pi n Z}{m_e v^2} \left(\frac{qe^2}{4\pi\varepsilon_0}\right)^2 \ln\left(\frac{2m_e v^2}{I}\right), \qquad (5.6)$$

which is valid for slow-moving/heavy charged particles. A correction for relativistic particles gives the relativistic Bethe formula, whose derivation we will gloss over:

$$-\frac{\mathrm{d}E}{\mathrm{d}x} = \frac{4\pi n Z}{m_e \beta^2 c^2} \left(\frac{qe^2}{4\pi\varepsilon_0}\right)^2 \left(\frac{1}{2}\ln\left(\frac{2m_e\,(\gamma\beta c)^2\,T_{\max}}{I}\right) - \beta^2\right),$$
$$(5.7)$$

where $\beta = v/c$, γ is the Lorentz factor, $\gamma = (1 - \beta^2)^{-1/2}$, and

$$T_{\max} = \frac{2m_e(\gamma\beta c)^2}{1 + 2\gamma\frac{m_e}{M} + \left(\frac{m_e}{M}\right)^2} \qquad (5.8)$$

is the maximum energy that can be transferred in a single interaction. Note that the mass, M, of the charged particle appears in the relativistic form via $T_{\max}$.

As a particle travels through a medium, then, it loses energy at a rate determined by several factors, including the type of particle, the medium, and the particle's instantaneous energy. The rate at which this energy is lost with respect to the distance traveled $(-\mathrm{d}E/\mathrm{d}x)$ is known as the material's *stopping power*. It should be noted that the x appearing in the Bethe formula is not a literal length, but really a measure of "amount of material traversed." That is, it is related to the actual path length, ℓ, by $x = \rho\ell$, where ρ is the density of the medium. As such, it is typically measured in SI units of g cm^{-2}.

There is a minimum in the Bethe formula at around $\beta\gamma \sim 3 - 3.5$ depending on the medium. Above this, relativistic effects dominate and the stopping power scales as roughly $\ln(\gamma^2)$, increasing slowly with energy. This increase is attributable to the fact that the electric field of the charged particle is highly Lorentz contracted and thereby concentrated, allowing for greater ionization. In reality, this increase eventually levels off, however, in the "Fermi plateau." This is because the maximum transferable energy, $T_{\max}$, appearing in the

Bethe formula grows without limit, which in turn would allow electrons to be liberated from the medium entirely. In practice, then, there is an energy cutoff that must be imposed, leading to the plateau.

Below the minimum at $\beta\gamma \sim 3$, the stopping power decreases with energy, scaling roughly as $\beta^{-2} \sim E^{-1}$. This is due to the fact that the energy of the particle at these scales is more directly related to its velocity, and a slower-moving particle has more time to interact and ionize atoms. As the energy of a charged particle drops lower still, other factors contribute to the stopping power, including the scattering of the particle by atomic nuclei in the medium and the particle's capture of electrons from the material. As such, the Bethe formula begins to lose its validity. In fact, the stopping power is observed to reach a maximum at around $\beta\gamma \sim 0.01$, and then begin to decrease at a rate for which the Bethe formula cannot account. Empirical models exist to describe this region, and at the lowest energy scales the stopping power is found to scale approximately linearly with particle energy. Ultimately, however, the question of how much energy the particle loses at these scales becomes irrelevant, since it will eventually reach the average thermal energy of the medium (thermalize), at which point no further energy dissipation will occur.

Photon Interactions with Matter

When a photon passes through a medium, it may ionize an atom of the material through the photoelectric effect. That is, if it has sufficient energy, it may be absorbed by an electron in one of the material's atoms, liberating that electron from its atomic orbital. For a given orbital, there is a minimum required energy, ϕ, for this process to occur. The difference between the photon energy and the ionization energy manifests itself as the kinetic energy of the ionized electron, through $E_{\text{kin}} = hf - \phi$, where h is the photon's frequency. The freed electron may then cause further ionization through the mechanism discussed in the previous section. At the same time, the atom from which the electron was liberated may also be left in an excited state, if the electron was removed from an inner orbital. This allows an electron from a higher orbital to fall into the lower-energy "hole" left by the photoelectric effect. As the electron makes this transition,

FIGURE 5.2 The Feynman diagram for Compton scattering along with a parametrization of the situation. The electron is assumed to be initially at rest, and the photon is scattered through an angle θ.

it emits the excess energy in the form of another photon, of lower energy than the original.

In addition to these ionization effects, photons may also interact with free or quasi-free electrons in the medium through inelastic scattering, known as Compton scattering. That is, the photon may be absorbed by a free electron, increasing the electron's kinetic energy. The electron then re-emits a photon, but one of a lower energy than the original. This can be summarized by the Feynman diagram in Figure 5.2. The electron must be free for such scattering processes, since the energy levels of bound electrons in atomic orbitals take only discrete values. So the absorption of a photon in this case would place the electron "between" energy levels, which is forbidden. By contrast, the free electrons in the conduction band of a material have a continuum of energy levels available to them. The energy redistribution of Compton scattering is achieved by producing a net increase in the electron kinetic energy, and hence also in its momentum. For momentum to be conserved, this requires that the re-emitted photon have a different trajectory from the photon that was absorbed. There is a direct relationship between this scattering angle and the energy lost to the electron, given by

$$\lambda_{\text{final}} - \lambda_{\text{initial}} = \frac{h}{m_e c}(1 - \cos\theta), \tag{5.9}$$

where λ is the photon wavelength, related to its energy through $E = hc/\lambda$. The reader is invited to derive this relationship in Exercise 4.

Through Compton scattering and the photoelectric, then, photons may dissipate their energy through direct interactions with the medium. As with the Bethe formula, these effects become less important at higher energies. In the case of both photons and charged particles, further effects dominate at higher energy scales. Whereas the low-energy behaviors are goverened by very direct interaction with the material, we will see that the high-energy effects are rather indirect, though they still require the medium in order to occur.

Bremsstrahlung and Pair-Production

When a high-energy electron passes an atom, it can interact with the atom via photon exchange. This allows the transferral of energy and momentum, in turn allowing the electron to emit one or more photons as in Figure 5.3. Note that the electron cannot simply emit a photon without such interaction, as this would necessarily violate momentum conservation. This process is known as Bremsstrahlung (a German word meaning "braking radiation"), as it results in the loss of a significant proportion of the electron's kinetic energy. Likewise, when a high-energy photon passes an atom, it can interact via virtual particle exchange to create an electron-positron pair, as in Figure 5.4. So an initial charged particle will tend to produce a shower of lower energy particles through a combination of Bremsstrahlung and pair-production, together referred to as radiative losses. The energy lost through Bremsstrahlung is found to be proportional to the total energy, and so follows an exponential decay. As such, we can define a radiation length, X_0, to be the length scale (in g cm^{-2}) over which E drops by a factor of e. So we have

$$E = E_0 e^{-x/X_0}. \qquad (5.10)$$

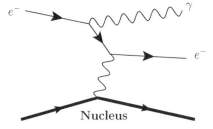

FIGURE 5.3 Feynman diagram contributing to radiative energy loss of a charged particle through braking radiation or Bremsstrahlung.

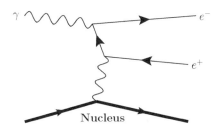

FIGURE 5.4 Feynman diagram contributing to electron-positron pair production in the presence of an electromagnetic field, in this case in the form of a nucleus.

Similarly, the characteristic length scale for photons is the mean free path, ℓ_{free}, with the photon intensity varying as $I = I_0 \exp(-x/\ell_{\text{free}})$. The mean free path tends to be of order $9X_0/7$.

While the total energy lost to a medium is the sum of the ionization and radiative losses, there is a critical energy, below which the ionization losses dominate, and above which the radiative losses dominate. The dissipation of a charged particle's energy, then, is characterized by a shower of exponentially increasing numbers of lower-energy particles, after which the charged particles in the shower dump their remaining energy via ionization, while the photons dissipate their energy through Compton scattering and the photoelectric effect. This leads to a fairly predictable number of final-state particles as a function of initial-particle energy. The value of the critical energy at which radiative losses begin to dominate depends strongly on the type of particle considered. For electrons, radiative losses play a big role, and since these losses scale with energy, it allows for very efficient energy dissipation. For heavy particles such as hadrons, the role of radiative losses is much less significant, since the critical energy is considerably higher. In fact, for heavy particles, even the ionization effects described by the Bethe formula play a smaller role than for lighter particles, with the majority of the stopping power coming instead from direct collisions with nuclei. For this reason, energy dissipation is typically a much slower process for heavy particles.

Čerenkov Radiation

Another way that charged particles can interact with their surroundings is via Čerenkov radiation. This is the electromagnetic analogue of a bow wave in water surfaces or a sonic boom in air. When an object capable of causing disturbances in a medium travels through that medium at a speed v, it causes waves to propagate outward at a speed c_{med}. If the object travels faster than the waves can propagate, then constructive interference of successive waves creates a plane wave that propagates at a particular angle away from the object's direction of motion.

In time t, a wave propagates a distance $c_{med}t$, while the object moves a distance vt. This gives an angle

$$\theta = \cos^{-1}\left(\frac{c_{med}}{v}\right), \tag{5.11}$$

at which this shock wave will develop.

In the case of electromagnetic radiation $c_{med}/c = 1/n$ where n is the refractive index of the material. This is useful for measuring β for a particle as

$$\beta = \frac{v}{c} \implies \theta = \cos^{-1}\left(\frac{1}{n\beta}\right), \tag{5.12}$$

which, since it depends on velocity, whereas most methods are momentum-dependent, can aid in determining the mass, and thereby the identity, of a particle. Since the refractive index is wavelength-

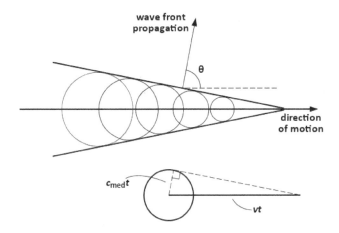

FIGURE 5.5 The geometry of wave-formation leading to Čerenkov radiation.

dependent, the number of Čerenkov photons also varies with wavelength, and is found to peak in the blue-UV range, making Čerenkov radiation visible as a pale blue glow.

Čerenkov radiation is produced when highly relativistic charged particles travel through a dielectric medium. If such a particle travels across a boundary between two media of different dielectric constants, there is yet another type of radiative loss known as "transition radiation." Unlike Čerenkov radiation, this is emitted over a range of angles. However, the peak intensity of the radiation is again characteristic of a particular velocity or, equivalently, Lorentz factor. Again, this is useful in identifying particles, and a transition radiation detector is designed to make use of this.

5.1.2 Early Detectors

This section is intended to give the reader an idea of the history of particle detection, and will look at some of the methods that were used in the early days of particle physics experiments. However, it is important to realize that these methods have not all been superseded. While the detectors used in the highest-energy experiments have generally moved on to other methods, some of the methods in this section are still used in other contexts, such as in specific low-energy experiments and day-to-day applications. All of the detectors mentioned here are used to determine the trajectories of charged particles, and work on the general principle of making those particles' tracks visible in some way. Particle trajectory by itself can reveal such information as the decay products of unstable particles and particle lifetime. Additionally, when these detectors are placed in magnetic fields, the curvature of particle trajectories provides additional information regarding momentum, charge, and energy. When a detector is used specifically to measure particle momentum in this way, it is referred to as a spectrometer.

Emulsions

One of the simplest types of detection apparatus is the nuclear emulsion. This is a particularly high-quality photographic emulsion that is applied in a thick layer to a photographic plate. The detection of charged particles then works in exactly the same way as light-detection in pre-digital photography. A charged particle causes a chemical reaction in the silver-halide crystals suspended in the emulsion. These tracks thus capture a permanent record of any charged particle passing through, and can be made visible at a later date through photographic development of the plate. The visible tracks are then observed with a microscope. The advantages of this method are that the spatial resolution is extremely high, and the emulsion can be installed and left *in situ* for the duration of an experimental run, since the recorded tracks are permanent. However, there are also significant disadvantages: first, the tracks have essentially no time resolution, as there is no record of *when* the particle associated with a particular track passed through. Second, since the plates require development and later study, there is no real-time output.

Cloud Chambers

While the emulsion is arguably the simplest type of detector, it is not, in fact, the oldest. As far back as the 1920s, physicists were using the cloud chamber to illuminate the tracks of charged particles. The principle behind this apparatus is that a vapor cannot spontaneously condense, even below its boiling point, without a condensation nucleus. This could be in the form of a dust particle, or an existing droplet of the condensed vapor, but can also be in the form of an ionized atom. A cloud chamber, then, consists of a saturated vapor cooled to below its condensation point, in a chamber that is sealed in order to avoid condensation around other types of nuclei such as dust motes. The cooling required for condensation is typically achieved by adiabatic expansion of the vapor inside via the rapid movement of a diaphragm or piston. Any charged particle moving through the vapor ionizes the atoms along its track, which then cause condensation of the vapor around them. The tracks are thus visible as small clouds of condensation, much like the contrails left by a passing aircraft. These tracks are then illuminated and captured on film, typically with cameras mounted at various angles to allow a three-dimensional picture of the tracks to be reconstructed.

A significant limitation of the cloud chamber is that one of the best types of condensation nucleus is existing droplets of the medium. This means that the cloud will rapidly increase in size until the tracks are difficult to observe. To combat this, the chamber is quickly reset to its supersaturated vapor state, ready for another expansion. The problem, then, is that there is a significant "dead time" during which the chamber is unable to show any tracks. A second disadvantage is that, as the energy scale of experiments was pushed higher, the low density of the gaseous medium required large chambers to achieve any significant reduction in particle energy. In smaller chambers, high-energy particles would simply exit the chamber before decaying to their final products, and so not allow for particle identification. Eventually, experimental energies surpassed the levels for practical and cost-efficient cloud chambers.

Bubble Chambers

In the 1950s, the bubble chamber began to supersede the cloud chamber. While the principle behind the bubble chamber is very similar to its predecessor, it solves one of the problems of the cloud chamber. Specifically, since the bubble chamber uses liquid as its medium rather than vapor, it has a much greater density, allowing the device to be physically smaller yet achieve the same stopping power. The tracks of charged particles in this case are revealed as trails of bubbles in a superheated liquid, forming at the nucleation sites, again provided by ionization of the medium. The superheated phase was achieved by a sudden decrease in pressure of the chamber by means of a piston. While the bubble chamber was an improvement, its design still suffers from some of the disadvantages of the cloud chamber: specifically, the dead time between superheated phases.

Spark Chambers

Used around the same time as the cloud and bubble chambers, there was another type of detector that also made visible the tracks of charged particles, but through a very different method. This was the spark chamber, and its operation relies on a high voltage being applied between parallel plates, greater than the electrical break-down voltage. The ions produced by the passage of a charged particle provide the ideal means for a visible spark to jump between the plates. This voltage could not be applied continuously, since even in the absence of an ionized path, the plates would discharge given the voltage between them. The spark chamber typically has a lower spatial resolution than the cloud or bubble chambers but has some advantages over these methods, depending on the aim of the experiment. In particular, if used in conjunction with a secondary trigger detector, capable of determining the possibility of an interesting event, the voltage could be applied in a very short time, ensuring the event would be captured. In contrast, the cloud and bubble chambers can only detect an event when they happen to be in the operational phase. In addition, spark chambers are typically more cost-effective. The spark chamber is now largely obsolete, but is mentioned here mainly as it is more closely related than the previous examples to more modern detector designs.

5.1.3 Modern Detectors

Modern detectors that monitor particle trajectories generally do not rely on making the ionized paths visible as the older detectors did. Instead, the electrons and ions liberated by passing charged particles are moved by electric fields to a circuit, where they can be detected directly as a signal or pulse. A great advantage of these devices over the older methods is that, since the signal is electronic, it may be input directly and automatically into computer storage without the need for additional recording equipment. An array of data lines then builds up a three-dimensional image of the particle trajectories in real time.

Despite the ubiquity of this approach, there is still a great deal of variation in the methods for detecting such signals. The behavior of ions in these detectors varies considerably with the applied voltage. If the voltage is too low, then ions and their liberated electrons will simply recombine without generating a signal at all. As the voltage is increased, we enter the ionization region, in which free electrons will drift toward the anode and cause a small signal. At very high voltages, the free electrons gain sufficient energy from the applied electric field that they cause secondary ionizations, which in turn cause further ionization, and so on. In this way, a cascade of electrons is produced, known as a Townsend avalanche, which causes a large signal at the anode. This is the principle behind a Geiger counter, in which each particle that enters will cause a large signal followed by a period of dead time while the medium undergoes recombination.

In between the ionization and Geiger regions, there is a range of voltages at which ionization produces a number of secondary ionizations proportional to the energy of the initial particle. This is the proportional region, in which many modern detectors operate. The multi-wire proportional chamber is one such detector, consisting of a plane of wire anodes sandwiched between two cathode planes, as in Figure 5.6. As a charged particle passes through the chamber, the ionization of the medium (often a noble gas) causes electrons to drift toward the nearest anode and ions to drift toward the cathode. Since the arrangement of electrodes leads to a field strength that is greatest close to the anodes, a drifting electron enters the proportional region as it approaches. This encourages secondary production allowing for a substantial signal in the form of a pulse of current. Surprisingly, however, the other anodes in the chamber also receive a pulse known as an induced pulse. As the avalanche appears in the vicinity of one anode, which then removes the electrons, a large number of positive ions is produced around this anode. These alter the arrangement of

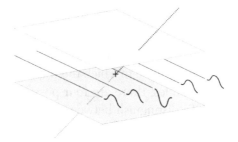

FIGURE 5.6 The signal generated by a MWPC as a particle passes through (dotted line). The planes at the top and bottom of the image are held at a negative potential while the wires running through the center are positive. The point at which the avalanche develops is marked by a cross.

field lines, since they also contribute to the field. Initially, the field lines from these ions all connect to the avalanche anode but, as the ions drift away from this point, the field lines distribute themselves across the remaining anodes. So a pulse is detected in the remaining anodes that is opposite in sign to that in the avalanche anode. In this way, we may determine which wire the particle passed most closely to. By separating one of the cathode planes into a series of strips, the particle's position along the avalanche anode can also be resolved.

Drift Chambers

The spatial resolution of wire chambers can be improved by taking into account the time required for the avalanche to reach the wire. Detectors that make use of this fact are known as drift chambers. For this to work, it must be possible to predict the time taken for an avalanche to develop from an ion at any point in the chamber. This is achieved, in the case of multi-wire chambers, with the aid of additional field-shaping wires of tuned potentials, such that the field in the chamber is very uniform. The simplest type of drift chamber, though, consists of a single anode wire running axially through the length of a hollow cylindrical cathode. Such an arrangement is known as a straw chamber, and is commonly used since it is an inexpensive design. A straw chamber by itself gives only limited information, but by measuring the timing of a pulse relative to other nearby straw chambers, the trajectory of the particle can be found very accurately. To see how this works, consider a charged particle passing through two straw chambers as in Figure 5.7. The relative time between pulses in the chambers is enough to determine the radial distance of the particle's trajectory from each straw. This identifies a circle around each anode to which the trajectory must be tangential. For two such circles, there are four possible trajectories as shown in the figure. To distinguish between these four possibilities requires a third chamber, and further chambers reduce the uncertainty in the measurement. A *straw tracker*, then, is a large collection of straw chambers used in parallel to track a charged particle. Notice that the tracks labeled C and D would remain indistinguishable if the third chamber were arranged in line with the first two. For this

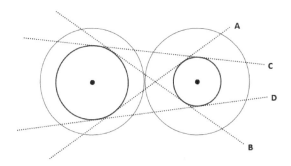

FIGURE 5.7 The possible paths of a charged particle leading to equivalent signals in a pair of straw chambers.

reason, neighboring chambers in a straw tracker are typically hexagonally close packed, as in Figure 5.8. Paths C and D are then easily distinguished by which chamber they pass through next.

Solid-State Detectors

The same concepts as employed in the *Multi-Wire Proportional Chamber* (MWPC) are used in the solid state as well as the gaseous state. Here, it is not electrons and ions that cause the signal but

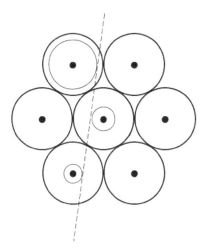

FIGURE 5.8 Hexagonally close packed straw chambers in a typical straw tracker. A charged particle trajectory is shown, along with the region to which each individual chamber can attribute that trajectory.

electrons and holes in a semiconductor, most commonly silicon. However, given the sensitivity of semiconductor devices, they typically operate in the ionization region, rather than the proportional region. Such devices allow for greater spatial resolution but at significantly greater cost. For this reason, modern experiments tend to use a combination of these methods, with the semiconductor devices only used close to the beam where spatial resolution is crucial, and more cost-effective gaseous-state detectors used outside.

Čerenkov Detectors

As the name suggests, Čerenkov detectors rely on detecting the Čerenkov radiation given off by energetic charged particles. The name is something of a catch-all term, however, for a number of different detector types that vary in purpose and design.

A threshold detector makes use of Equation 5.11 to aid in particle identification. For a particle of known momentum (measured in some type of spectrometer), we may not necessarily know the mass or velocity of the particle, just the product of the two. If a particle has a sufficiently high velocity β, it will radiate Čerenkov radiation. However, notice that Equation 5.11 has no solution if β is too small. A lightweight particle may thus emit Čerenkov radiation where a heavier particle of equal momentum may not, and a threshold detector distinguishes between particle species in this way. More elaborate threshold detectors may consist of two or more materials of different refractive indices to allow for further distinctions to be made.

A ring-imaging Čerenkov (or RICH) detector allows the emitted cone of Čerenkov photons to propagate freely for some distance before striking a plane of photon detectors perpendicular to the cone's symmetry axis. In this way, the photons imaged by the detectors form a ring whose radius can be used to calculate β, again using Equation 5.11. In this way, the velocity of the initial charged particle may be calculated in addition to its momentum, allowing for calculation of the mass and again aiding in particle identification. In addition, the measured energy of detected photons in a RICH detector can be used to extrapolate backward and calculate the energy of the initial particle. A contemporary example of this type of detector is its use in the LHCb experiment, which aims to study $\mathbb{CP}$ violation in the B mesons.

Calorimeters

All of the methods considered so far have been non-destructive. Although some energy is lost by the particle in ionizing the medium or emitting radiation, this amount is very small in comparison to the particle's total energy. In order to measure particle energy directly, however, the methods tend to be destructive, relying on the total or near-total absorption of that energy. This is achieved with a calorimeter, of which there are two major categories: electromagnetic and hadronic.

An electromagnetic calorimeter is designed to absorb the energy of a particle that interacts primarily through the electromagnetic force, whereas a hadronic calorimeter is designed for particles that lose the majority of their energy via strong nuclear interactions. We will consider the electromagnetic type first. As we saw in Section 5.1.1, a high-energy charged particle will produce a shower of other particles with lower energies, via radiative losses. Below a critical energy, these processes become negligible so a particle with a particular energy will lead to a characteristic number of secondary electrons and photons being emitted, which then lose their energy through ionization losses. Calculating the initial particle's energy, then, is achieved essentially by counting the number of particles in the final shower. This can be achieved in a number of ways depending on the design of the calorimeter. Although it is possible to construct a homogeneous calorimeter from a single material, this tends to be expensive. More commonly, calorimeters sample the energy produced in showers by layering two different materials. The first is a high-density material intended to encourage shower production and the second is the sampling material. The nature of the sampling material depends on the design of the detector; one possibility is to have a secondary set of ionization chambers to measure the passage of charged particles in the shower. Since the positions of these secondary particles is unimportant, these ionization chambers can be of a very simple design.

Another possibility is to use a scintillating material: that is, a material which radiates photons in the visible range when excited by interactions with ionizing radiation. The photons emitted by the scintillator may then be detected via devices such as photomultiplier tubes or, more recently, single-photon detectors such as silicon photomultipliers. By amplifying the signal using these methods, both the intensity and the energy of the scintillation radiation may be accurately measured. In yet other calorimeter designs, the sampling may be achieved with Čerenkov detectors. Although these sampling techniques will inevitably not capture all of the energy dissipated by the initial particle, the energy is distributed in a predictable manner allowing for accurate reconstruction of the total energy.

A hadronic calorimeter works in much the same way as an electromagnetic calorimeter. The primary difference is in the nature of the interactions, rather than the design. While hadrons will interact with their surroundings electromagnetically if charged, hadrons being of a much greater energy than leptons or photons, such interactions do little to dissipate their energy. The majority of a hadron's energy is dissipated instead through collisions with the nuclei of the calorimeter medium, leading to the production of secondary hadrons. A hadronic shower, then, is formed in a similar fashion to the electromagnetic shower produced by light charged particles. However, since nuclear collisions are far less likely to occur than electromagnetic interactions, the mean free path for a hadron is much greater than for a light charged particle, leading to showers developing over greater distances. As such, hadronic calorimeters must be physically larger than their electromagnetic counterparts to ensure the full dissipation of energy. Additionally, a variable proportion of the energy of nuclear collisions tends to be carried away in the form of neutrinos and neutral pions, both of which can escape detection. For this reason, there tends to be considerable statistical fluctuation in the measurement of hadron energies.

Layered Designs

All of the types of particle detection that we have looked at in the preceding sections are generally capable of determining just one of the properties of a passing particle. In order to build up a clear picture of an interaction, it is necessary to layer different detector types on top of one another. The ATLAS and CMS detectors at CERN's Large Hadron Collider are both prime examples of this principle. Both detectors have silicon detectors as their innermost layers, close to the beam for precise measurement of the positions of interactions. Beyond this are additional layers of trackers for measuring the momentum of particles. Intense magnetic fields are used to bend the paths of particles throughout the detectors to aid in identification. Further out still, there are electromagnetic and hadronic calorimeters to measure the energy of the particles through absorption. Finally, beyond the calorimeters there are further sets of trackers for the precise measurement of the momentum of muons. Muons are by far the most penetrative of charged particles and will escape even the dense material of the calorimeters. Therefore, for an accurate measurement of the energy and momentum carried away from an interaction by muons, it is necessary to track the muons' trajectories for as far as is feasible. The high penetration of muons arises from a combination of the particle's properties. First, the muon is a lepton and so does not interact through the nuclear processes that serve to slow hadrons in a calorimeter. Second, the mass of the muon is sufficiently high that the stopping power due to radiative losses is weak; however, the mass is not so high as to cause the muon to decay before leaving the detector. This is in contrast to the muon's heavier cousin, the τ, whose decay width is such that it typically decays close to its point of creation, at least at current collision energies.

5.2   ACCELERATORS

In the early days of particle physics, discoveries relied on high-energy cosmic rays. For example, this is how the muon was discovered. However, for higher-energy experiments, particles must be accelerated artificially. There are two main types of accelerator experiment: fixed-target beams and colliders. In a fixed-target experiment, a single particle beam is accelerated to high energy and then directed at a target constructed from high-density material. The collisions of accelerated particles with the nuclei of the target result in a high concentration of energy in a localized region, leading to the production of new particles that may be studied using the methods of the preceding sections. While fixed-target experiments typically produce a large number of events due to the high likelihood of collisions, there is a significant downside to the design. This is that conservation of momentum requires the products of any collision to be moving in the direction of the beam. This essentially guarantees that some of the energy of the beam will be lost as kinetic energy of the products. The alternative is to aim two beams at each other in a collider experiment. In this way, there is no net momentum in the system and all of the beam energy is potentially available for particle production. The technical difficulty of this approach, however, is considerably increased, since each particle beam is typically a few nanometers in diameter, and two of these must collide head-on. An often-quoted analogy is the point-to-point collision of a pair of knitting needles launched from either side of the Atlantic Ocean. Needless to say, the focusing system must be extraordinarily precise.

Acceleration is achieved through the manipulation of charged particles with electric and magnetic fields. Since the force due to a magnetic field is given by $\mathbf{F} = q\mathbf{v} \times \mathbf{B}$, this force is always perpendicular to the direction of motion. Therefore, a magnetic field can do no work, and it is only electric fields that can increase the energy of a charged particle. The simplest accelerator, then, consists of a static electric field produced by a pair of plate-like electrodes with a hole to allow the passage of any accelerated particles. Such designs are indeed used for very low energy experiments, but there is a strict limit on the energy attainable in this way. If we accelerate a particle with a single constant electric field, then the kinetic energy that can be achieved is directly proportional to the applied potential difference. So this method is impractical for high-energy experiments since the required potential would be immense, and the system would suffer electrical breakdown. There are numerous accelerator designs that overcome this problem.

5.2.1 Linear Accelerators

To avoid the issues with static electric fields, accelerators instead use alternating fields. The simplest example of such a device is a linear accelerator or linac. Linac is a generic term for any accelerator in which each particle passes only once through the machine, and so has only one chance to be accelerated. As an example of the operation of a linac, consider an accelerator in which the electric field is produced by a set of hollow tube-shaped electrodes of alternating potentials. An electric field between these plates accelerates the particle. If the fields were static, however, the particle would accelerate between one pair of plates and then decelerate between the subsequent pair. So the voltages on the plates constantly switch back and forth so that, as the particle moves into the next section, the field is still in the correct orientation to accelerate the particle further. Since the speed of the particle increases throughout the length of the linac, but the frequency of the alternating voltage remains constant, the distance between electrodes, as well as the length of the electrodes themselves, must increase along the tube.

Accelerators do not accelerate particles in isolation but as beams. Of course, most particles in the beam will not be perfectly in phase with the oscillating field, and will not be given the maximum amount of energy in each section; others will be more than 90° out of phase and actually decelerate in some sections. This may seem to be a problem, but it is actually rather useful as it causes the particles to bunch together. A particle exactly at the mean position of a bunch will receive a fraction of the maximum possible energy-gain. Any particle lagging behind this mean position will be accelerated more and catch up with the main bunch, while any particle ahead of the bunch will not be accelerated as much and so the bunch will catch it up. This

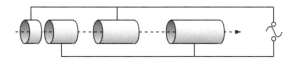

FIGURE 5.9 Schematic representation of a linear accelerator using alternating voltages.

bunching helps to maximize the possibility of a collision when the beam of particles meets its target. Indeed, the second key quantity of interest (after energy) for any accelerator is its *luminosity*: essentially a measure of the average rate at which events occur.

5.2.2 Cyclotrons

A cyclotron is also based on the principle of alternating electric fields but is built in a very different way from a linac. Rather than a large number of electrodes that each particle traverses only once, a cyclotron consists of just two electrodes, which each particle visits multiple times in a spiral trajectory. This is achieved through two flat hollow semi-cylindrical electrodes, named "Ds" in reference to their shape. The electrodes are separated by a gap on their straight sides as shown in Figure 5.10. It is when crossing this gap that the particle will receive a boost in energy. In order that each particle may traverse the accelerating gap more than once, a static magnetic field is applied in the axial direction. Due to the Lorentz force, the charged particle then travels in a circular path between accelerations, gradually spiraling outward as its energy increases. Once the particle reaches the outer edge of the D, it is fired at its target. The beauty of this design is hidden in the mathematics describing its behavior. In the non-relativistic regime, the Lorentz force acting on the particle

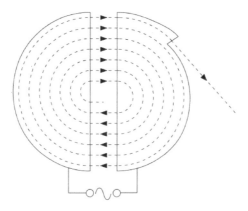

FIGURE 5.10 Schematic representation of a cyclotron demonstrating a typical accelerated particle's trajectory through the device.

is given by

$$\mathbf{F} = q e \mathbf{v} \times \mathbf{B}, \tag{5.13}$$

where **v** is the particle's velocity and **B** is the applied magnetic field. So the particle experiences a centripetal acceleration of magnitude

$$a = \frac{q e v B}{m}. \tag{5.14}$$

Recall that centripetal acceleration is also given by

$$a = \frac{v^2}{r}, \tag{5.15}$$

and equating these gives

$$\frac{v}{r} = \frac{q e v B}{m}. \tag{5.16}$$

We now calculate the time period of the particle's orbit as

$$T = \frac{2\pi r}{v} = \frac{2\pi m}{q e B}, \tag{5.17}$$

which is independent of the particle's speed! Despite the particle gaining energy and traveling at ever-increasing speeds around the cyclotron, this increase is matched perfectly by the increasing distance the particle must travel as it works its way out on its spiral trajectory. This makes the operation of a cyclotron wonderfully simple, since the time between successive accelerations remains constant throughout the time the particle is in the machine. As such, the frequency of the alternating voltage applied to the Ds can also remain constant at

$$f = \frac{1}{T} = \frac{q e B}{2\pi m}. \tag{5.18}$$

From the 1930s to the 1950s, the cyclotron was the design utilized for the highest-energy experiments of the day. However, the design has its limitations, and is not suitable for acceleration to extreme high energies. There are two problems: first, the maximum energy attainable depends on both the magnetic field strength and the radius of the Ds through

$$E = \frac{1}{2}mv^2 = \frac{(qeBr_{\text{max}})^2}{2m}, \tag{5.19}$$

so high-energy experiments require either very high magnetic field strengths or prohibitively large Ds. The second issue is the appearance of m in the previous derivation for the time period. This derivation assumes a constant m and so in fact applies only in the non-relativistic limit. At higher energies, relativistic effects become important and the constant-frequency relationship breaks down. This second problem can be solved with the use of a *synchrocyclotron*: essentially the same design with a variable-frequency voltage, allowing the voltage frequency to be adjusted in synchrony with the natural frequency of the particle's orbit. An immediate problem with this solution is that the frequency cannot be simultaneously tuned to suit particles at different stages of acceleration.

Radio Frequency Acceleration

As relativistic effects become important, one consequence is that high-energy particles all travel essentially at c. Increasing the energy of such a particle leads to an increase in relativistic mass rather than in speed. This allows for an efficient acceleration mechanism in which a particle in effect "surfs" an electromagnetic wave along the beam line, gaining energy as it does so. At first sight, this may seem counterintuitive, since the electric field in an electromagnetic wave is orthogonal to the wave's propagation, and would apparently serve to knock the particle off course. However, it is important to realize that this is only true in free space. While electromagnetic waves in free space are transverse to the direction of propagation, the fields in a cylindrical waveguide are constrained by the boundary conditions to lie along the direction of travel. This is because the guide is made of a conducting material, so the electric field must meet it perpendicularly. Equivalently, the magnetic field cannot pass through the

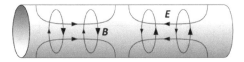

FIGURE 5.11 The wave mode excited in a cylindrical waveguide.

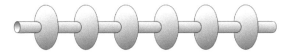

FIGURE 5.12 A radio-frequency accelerating cavity. The baffles reduce the lateral components of the electric field, leaving an almost purely longitudinal field for acceleration.

material. This gives a wave mode similar to that in Figure 5.11, in which the electric field lies parallel to the beam.

The effectiveness of this method is maximized by introducing a set of baffles to the waveguide, as in Figure 5.12, such that the length of each section is equal to the wavelength. Resonance then sets up standing waves in the guide with minimal fields in the directions perpendicular to the beam axis.

Klystrons

For radio frequency acceleration to be of use, the waves must have a large amplitude, and the typical means of producing such waves is with a klystron. A klystron is essentially a small electrostatic linear accelerator. Electrons are emitted from a cathode and accelerated under high voltage toward an anode. They then pass through a cavity resonator which is fed by a low-amplitude signal at the desired output frequency. This causes the electrons to begin to bunch: electrons out of phase with the input field accelerate or decelerate. The electrons then pass through a drift tube. This is a tube of precise length such that the bunching of the electrons is maximized by the time they reach the far end. They then pass through a second resonator, building up a strong standing wave as successive bunches pass through. Since the bunches arrive at the frequency determined by the input signal, the output is a wave of the same

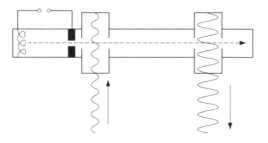

FIGURE 5.13 A klystron. Electrons are accelerated from the cathode at the left, through the anode and into the drift tube. They are bunched by the low-amplitude waves in the first resonating cavity and produce high-amplitude waves in the second.

frequency. However, since the kinetic energy of the electrons is determined by the high voltage across the electrostatic accelerator, the signal is greatly amplified. In order to ensure that the electron beam does not spread out along the klystron's length, the beam must be focused. This can be achieved quite simply with a longitudinal magnetic field. Any component of the electron's velocity that is perpendicular to the length of the klystron then results in a Lorentz force such that the electron travels in a helical path through the drift tube.

5.2.3 Synchrotrons

A synchrotron is a ring-shaped vacuum, in which acceleration occurs in radio-frequency cavities at one or more stations around the ring. The particles are then shepherded around the ring by magnetic fields to be accelerated again. We can think of it essentially as a modified linac, in which the particles leaving one end are brought back to the beginning. The name synchrotron comes from the fact that the magnetic field must again be tuned as the particles increase in energy. The magnetic fields used to bend the beam around the ring are produced by dipole magnets. In practice, to keep the circumference of the ring as small as possible, these magnets must be extremely powerful and so superconducting electromagnets are typically used. For example, at the time of writing, the most powerful accelerator is CERN's Large Hadron Collider, at a collision energy of 13 TeV. The dipole magnets used in this collider produce a maximum field

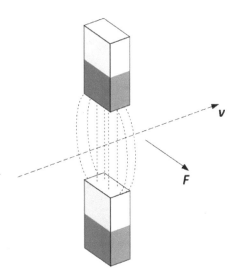

FIGURE 5.14 Schematic representation of a dipole magnet used to bend charged-particle beams. The dipoles typically employed are of a much more sophisticated design to reduce lateral components of the magnetic field.

strength of around 8 T, on the order of 10^6 times the Earth's geomagnetic field. Despite this, the ring is still over 8 km in diameter.

Dipole magnets have the fortunate side effect of correcting small deviations of particle trajectories from the mean path. In particular, a particle moving too far toward the inner edge of the ring will experience a weaker force than particles on the mean path, and so the diameter of such a particle's orbit will begin to increase. Similarly, a particle drifting toward the outer edge of the ring will experience a greater force, again correcting the deviation. The net effect is that every particle in the ring travels in a circular orbit with the same diameter as the ring, but different particles' orbits are not concentric. Without acceleration, relative to the beam average, each particle experiences a periodic variation in its momentum as it traverses the ring, known as betatron oscillation. So dipole magnets by themselves are capable of focusing the beam to some extent. However, we require something more precise than this "weak focusing," and for this we turn to quadrupole magnets and "strong focusing."

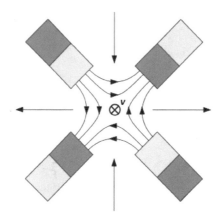

FIGURE 5.15 Schematic representation of a quadrupole magnet used to focus a beam of charged particles.

Magnetic Focusing

Strong focusing works on the principle of using a repeating lattice of focusing magnets. If a beam of (positively charged) particles passes through a quadrupole magnet as in Figure 5.15, then it will be focused in the vertical plane by the Lorentz force. At the same time, though, it will be defocused in the horizontal plane.

The horizontal defocusing can be corrected by a quadrupole in the opposite orientation, though clearly this will again defocus in the vertical plane. Indeed, if a vertically focusing quadrupole is followed immediately by a horizontally focusing quadrupole, there is no net effect. However, if the beam is allowed to drift a distance between the two quadrupoles, then overall it is focused in both planes. This arrangement is known as a FODO lattice as (in one plane) it consists of a focusing magnet (F), then nothing (0), then a defocusing magnet (D) and finally another stretch of nothing (0).

To see why this works, consider a particle that is perfectly aligned with the beam horizontally but which has a small vertical displacement from the beam. Suppose that its velocity is also perfectly aligned with the beam, that is, it has a longitudinal velocity. If the particle passes through a horizontally focusing magnet, this will have no effect on its horizontal velocity since the field at its location is itself horizontal. The particle will be defocused vertically, however, which is to say that it will acquire a vertical component to its velocity, away from the beam. Since the field strength of a quadrupole is weakest at the center, the vertical component of the particle's velocity is small. However, by the time the particle reaches the next (vertically focusing) magnet, it has now drifted further from the beam. As such, it is in a region with a stronger field and so feels a greater restoring force than the defocusing force experienced previously. The net effect is that the particle is focused in the vertical plane.

A quadrupole lattice cannot, by itself, correct for all variations in a beam. In particular, since each bunch will contain particles of a range of energies, some particles will experience a greater deflection than others in the quadrupole field. In this context, the range of energies present is known as the beam's *chromaticity*. The appropriate corrections that must be made to each particle's trajectory can be treated as a power series in energy. The dipole and quadrupole magnets are then capable of correcting particle paths to lowest order. To correct for higher-order deviations requires higher-order magnetic fields, and so sextupole and octopole magnets are also used in a focusing lattice. The number of magnets required at each order diminishes as the pole order increases, since the necessary higher-order corrections are much smaller in magnitude than the lower-order corrections.

Synchrotron Radiation and Future Accelerators

Synchrotrons are not without their own disadvantages. Key among these is the problem of energy losses. As a charged particle moves in a circular orbit, it is constantly accelerating, and accelerating charges emit electromagnetic radiation and lose energy in the process. In this context, this radiation is known as synchrotron radiation. This puts an upper limit on the energy attainable by any given synchrotron, since the energy lost through radiation increases with the total energy of the particle. At some point, the energy lost through synchrotron radiation during one lap of the ring will equal the energy input during the same period. At this point, further acceleration is impossible. The amount of synchrotron radiation emitted by a particle depends on the particle's mass and energy, and the radius of the ring. This is one reason for the large scale of modern accelerators. It also means that electrons, with their small mass, are far more susceptible to radiative energy losses than hadrons since synchrotron radiation scales with γ, which is velocity-dependent, rather than energy-dependent. This is partly the reason that protons were the particle of choice for acceleration to the high energies of the Large Hadron Collider. Another reason for this choice is the composite nature of protons. Since each event is caused by the interaction of *some part* of each proton, the amount of energy involved in each collision is variable. This is ideal for an all-purpose machine designed to probe a range of energies. The downside is that the effective collision energy per event is much less than the total proton energy. For this reason, coupled with the issue of synchrotron radiation, many believe that the next generation of accelerators should return to a linear design using leptons. A particularly intriguing concept is that of the Compact Linear Collider (CLIC). The proposed CLIC design allows for acceleration to high energy over a shorter distance than "conventional" linear accelerators, and would utilize a secondary drive beam to produce very high field strengths for the primary cavity resonators. In essence, it is a linac with a secondary linac in place of conventional klystrons.

The preceding sections have hopefully given the reader a flavor of some of the exciting physics involved in both the acceleration and detection of particles. To call this an overview, however, would be generous, since we have barely scratched the surface of these topics. The interested reader will find whole books and indeed whole journals dedicated to the topics of individual paragraphs of this chapter.

5.3 MEASURABLE QUANTITIES IN PARTICLE PHYSICS: MATCHING THEORY TO EXPERIMENT

We finish this chapter with a look at the quantities that experiments aim to measure and how we might hope to predict their values theoretically. We will consider two important quantities: the decay rate and the cross-section. From an experimental point of view, these are very different concepts, but we will see that they are closely connected from a computational standpoint.

5.3.1 Cross-Sections

The cross-section is a measure of the likelihood of an interaction between two particles when one is fired toward the other. It behaves as an effective area through which the incident particle must travel in order to effect a collision. In order to understand the cross-section as it is used in particle physics, let us first consider the simpler case of the interaction of two *classical* objects.

Classical Scattering

Suppose a sphere of radius r is fired toward a second stationary sphere of radius R and much greater mass from which it scatters elastically. What can we deduce about the resulting collision?

First, we must parametrize our problem. An imaginary line drawn parallel to the moving sphere's trajectory that passes through the center of the target sphere will be referred to as the collision axis. Since the situation is rotationally symmetric about this collision axis, the only variations we can make to the setup of this hypothetical experiment are the speed of the projectile and its perpendicular distance from the collision axis. This latter quantity is known as the impact parameter and is conventionally denoted b. What might the outcome of this experiment be? Well, assuming that the spheres collide elastically, the projectile must be deflected in some direction which is best characterized by spherical polar coordinates originating at the target as in Figure 5.16.

Since the situation has radial symmetry, the azimuthal angle ϕ will be unchanged before and after the collision, so the angle of interest is θ. Consider a plane perpendicular to the collision axis placed just before the target. The projectile must pass through some part of this plane. Let's call the small area through which it passes $d\sigma$. After the

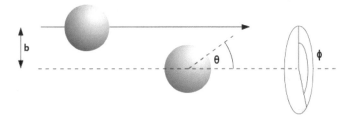

FIGURE 5.16 A collision between two classical objects, showing the impact parameter and the angular coordinates used to parametrize the subsequent trajectory of the scattered objects.

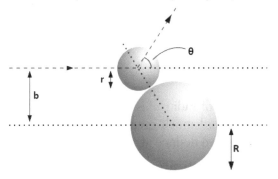

FIGURE 5.17 The point of contact between two classical colliding objects.

collision, the projectile is scattered into some solid angle characterized by θ and ϕ. Let's call the small region of solid angle around the projectile's trajectory $d\Omega$. How, then, does $d\Omega$ depend on $d\sigma$? To answer this, first notice that

$$d\sigma = b \, db \, d\phi \quad \text{and} \quad d\Omega = \sin\theta \, d\theta \, d\phi, \tag{5.20}$$

so we find that

$$\frac{d\sigma}{d\Omega} = \frac{b}{\sin\theta} \left| \frac{db}{d\theta} \right|. \tag{5.21}$$

This quantity is known as the differential cross-section.

Zooming in to the collision itself as in Figure 5.17, we see that a little geometry gives

$$b = (R + r) \sin\left(\frac{\pi - \theta}{2}\right) = (R + r) \cos\left(\frac{\theta}{2}\right), \qquad (5.22)$$

since in an elastic collision, the angle of incidence of the lighter sphere must equal its angle of reflection. Therefore, the differential cross-section becomes

$$\frac{d\sigma}{d\Omega} = \frac{(R + r) \cos\left(\frac{\theta}{2}\right)}{\sin\theta} \times \frac{(R + r) \sin\left(\frac{\theta}{2}\right)}{2}$$
$$= \frac{(R + r)^2}{4}. \qquad (5.23)$$

This is a measure of how much of the cross-sectional area of the collision scatters into each solid angle. In more general situations, the differential cross-section will typically depend on other factors such as the projectile's energy. To find the total cross-sectional area of the plane through which the projectile can pass and experience a deflection, we integrate the differential cross-section over all solid angles:

$$\sigma = \int \frac{d\sigma}{d\Omega}\, d\Omega$$
$$= \int_0^{2\pi} d\phi \int_0^{\pi} d\theta\, \frac{(R + r)^2}{4} \sin\theta \qquad (5.24)$$
$$= \pi(R + r)^2.$$

This expression is as we would expect, since it is the area of the region through which the projectile must pass to collide with the target at all. This is why such quantities are known as the cross-sections for the collision. Notice that another way to think of the cross-section is as the likelihood of a collision occurring at all. If we fire the projectile randomly, without aiming, but we know that it is constrained to travel along a tube of unit cross-sectional area, then the probability that a collision will take place is given by σ.

It was through these ideas that Rutherford derived his scattering formula, which we discussed in Section 1.2. By considering a classical electromagnetic interaction between two particles, one may show that the relationship between the impact parameter, b, and the scattering angle, θ, is given by

$$b = \frac{kq_1 q_2 e^2}{2E} \cot\left(\frac{\theta}{2}\right), \tag{5.25}$$

where k is the Coulomb constant, E is the energy of the scattered particle, and q_1, q_2 are the particles' charges. By substituting this into Equation 5.21, it is straightforward to arrive at the differential cross-section for classical electromagnetic scattering from a stationary charged object,

$$\frac{d\sigma}{d\Omega} = \frac{k^2 q_1^2 q_2^2 e^4}{16E^2 \sin^4(\theta/2)}, \tag{5.26}$$

as the reader is invited to show in Exercise 5.

Inelastic Scattering

A much more complicated situation that may occur is one in which the scattering is inelastic. For instance, consider a situation in which the target and projectile are both shattered by the collision. There are now many more possibilities to consider for the final state of the system. Let's assume that there are N particles (still classical objects) in the final state. In principle, the momentum of each of these particles can take any value as long as the total momentum in the system is conserved. To parametrize the system, then, it is no longer enough to use just the angle of the final trajectory; we must also specify the speed. The simplest way to do this is to consider for each particle a small range of final momenta d^3p. In this way, $d\Omega$ is replaced by N factors of d^3p. Since this could lead to a very unwieldy notation, it is typical to continue to use the symbol $\frac{d\sigma}{d\Omega}$ to refer to the differential cross-section, even though it is now nothing more than shorthand for

$$\frac{d\sigma}{d\Omega} \equiv \frac{d\sigma}{d^3p_1 \ldots d^3p_N}. \tag{5.27}$$

Since this differential cross-section must conserve momentum, the one factor that we know it must contain, before even performing any calculations, is a δ-function. That is, we must have

$$\frac{d\sigma}{d\Omega} \propto \delta^3 \left(\mathbf{p}_i - \mathbf{p}_1 - \cdots - \mathbf{p}_N \right), \qquad (5.28)$$

where $\mathbf{p}_i$ is the initial momentum in the system.

We could now, in principle, calculate the likelihood of a collision of this type with particular final momenta by analyzing the mechanics of the specific system we are interested in. We could then find the likelihood of this type of collision occurring at all by integrating over all possible final momenta, to arrive at the cross-section. Notice, though, that this is the cross-section specifically for a collision with N final particles. If we wish to know the probability of *any* kind of collision, we must also sum over all values of N.

We can see, at least in principle, how to calculate a cross-section theoretically, then, but how might it be measured experimentally? Well, assuming that we fire multiple projectiles at the target, and that we know the number of projectiles per unit time, or luminosity, is L, then the total number of events should be $N = L\sigma$. This is what makes the cross-section such a useful quantity. Experiments will vary in luminosity because of design differences, but as long as the luminosity is known, the cross-section can be calculated, and its value is independent of the experimental setup. Similarly, we can find the differential cross-section experimentally by measuring the total number of events with a particular arrangement of final states, $dN/d\Omega$, and then factoring out the luminosity.

Quantum Scattering

We are now ready to consider the quantum-mechanical version of the cross-section. When dealing with elementary particles, the final state particles are distinguishable if they are of different kinds and indistinguishable if they are of the same kind. For this reason, we must separate collision outcomes not just by the number of final-state particles but also by the particular set of particles produced. In this way, we find that there is a separate differential cross-section for each process that may take place. The total cross-section for the interaction of two elementary particles is the sum of the cross-sections for each possible interaction, as determined by conservation laws.

To calculate the cross-section, we imagine that the system is restricted to a finite volume, V, with sides of length L. We fire a beam of particles of type A into a target of particles of type B for a set time T. We are interested in some particular reaction $A + B \rightarrow F_1 + \ldots F_N$, with some particular set of final-state momenta $\mathbf{p}_1, \ldots, \mathbf{p}_N$. However, since the probability of achieving *exactly* these momenta is zero, we again consider a small momentum range d^3p around each chosen value. Since the cross-section, σ, behaves as an area, the number of events of the correct type per unit volume per unit time is given by

$$P = \Phi n_B \sigma, \tag{5.29}$$

where n_B is the number of target particles per unit volume, and Φ is the incident flux, or the number of incident particles passing through a unit area in unit time. So the number of events with the chosen final-state momenta is

$$dP = \Phi n_B d\sigma. \tag{5.30}$$

If we imagine a plane perpendicular to the beam, we can see that the flux is equal to the density of A-type particles multiplied by the length of a section of beam that can pass that plane in unit time. In other words,

$$\Phi = n_A |\mathbf{v}_A|, \tag{5.31}$$

where $\mathbf{v}_A$ is the velocity of the particles in the beam. So we have

$$d\sigma = \frac{dP}{\Phi n_B} = \frac{dP}{n_A n_B |\mathbf{v}_A|}. \tag{5.32}$$

For relativistic particles of energy E, the conventional normalization is for the state to consist of $2E$ particles per unit volume, where E is the particle's energy. This seems like a strange normalization condition at first sight, but it is used with good reason. We cannot use the non-relativistic normalization of one particle per unit volume, since volume is not a Lorentz-invariant quantity. That is, observers in different reference frames will disagree on the volume occupied by the particle. In particular, if one observer measures a volume as V, then a second observer moving relative to the first at a speed v, will see a volume that is Lorentz contracted along the direction of relative motion by a factor $\gamma = 1/\sqrt{1 - v^2}$. A consistent normalization, then, must have a number of particles that also scales as γ. Since the energy scales in the correct way, $2E$ particles per unit volume is a Lorentz-invariant normalization. The factor of 2 is not so well justified and is merely convention. If this normalization condition still feels strange given that we only want to describe one particle, we can look at it from some other points of view. First, a relativistic treatment, as we will see later, necessarily involves antiparticles. So the notion of having only one particle is somewhat flawed anyway, as a particle will be accompanied by a host of virtual particle and antiparticle pairs. Second, we are in fact rarely interested in describing a single particle, since we typically prepare samples as a beam of many similar or identical states. A third way to consider the condition is as a variable volume, rather than a variable particle number. That is, the condition is equivalent to the condition that we have one particle per $1/(2E)$'s-worth of a unit volume. Whichever interpretation you choose to go with, imposing the condition, we find

$$d\sigma = \frac{dP}{2E_A \cdot 2E_B \, |\mathbf{v}_A|}. \tag{5.33}$$

The only thing left to do is to find dP. Since this is related to the probability of a transition from the initial (two-particle) state to the final (many-particle) state, we can write it as

$$dP = \frac{\left| \left\langle \text{final} \left| \widehat{T} \right| \text{initial} \right\rangle \right|^2}{VT} d(\text{Phase}), \tag{5.34}$$

where $\widehat{T}$ is the transfer operator from initial to final state, and d(Phase) is a Lorentz-invariant phase space factor to account for the number of final states with the appropriate momenta.

To calculate d(Phase), notice that, confined as we are to our finite volume V, the allowed momentum states for a final-state particle are those with an i-th momentum component given by $(2\pi/L)\,k_i$, for $k_i \in \mathbb{Z}$. This gives a density of states of $(L/2\pi)^3 = V/(2\pi)^3$, so the number of final states for a particle to scatter into within a small region of phase space around some particular momentum is given by

$$\frac{\mathrm{d}^3 p}{2E(2\pi)^3},\tag{5.35}$$

where $\mathrm{d}^3 p$ is the volume of the small phase space region, and E is the energy associated with this final-state momentum. Again, the $2E$ arises from the conventional normalization for relativistic particles, that there are $2E$ particles in the volume V. The differential cross-section must contain one factor like the previous one for each particle in the final state.

Finally, the transition amplitude $\left\langle \text{final} \left| \widehat{T} \right| \text{initial} \right\rangle$ must necessarily contain a δ-function to constrain the final momenta and ensure overall momentum conservation. For this reason, we write it in the form

$$\left\langle \text{final} \left| \widehat{T} \right| \text{initial} \right\rangle = (2\pi)^4 \mathcal{M} \delta^4 \left(p_A + p_B - \sum_f p_f \right).\tag{5.36}$$

Since this amplitude is squared, we find that one δ-function constrains the momenta of the second so we end up with

$$\left| \left\langle \text{final} \left| \widehat{T} \right| \text{initial} \right\rangle \right|^2 (2\pi)^8 \, |\mathcal{M}|^2 \, \delta^4(0) \delta^4 \left(p_A + p_B - \sum_f p_f \right).$$
$$\tag{5.37}$$

The quantity $\mathcal{M}$ is known as the invariant amplitude, and the evaluation of this quantity is specific to the process under consideration. For this reason, we postpone discussion of the invariant amplitude to later chapters. The $\delta^4(0)$ looks rather worrying, since we would expect it to evaluate to an infinite constant. Fortunately, since we are considering a finite universe of volume V and duration T, the

δ-function evaluates to $VT/(2\pi)^4$. Putting everything together, we find that the time and volume cancel out, and we are left with

$$d\sigma = \frac{|\mathcal{M}|^2}{2E_A \cdot 2E_B \, |\mathbf{v}_A|} (2\pi)^4 \delta^4 \left(p_A + p_B - \sum_f p_f \right)$$
$$\times \prod_f \left(\frac{d^3 p_f}{2E_f (2\pi)^3} \right). \tag{5.38}$$

This is not quite our final expression, however, since we have not yet taken into account all scenarios. First, we must consider a statistical factor. If two particles in the final state are indistinguishable, then the expression in Equation 5.38 will double count each possible arrangement of momenta. For example, the state "electron 1 with momentum p_1 and electron 2 with momentum p_2" is the *same state* as "electron 1 with momentum p_2 and electron 2 with momentum p_1." In this situation, then, we must halve the result of Equation 5.38. Similarly, if there are j indistinguishable particles, there are $j!$ ways to label them, and we will over-count by this factor. So the first amendment we wish to make to the previous expression is to include a statistical factor of

$$S = \prod_{\substack{\text{particle} \\ \text{types}}} \frac{1}{(\# \text{ particles})!}. \tag{5.39}$$

The second amendment we wish to make is to allow for both initial particle types to be in motion (as in a collider). One way to do this is simply to take the relative velocity $|\mathbf{v}_A - \mathbf{v}_B|$. While this is a legitimate solution, it is rather unsatisfying, since it is still not manifestly Lorentz-covariant. A neater solution is to replace the denominator with

$$E_A E_B \, |\mathbf{v}_A - \mathbf{v}_B| = \sqrt{(p_A \cdot p_B)^2 - (m_A m_B)^2}, \tag{5.40}$$

where the proof of this equality is left as an exercise for the reader (Exercise 7).

This gives us a complete and manifestly Lorentz-covariant expression for the differential cross-section:

$$d\sigma = \frac{S\,|\mathcal{M}|^2}{4\sqrt{(p_A \cdot p_B)^2 - (m_A m_B)^2}} \prod_f \left(\frac{d^3 p_f}{2E_f (2\pi)^3} \right) \times$$

$$\times (2\pi)^4 \delta^4 \left(p_A + p_B - \sum_f p_f \right). \tag{5.41}$$

It is worth noting that this expression is spin-polarized. That is, the calculation of $\mathcal{M}$ requires us to assign specific spin orientations or helicities to all initial- and final-state particles. If we do not know these values, then we must sum over the cross-sections for each possibility. More precisely, since the particles in the beam and the target will have a mixture of spins, of which we are unaware, we must average over the possible initial spin states. On the other hand, if the final-state particles are detected but their spins are unmeasured, then all possibilities will count toward the measured cross-section. In this case, then, we must *average* over initial spins but *sum* over final spins.

A final point on the cross-section that we have derived is to emphasize that this quantity is process-dependent. That is, there is a cross-section for, say, $e^+ + e^- \to \mu + +\mu^-$, but another, independent cross-section for $e^+ + e^- \to \gamma + \gamma$. The likelihood of the incident particles interacting *at all*, then, is the sum of the cross-sections for all of the possible individual processes that could occur. As such, we define a total cross-section, $\sigma_T = \sum \sigma$, to be the cross-section for any type of interaction, regardless of the products. Notice that one of the possible outcomes is for the products to be the same as the incident particles, so the total cross-section necessarily includes elastic scattering. Some detectors are specifically built with the aim of measuring total cross-sections. A contemporary example of this approach is the TOTEM (or TOTal cross section, Elastic scattering and diffraction dissociation Measurement) experiment, which shares an interaction point with the Compact Muon Solenoid (CMS) at the LHC. TOTEM is designed to lie close to the beam line a long way from the interaction point, so as to detect anything scattered with small angles from the beam that other detectors would miss. In this way, the experiment aims, among other things, to take precision measurements of the total proton-proton cross-section.

5.3.2 Lifetimes

The second measurable quantity we will consider is the lifetime, or equivalently the decay rate, of unstable particle species. An unstable particle has a finite probability of decaying during any given time interval, and the decay rate, Γ, is the probability of decay in a given unit time, just as in the case of radioactive decay. The lifetime, τ, of a particle species is a measure of the average lifetime of a particle of this type before it decays, and is related to the decay rate by $\tau = 1/\Gamma$. Equivalently, the lifetime is the time it takes for the size of a sample of such particles to decrease by a factor of e. A related quantity is the half-life: the time it would take a similar sample to halve in size. The half-life, $t_{1/2}$ is related to the lifetime by $t_{1/2} = \tau \ln 2$. These quantities may be measured experimentally in one of two ways. First, if the lifetime is sufficiently long for the particle to be observed before decay, and if the decay products are observable, then the decay rate may be measured directly by counting the number of decays per unit time. More commonly in particle physics, however, the lifetime of an unstable particle is so short that the particle is not directly observed before decay. Instead, only the products of its decay are observed. Such a particle is known as a resonance. In this case, the decay rate can be measured experimentally through statistical analysis of the invariant mass for groups of particles. If the same final states appear many times in collisions, and the invariant masses are found to peak around a particular value, this is evidence of a resonance with that mass (Figure 5.18).

One may expect that the invariant mass measurements should form an infinitesimally thin spike around the resonance mass rather than a broad peak, but remember that, since the particle is short-lived, there is considerable uncertainty in its energy. The width of the resonance peak at half maximum height is equal to the decay rate of the resonance (in natural units). This can be seen by looking at the wavefunction for an unstable particle in its rest-frame, which

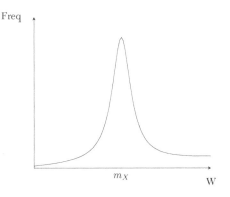

FIGURE 5.18 A typical resonance peak in invariant-mass data.

takes the form:

$$|\psi(t)\rangle = |\psi(0)\rangle \, e^{-(\frac{\Gamma}{2}+im)t}, \qquad (5.42)$$

where Γ is the decay rate and m the particle mass. Note that this is simply a plane-wave type state of energy m and no momentum, which is also exponentially decaying at a rate $\Gamma/2$. The $\Gamma/2$ exponent ensures that the probability of measuring the particle decays at a rate Γ, since $P(t) \propto \langle \psi(t) \, | \, \psi(t) \rangle$.

A Fourier transform to the energy domain gives the Breit-Wigner formula

$$\rho(W) = \frac{\kappa}{(W - m)^2 + \frac{\Gamma^2}{4}}, \qquad (5.43)$$

where $\rho(W)$ is the frequency density of measurements taken at a particular invariant mass W, and κ is a constant. This is in excellent agreement with the measured distributions of invariant mass, with Γ equal to the width of the peak. The Breit-Wigner peak is the characteristic shape to look for in experimental data that may indicate the existence of a previously unknown particle. Of course, its discovery requires many scattering events for a particular set of final state particles over a range of values of the invariant mass. Trial peaks are then fitted against the data and the accuracy of the fit assessed to determine the most likely properties of any potential discoveries. As the fit becomes better with increased data points, the probability that the peak is nothing more than a random statistical fluctuation decreases and it is more likely that it represents a genuine discovery. Potential discoveries are generally announced when the fit is sufficiently accurate that the likelihood of error is only 0.0027. This probability is equivalent to being three standard deviations from the mean in a normal distribution, so such events are said to occur "at three sigma." This is not generally considered strong enough evidence for a discovery, however, which is reserved for events at five sigma. This translates to a probability of only around 5×10^{-7} that the discovery is erroneous and due simply to statistical noise. That the bar is set so high is merely evidence that noise does lead to spurious results from time to time. In fact, a recent example of this arose in the LHC data in late 2015. An excess of two-photon final states was discovered in gluon-gluon interactions, and fitting the resonance peak to the data appeared to point to the existence of a particle with a mass around 750 GeV. This fit was accurate to almost four sigma, suggesting that the likelihood of it being merely a statistical fluctuation was around 1 in 15,000. This apparent resonance was even given a name: the digamma. However, as more data accrued, the ratio of peak to noise diminished and the significance of this fit decreased and vanished. The peak really was just an accident of the data. This goes to show that when the number of events is large enough, even very unlikely artifacts can appear in the noise. In other words, 15,000 isn't such a large number when hundreds of millions of collision events occur each second.

From a theoretical standpoint, the decay rate is similar to the total cross-section, in that it is indiscriminate of the final products of the decay. When calculating a decay, we must consider a single *decay mode* at a time. That is, we must calculate the rate at which the initial unstable particle species will transition to a particular final state. To find the experimentally determined decay rate, as for the total cross-section, we must then sum over the individual modes' partial decay rates: $\Gamma = \sum_k \Gamma_k$, where k is a particular decay mode. The ratio of a mode's individual decay rate to the total decay rate is a measure of the proportion of initial particles that decay via that mode, and is known as the *branching fraction* or *branching ratio*: $BF_k = \Gamma_k/\Gamma$.

How are we to calculate the decay rate for a particular mode, then? Well, one way to approach this problem is to think of a decay as a one-particle "collision." The decay rate is then equivalent to the cross-section for this process. That is, the decay rate will be proportional to the square of the transition amplitude and the available phase space, so these factors take the same form as they did for the cross-section. Similarly, the decay rate will also include a delta-function to ensure momentum conservation, though this will now have only one initial-state particle in it. The only other part of the cross-section formula that we must modify is the pre-factor relating to the initial state. Before rewriting the cross-section formula to make it manifestly Lorentz-covariant, the pre-factor took the form:

$$\frac{1}{2E_A \cdot 2E_B \,|\mathbf{v}_A|}. \tag{5.44}$$

In the case of a decay rate, there is only one particle to consider in its rest frame, and no relative velocity to worry about. So this factor becomes simply $1/2m_A$, where m_A is the mass of the decaying particle. So the partial decay rate is given by

$$\Gamma_k = \frac{S\,|\mathcal{M}|^2}{2m_A} \prod_f \left(\frac{\mathrm{d}^3 p_f}{2E_f (2\pi)^3} \right) \times (2\pi)^4 \delta^4 \left(p_A - \sum_f p_f \right). \tag{5.45}$$

The amplitudes that will allow us to calculate these measurable quantities will be the subject of later chapters.

EXERCISES

1. Consider a classical object of mass m and initial velocity $\mathbf{v}$ colliding with a second stationary classical target of mass m_t. After the collision, the mass m has a velocity $\mathbf{v}_1$ and the target has a velocity $\mathbf{v}_t$.

(a) By considering the conservation of (non-relativistic) energy and momentum, show that

$$\left(1 + \frac{m_t}{m}\right) |\mathbf{v}_t|^2 = 2\mathbf{v} \cdot \mathbf{v}_t.$$

(b) Hence find the maximum speed that may be transferred to the target during the collision and find a necessary condition on the masses for this maximum to be achieved.

2. **(a)** Show that the non-relativistic Bethe formula may be written as $-dE/dx = A\ln(BE)/E$ where A and B are constants.

(b) Hence show that the Bethe formula predicts a maximum stopping power at low energy that is independent of particle type. Find the value of $\beta\gamma$ at which this occurs. You may assume a mean ionization energy equal to that of lead: 823 eV.

(c) Below this maximum, the Bethe formula is no longer valid and the stopping power is found to scale approximately linearly with particle energy. By considering the behavior and interdependence of stopping power, particle energy, and path length, at values of $BE \gg 1$, $BE \sim 1$, and $BE \ll 1$, sketch a rough plot of how the stopping power varies with path length. (The characteristic shape of this plot is known as the Bragg curve.)

(d) Use the (non-relativistic) Bethe formula to estimate the maximum stopping power due to ionization for α particles in lead, given that lead has atomic number 82 and a density of 11.35 g cm^{-2}.

3. Lead (atomic number 82) has a radiation length of 6.37 g cm^{-2} and electrons have a critical energy of around 10 Mev, below which they lose their energy predominantly through ionization losses. A 300 GeV electron enters a calorimeter made of lead and produces a shower.

(a) Assuming that the mean-free path of photons in lead is of the same order as the radiation length, estimate the total final number of particles in the shower.
(b) Estimate also the shower penetration depth given that the density of lead is 11.35 g cm^{-3}.

4. The energy and momentum of a photon are given by $E = p = hc/\lambda$, where λ is the photon wavelength. Assuming that an electron is initially at rest, show that the scattering angle of the photon in Compton scattering is directly related to the transferred energy (Equation 5.9).

5. When a non-relativistic charged particle is scattered from a heavy charged center via a Coulomb interaction, it follows a hyperbolic trajectory as follows.

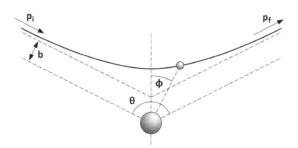

(a) Show that the only net momentum change is in the direction of the line of symmetry of the hyperbola (vertical axis).
(b) Show that the initial and final momenta, $\mathbf{p}_i$, $\mathbf{p}_f$ are related by

$$\frac{|\mathbf{p}_f - \mathbf{p}_i|}{|\mathbf{p}_i|} = 2\sin\left(\frac{\theta}{2}\right),$$

where θ is the scattering angle.

(c) The hyperbola may be parametrized by the variable angle ϕ shown in the figure. By considering the vertical component of the Coulomb force, show that

$$|\mathbf{p}_f - \mathbf{p}_i| = \int d\phi \, \frac{kq_1q_2e^2}{bv} \cos\phi,$$

where q_1e and q_2e are the charges of the particles and k is the Coulomb constant. (Hint: you may find it helpful to consider the angular momentum of the particle about the scattering center at a time in the long-distant past or future.)

(d) Using your results to the previous parts, derive Equation 5.25.

(e) Hence derive the Rutherford scattering formula.

6. What assumptions are made in the above derivation of the Rutherford scattering formula? Why is a departure from the formula expected for high-energy particles if the target is an atomic nucleus?

7. Show that

$$E_A E_B \, |\mathbf{v}_A - \mathbf{v}_B| = \sqrt{(p_A \cdot p_B)^2 - (m_A m_B)^2},$$

as in Equation 5.40.

8. Perform a Fourier transform on Equation 5.42 to arrive at the Breit-Wigner resonance formula (Equation 5.43), and find the value of the constant κ.

PARTICLE CLASSIFICATION

The first step toward understanding the different particles in nature was to classify them. By cataloguing what particles there were, physicists began to understand the relationships between them. There are several different ways to classify particles, and we consider some of these here. We will find that much of our attention will be given to understanding the hadrons, since the composite nature of these particles allows for a very rich structure.

6.1   THE SPIN-STATISTICS THEOREM

Arguably the most important property of a particle is its spin or intrinsic angular momentum. To be clear, by this we do not mean the magnetic spin quantum number m_s, which can take on a number of different values for any given particle, based on the orientation of its spin. Nor, strictly, do we mean the overall magnitude of the particle's spin; instead when we talk of a particle's spin, we generally use the term as a shorthand for "spin quantum number." So a spin-$\frac{1}{2}$ particle has a spin quantum number of $\frac{1}{2}$ and therefore an intrinsic angular momentum (spin in the true sense) of $\sqrt{\frac{1}{2}(\frac{1}{2}+1)} = \sqrt{\frac{3}{4}}$. This number is a fundamental part of the identity of a particle: electrons are spin-$\frac{1}{2}$, and nothing can change that, any more than it could change the electron's mass. The reason that this is of such importance is that it is the spin of a particle that determines which type of statistics it obeys.

Consider a quantum system consisting of two indistinguishable particles. By indistinguishable, we mean that any physical measurement of the system does not change if the two particles are swapped. If the particles are labeled A and B, then the state vector for this system can be written as $|\psi(A, B)\rangle$, and swapping the two particles then gives a state vector $|\psi(B, A)\rangle$. In order to satisfy the requirement that the system is physically identical before and after such an exchange, it may seem obvious that the state vector itself be equal before and after: $|\psi(A, B)\rangle = |\psi(B, A)\rangle$. However, this is in fact too restrictive a condition, since the state vector is not itself a physical quantity. All that we require is that the associated probability density remain the same, or that $\langle\psi(A, B) | \psi(A, B)\rangle = \langle\psi(B, A) | \psi(B, A)\rangle$, so in fact the state vectors can be related by a complex phase $|\psi(B, A)\rangle = e^{i\alpha} |\psi(A, B)\rangle$, where $\alpha \in \mathbb{R}$, since such a phase will cancel out of any physically meaningful quantities. By a similar argument to that used for the eigenstates of parity, we can now restrict the values of this phase. In particular, two successive exchanges of the indistinguishable particles clearly returns the system to its original state. If we let the operator $\widehat{E}$ have the effect of exchanging the two particles, we find

$$\widehat{E}^2 |\psi(A, B)\rangle = e^{2i\alpha} |\psi(A, B)\rangle = |\psi(A, B)\rangle, \qquad (6.1)$$

so $e^{2i\alpha} = 1$, which restricts the value of the phase to ± 1. That is, the state vector is either unchanged under exchange of particles, or it picks up a relative negative sign.[1]

Consider the case that the state vector changes sign under particle exchange. If we construct the state vector out of one-particle states,

---

[1] An interesting exception to this argument occurs when we restrict ourselves to two spatial dimensions. In this case, two successive exchanges of two particles—equivalent to a full rotation of one about the other—does not necessarily return the system to its original state. This allows two-dimensional particles to have any phase under exchange. Such particles are known as anyons.

then it must in this case be antisymmetric:

$$|\psi(A, B)\rangle = |\psi_A\rangle \otimes |\psi_B\rangle - |\psi_B\rangle \otimes |\psi_A\rangle, \qquad (6.2)$$

where $|\psi_A\rangle$, $|\psi_B\rangle$ are the states of particles A and B respectively. If both particles are now placed into the same one-particle state $|\psi_1\rangle$, we find

$$|\psi(A, B)\rangle = |\psi_1\rangle \otimes |\psi_1\rangle - |\psi_1\rangle \otimes |\psi_1\rangle = 0. \qquad (6.3)$$

Since this state is identically zero, it has no probability density associated with it. This demonstrates, then, that such a combination of particles has no likelihood of occurring, or is forbidden. This is the basis of the Pauli Exclusion Principle, which states that no two particles (that change sign under particle exchange) can occupy the same state. Notice that no such restriction applies in the case that the state vector is unchanged under particle exchange, since in this case, the expression for the overall state would be symmetric in one-particle states. This greatly affects the behavior of the particles: those that change sign under exchange obey Fermi-Dirac statistics, and are termed fermions, while those that do not change sign obey Bose-Einstein statistics, and so are bosons.

The spin-statistics theorem is a powerful result that demonstrates a one-to-one correspondence between the spin of a particle and the statistics that it obeys. In particular, it states that all integer-spin particles are bosons, while all particles with half-odd-integer spin are fermions. In the Standard Model, there is one spin-0 boson (the Higgs) and several spin-1 bosons that mediate the three non-gravitational forces. These are the photon for electromagnetism, eight gluons for the strong nuclear force, and the $W^\pm$ and Z^0 for the weak force. The fermions will be explored further in the following sections.

6.2   THE STRONG FORCE

Another top-level distinction to be made between particle species is whether or not the particle interacts through the strong nuclear force. The reason this makes such a key difference to the particle's behavior is primarily due to the concept of confinement. The strong force is responsible for the existence of bound states of strongly interacting particles, and confinement is a property of this force that leads to the complete isolation of such bound particles from the outside world. That is, the internal dynamics of the bound state, along with its constituents, are unobservable from outside the confines of that state at normal energy scales. While the internal structure may be probed with very high-energy experiments, at normal scales the bound state behaves almost as though it were a fundamental particle itself. For this reason, it is necessary to catalog both the fundamental particles and the strongly-bound states. Indeed, it is the existence of a multitude of such composite particles that led to the particle explosion in the 1950s. We thus make a distinction between those fermions that do not interact through the strong force, the leptons, and those that do. The classification of leptons is straightforward, as we have already seen in Chapter 1. But of those strongly interacting particles, we further distinguish those that are fundamental, the quarks, and those composite particles that consist of bound states of quarks, the hadrons.

The sum of angular momenta of the constituents of a hadron contributes to its *internal* angular momentum. Since we see the bound state as a particle in its own right, we see this internal angular momentum as the particle's spin. The hadron as a whole can have, in addition, orbital angular momentum, but both the orbital and spin angular momenta of the constituents contribute to the spin of the hadron. This means that hadrons can, in principle, have any integer or half-integer value of spin. However, since orbital angular momentum is restricted to integer values, the type of statistics obeyed by the hadron is determined by the total spin of the constituent quarks, which in turn depends solely on the number of quarks in the hadron. Ignoring the possibility for now of more complicated bound states (though we will return to this in Section 10.6), we consider only two types. There are the baryons—fermionic states consisting of three quarks—and the mesons—bosonic states comprising a quark and an antiquark.

If we include composite particles in our classification process, then the number of particles we must consider increases dramatically. For this reason, the remainder of this chapter is dedicated to understanding how the hadrons are grouped, as well as the underlying structure behind their classification.

6.2.1 Isospin

When a multitude of new particles was being discovered in the 1950s, physicists needed a way to catalog them. They chose to do this with the introduction of two new quantum numbers: isospin and strangeness. We have already seen in Chapter 1 how strangeness was defined in order to explain the strangely long lifetimes of some particles. Now we must consider isospin. The origin of this quantum number (or more accurately, pair of quantum numbers) is the observation that the proton and neutron have similar masses, and may inter-convert during β decay. The idea was advanced that these two might be simply two states of the same particle: the nucleon. If the nucleon has a property similar to spin with an $SU(2)$ structure and an overall magnitude of $\frac{1}{2}$ for this property, then there would be two "orientations" available to it to account for the existence of the proton and neutron. Since this new spin-like property accounts for the existence of different isotopes, it was termed "isotopic spin," or isospin for short, and given the symbol I.[2] By convention, the orientation is measured in the "z" direction of isospin space, so the proton and neutron are distinguished by the third component of isospin I_3. The proton and neutron are thus assumed to form a doublet under an isospin $SU(2)$ symmetry group. While this idea is not taken literally anymore, the terminology has stuck and is now used to describe all hadrons.

The next important observation for this story was that, just like the proton and neutron, many of the new particles being discovered could also be grouped with others of similar mass. For example, the

---

[2] Strictly, isospin accounts for different *isobars* rather than isotopes, since a change in the isospin of a nucleus changes its atomic number and not its atomic mass number, but let's not split atoms.

Σ baryon masses are $m_{\Sigma^-} \approx 1197$ MeV, $m_{\Sigma^0} \approx 1192$ MeV, and $m_{\Sigma^+} \approx 1189$ MeV. These were grouped into an isospin triplet, with an isospin of $I = 1$. By convention, the member of an isospin multiplet with the highest charge was also assigned the highest value of I_3. So the Σ baryons have $I_3(\Sigma^-) = -1$, $I_3(\Sigma^0) = 0$ and $I_3(\Sigma^+) = +1$, while the proton and neutron have third components of isospin $+\frac{1}{2}$ and $-\frac{1}{2}$ respectively.

6.2.2 Flavor $SU(3)$

If we arrange hadrons according to several properties, an interesting pattern emerges. First, we categorize according to spin and parity. When listing the properties of a hadron, it is customary to denote the spin and parity with a single symbol, in which the parity $(+$ or $-)$ is written as a superscript on the value of the spin. For example, the proton is said to have a spin-parity of $J = \frac{1}{2}^+$. If the particle is also a $\mathbb{C}$ eigenstate, its C eigenvalue is included as a second superscript. For example, the π^0 has $J^{PC} = 0^{-+}$. Within each spin-parity family, we then arrange the hadrons on a graph of strangeness plotted against isospin. The result is that the lightest hadrons form a spin-$\frac{1}{2}$ octet, a spin-$\frac{3}{2}$ decuplet, and two integer-spin nonets that look suspiciously like $SU(3)$ representations (Figures 6.1–6.4). In fact, by rescaling the vertical (strangeness) axis, we can identify I_3 and S as the compatible generators of the symmetry group. More precisely, the T_8 generator is identified as $T_8 = \frac{2}{\sqrt{3}}(S + B)$ where B is the baryon number. The quantity $S + B$, or more generally $S + C + \widetilde{B} + T + B$, is known as the hypercharge, but this is not to be confused with the weak hypercharge, which we will introduce in Chapter 12.

The baryons, then, appear to fall into the **8** and **10** representations of $SU(3)$, but why should this be the case? Gell-Mann's hypothesis was that there is an underlying structure to the hadrons: they are built out of smaller spin-$\frac{1}{2}$ units—quarks—which transform as the fundamental representation. The relevant quantum numbers of these objects are:

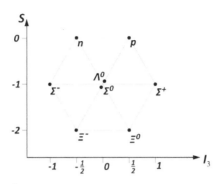

FIGURE 6.1 The spin-$\frac{1}{2}$ baryon octet.

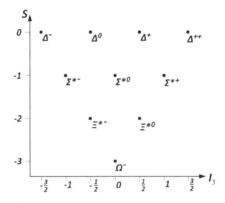

FIGURE 6.2 The spin-$\frac{3}{2}$ baryon decuplet.

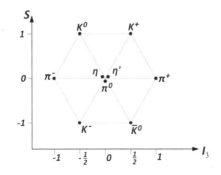

FIGURE 6.3 The spin-0 scalar meson nonet.

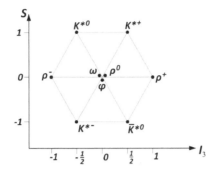

FIGURE 6.4 The spin-1 vector meson nonet.

Quark	I	I_3	S
u	$\frac{1}{2}$	$+\frac{1}{2}$	0
d	$\frac{1}{2}$	$-\frac{1}{2}$	0
s	0	0	-1

So it is the number of u and d quarks that determines isospin, and the presence of strange quarks that determines strangeness. Recall, though, that strangeness was defined by an unusually low decay rate. When we consider weak interactions in detail in Chapter 12, we will see exactly why it is that the presence of a strange quark vastly increases the lifetime of a particle.

As we have seen in Section 4.3.2, a combination of three fundamental representations gives both a decuplet and an octet (along with two additional representations, but we will deal with these later). Similarly, the combination of a fundamental with an anti-fundamental gives an octet and a singlet, or, put another way, a nonet. The proposal, then, was that baryons are formed from three quarks, while a meson is formed from a quark and an antiquark. As an example, the quark content of the spin-$\frac{1}{2}$ octet is shown in Figure 6.5, and a quick calculation or two will show that the quantum numbers appear to sum correctly to give the correct values of I_3 and S for each hadron. However, as we will see in Section 6.4, the details of this process are a little more complicated. Also, in order to account for the charges

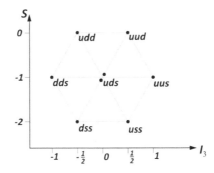

FIGURE 6.5 The spin-$\frac{1}{2}$ octet, showing the quark content of each baryon.

of the hadrons, we find that the d and s quarks have charges of $-\frac{1}{3}$, while the u quark has $+\frac{2}{3}$.

6.3   COLOR

Historically, the need for three quarks per hadron was incorporated into the quark model simply to fit the data. Three quarks was sufficient to explain the known baryons. In hindsight, the underlying reason for three-quark states is the existence of color. Color is an additional degree of freedom for quarks: each quark flavor (up, down, strange, etc.) comes in three varieties, known as colors. These have nothing at all to do with color in the everyday sense but are an arbitrary set of labels to distinguish the three types of quark. This terminology does allow for a useful analogy, however, in that the colors, labeled red, green, and blue, can be considered to mix in a baryon to form a neutral white. Just as an atom, with equal numbers of positive and negative charges, appears neutral at large distance scales, a baryon with all three colors present in a neutral white mixture appears to have no color from the outside. Similarly, antiquarks carry and anticolor and mesons consist of a neutral color-anticolor combination.

This color degree of freedom cannot be directly detected, again due to confinement; so why should it be included on theoretical grounds? The answer is that hadronic bound states without color appear to disobey the Pauli Exclusion Principle. We will see in Section 6.4 that this is true of all the hadrons, since their wavefunctions without color must be symmetric under quark exchange in order to match correctly the observed multiplets. The simple way to see the problem, however, is to look just at the Ω^- baryon in an $S_z = +\frac{3}{2}$ spin state. With this spin and a strangeness of -3, the baryon must consist of three strange quarks whose spins are aligned. In this configuration, each quark is in the same one-particle state, which contradicts the exclusion principle. Since quarks are fermions, the overall wavefunctions must be antisymmetric under exchange of any two quarks. Color was introduced as an additional $SU(3)$ degree of freedom, in which the three quarks of a baryon are completely antisymmetric, to allow for an overall antisymmetric wavefunction for the hadron.

Additional evidence for the existence of three colors comes from the production of hadronic showers in electron-positron collisions via processes such as

$$e^+ + e^- \rightarrow q + \bar{q} \rightarrow \text{hadrons}, \tag{6.4}$$

where q and $\bar{q}$ are a quark-antiquark pair. Since the electron does not participate in strong interactions (and weak interactions are negligible), we can see that the cross-section for q-$\bar{q}$ production is calculated from a lone Feynman diagram:

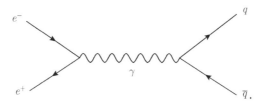

Although we have not yet delved into the full calculation of cross-sections from a theoretical viewpoint, we can see that the above process is essentially identical to μ^+-μ^- production: $e^+ + e^- \rightarrow \mu^+ + \mu^-$. As such the cross-sections should be essentially identical, up to the strength of the coupling constant at the vertices. That is, since the quarks have fractional charges, the hadronic cross-section should be scaled down relative to muon production by a factor of q^2, where q is the fractional charge of the quark in question. Recall that the cross-section is proportional to the square of the invariant amplitude and hence the squared charge. So at low energies, where strange hadron production is possible, but charmed hadron production is not, we would expect to find that the hadronic cross-section is related to the muonic cross-section by

$$\frac{\sigma(e^+ + e^- \to \text{hadrons})}{\sigma(e^+ + e^- \to \mu^+ + \mu^-)} = \frac{(q_u e)^2 + (q_d e)^2 + (q_s e)^2}{(q_\mu e)^2}$$

$$= \frac{(+\frac{2}{3})^2 + (-\frac{1}{3})^2 + (-\frac{1}{3})^2}{(-1)^2} = \frac{2}{3}.$$

(6.5)

Experimentally, however, this ratio is found to be 2, not $\frac{2}{3}$. As energies are increased above the threshold for charm production, this ratio changes both experimentally and theoretically, but again the experimental result is higher than the theoretical prediction by a factor of 3. This trend continues above the threshold for bottom-quark production. So production rates in such processes are consistently higher than theoretical predictions by a factor of 3. However, if the process $e^+ + e^- \to q + \bar{q}$ is seen as a shorthand for three different processes:

$$e^+ + e^- \to q_r + \bar{q}_{\bar{r}},$$
$$e^+ + e^- \to q_g + \bar{q}_{\bar{g}},$$
$$e^+ + e^- \to q_b + \bar{q}_{\bar{b}},$$

(6.6)

each of which can produce the detectable final-state hadrons, then the source of the discrepancy becomes clear.

While color was originally introduced just to clear up the symmetry issue, it is now known to be of fundamental importance to the strong nuclear force. Color is to strong interactions as charge is to electromagnetic interactions. The difference is that, where electric charge is, in a sense, a one-dimensional quantity (positive and negative may be opposites but can still be plotted on the same axis), color is three-dimensional. It comes in red, green, and blue varieties, each of which can be positive or negative (negative red is known as antired). This, then, is the underlying reason that baryons have three quarks. In a colorless system, such as a combination of red, green, and blue quarks, there is no net "color charge" to attract further quarks. The structure of strong interactions will be examined in greater detail in Chapter 10. For now, the only part of this discussion that we must bear in mind for the remainder of this chapter is that quarks are bound in color-antisymmetric combinations.

6.4 BUILDING HADRONS

The state vector for a hadron can be broken down into four components: the flavor state, the color state, the spin state, and the spatial dependence, $|\psi\rangle = |\psi_{\text{flavor}}\rangle |\psi_{\text{color}}\rangle |\psi_{\text{spin}}\rangle |\psi_{\text{space}}\rangle$. Of these, the spatial component has the potential to be the most complicated, involving the spherical harmonic functions that we will consider in Section 6.4.3. However, if we restrict our attention for now to those hadrons with no internal orbital angular momentum, in which all quarks are in the ground state with respect to their spatial dependence, then we can say that the spatial component is symmetric under quark exchange. In order to understand the existence of the different hadrons, it is enough to consider just the symmetry properties of these components, so this allows us to neglect the spatial component. We can further deconstruct the remaining components and write them in terms of combinations of one-particle states for the individual component quarks. For baryons, we have already seen that the color state must be fully antisymmetric. We can write this as

$$|\psi_{\text{color}}\rangle = \frac{1}{\sqrt{6}} \left(|rgb\rangle - |rbg\rangle + |brg\rangle - |bgr\rangle + |gbr\rangle - |grb\rangle \right).$$

(6.7)

We can see from this color wavefunction that hadrons are color *singlets*. That is, they form the **1** representation of color $SU(3)$. It is not enough for a hadron to be "colorless" in the sense of having no net color: they must be an antisymmetric combination of color states. In fact, if we could construct, for example, a meson consisting of just red and antired states, it would be strongly interacting, since it is not a singlet. This is analogous to the non-zero spin of the symmetric state $(|\uparrow\downarrow\rangle + |\downarrow\uparrow\rangle)$ as in the next section. For the flavor and spin components, things aren't quite so straightforward, and we will return to them shortly.

Since a meson is composed of one quark and one anti-quark, quark exchange is not an option. This makes meson state vectors rather more simple to construct than baryon states. The color component must clearly consist of a color and its anti-color. However, strong interactions within the meson mean that this color and anti-color are not fixed. The color component must, therefore, be a symmetric combination of color-anticolor states:

$$|\psi_{\text{color}}\rangle = \frac{1}{\sqrt{3}} \left(|r\bar{r}\rangle + |g\bar{g}\rangle + |b\bar{b}\rangle \right). \tag{6.8}$$

6.4.1   Quark Content

We have two more parts of the wavefunction to consider, and here there is more freedom and more subtlety than in the color part we have already considered. The guiding principle in constructing the spin and flavor parts of the wavefunction will be that the combined result must be fully symmetric under quark exchange. This is because the overall wavefunction must be fully antisymmetric, since quarks are fermions. This antisymmetry is already accounted for by the color part of the wavefunction, and the spatial part is fully symmetric for the hadrons we wish to consider.

Combining Spins

Let us begin by looking at the spin state of quarks in a baryon. We will need to consider a combination of three quarks, but for now, let's consider a simpler system of just two quarks. If a two-particle state $|\psi\rangle$ is constructed out of one-particle states $|\psi_A\rangle$ and $|\psi_B\rangle$ for the individual particles A and B, then the z component of spin for the system is given by

$$S_z = S_z^{(A)} + S_z^{(B)}, \tag{6.9}$$

where $S_z^{(A)}$ and $S_z^{(B)}$ act only on their respective particles. In other words

$$S_z |\psi\rangle = S_z^{(A)} |\psi_A\rangle + S_z^{(B)} |\psi_B\rangle . \tag{6.10}$$

It is straightforward to show that the states $|\uparrow\uparrow\rangle$ and $|\downarrow\downarrow\rangle$ have a z component of spin of $+1$ and -1 respectively. Similarly, the mixed states $|\uparrow\downarrow\rangle$ and $|\downarrow\uparrow\rangle$ are both S_z states with eigenvalue 0.

We can also form the total spin operator for the two-particle system:

$$\mathbf{S}^2 = S_x^2 + S_y^2 + S_z^2, \tag{6.11}$$

and while $|\uparrow\uparrow\rangle$ and $|\downarrow\downarrow\rangle$ are $\mathbf{S}^2$ eigenstates, the $S_z = 0$ mixed-spin states are not. In fact, we find

$$\mathbf{S}^2 |\uparrow\downarrow\rangle = \mathbf{S}^2 |\downarrow\uparrow\rangle = |\uparrow\downarrow\rangle + |\downarrow\uparrow\rangle . \tag{6.12}$$

From this, it quickly follows that the symmetric combination $|\psi_S\rangle = \frac{1}{\sqrt{2}} (|\uparrow\downarrow\rangle + |\downarrow\uparrow\rangle)$ *is* an eigenstate of $\mathbf{S}^2$ with eigenvalue 1. Similarly, the antisymmetric combination $|\psi_A\rangle = \frac{1}{\sqrt{2}} (|\uparrow\downarrow\rangle - |\downarrow\uparrow\rangle)$ is an $\mathbf{S}^2$ state with an overall spin of 0.

This suggests that the symmetric state should form part of a triplet, while the antisymmetric state is a singlet. This is indeed the case: the remaining two states of the triplet are $|\uparrow\uparrow\rangle$ and $|\downarrow\downarrow\rangle$, as can be demonstrated by forming the spin raising and lowering operators S_+, S_- for the two-particle system. We have

$$\begin{aligned} S_+ |\downarrow\downarrow\rangle &= |\psi_S\rangle \\ S_+ |\psi_S\rangle &= |\uparrow\uparrow\rangle \\ S_+ |\uparrow\uparrow\rangle &= 0, \end{aligned} \tag{6.13}$$

with a similar result for S_-.

Combining a third spin works in a similar fashion to produce a total of 10 three-particle spin states:

$$
\left.
\begin{array}{r}
|\uparrow\uparrow\uparrow\rangle \\
\frac{1}{\sqrt{3}}\left(|\uparrow\uparrow\downarrow\rangle + |\uparrow\downarrow\uparrow\rangle + |\downarrow\uparrow\uparrow\rangle\right) \\
\frac{1}{\sqrt{3}}\left(|\uparrow\downarrow\downarrow\rangle + |\downarrow\uparrow\downarrow\rangle + |\downarrow\downarrow\uparrow\rangle\right) \\
|\downarrow\downarrow\downarrow\rangle
\end{array}
\right\} S = \frac{3}{2}
$$

$$
\left.
\begin{array}{r}
\frac{1}{\sqrt{2}}\left(|\uparrow\downarrow\uparrow\rangle - |\downarrow\uparrow\uparrow\rangle\right) \\
\frac{1}{\sqrt{2}}\left(|\uparrow\downarrow\downarrow\rangle - |\downarrow\uparrow\downarrow\rangle\right)
\end{array}
\right\} S = \frac{1}{2}
\tag{6.14}
$$

$$
\left.
\begin{array}{r}
\frac{1}{\sqrt{2}}\left(|\uparrow\uparrow\downarrow\rangle - |\downarrow\uparrow\uparrow\rangle\right) \\
\frac{1}{\sqrt{2}}\left(|\uparrow\downarrow\downarrow\rangle - |\downarrow\downarrow\uparrow\rangle\right)
\end{array}
\right\} S = \frac{1}{2}
$$

$$
\left.
\begin{array}{r}
\frac{1}{\sqrt{2}}\left(|\uparrow\uparrow\downarrow\rangle - |\uparrow\downarrow\uparrow\rangle\right) \\
\frac{1}{\sqrt{2}}\left(|\downarrow\uparrow\downarrow\rangle - |\downarrow\downarrow\uparrow\rangle\right)
\end{array}
\right\} S = \frac{1}{2}.
$$

The spin-$\frac{3}{2}$ quadruplet is fully symmetric in all three spins, while the two spin-$\frac{1}{2}$ doublets are each partially antisymmetric: the first in spins 1 and 2, the second in spins 1 and 3, and the third in spins 2 and 3. In fact, only eight of these 10 states are linearly independent since any one of the doublets may be formed as a linear combination of the other two.

Combining Flavors

Since the combination of spin and flavor wavefunctions must be symmetric overall, flavor states must share the same symmetry properties as the spin states we wish to combine them with. So the spin-$\frac{3}{2}$ states must be fully symmetric in flavor, while the spin-$\frac{1}{2}$ states must be of mixed symmetry. Simply writing down all 27 possible combinations of three quark flavors is not sufficient, however, since the majority have no symmetry. For example, $|uds\rangle$ is neither symmetric nor antisymmetric. So what combinations should we be using? To answer this, notice that, in taking the quarks to be interchangeable, we have made an implicit assumption that the quarks are identical apart from their flavor. That is, we are treating them as if they have the same properties, such as mass and charge. Of course, we know this not to be the case but here the hierarchy of forces comes into play. Since the strong force utterly dwarfs the weak and electromagnetic interactions, and since there is no difference between the quarks in terms of their strong interactions, to a good approximation, we can treat them as the same particle. We say that there is an "approximate $SU(3)$ flavor symmetry." So, just as we constructed our spin states as irreducible $SU(2)$ representations in order that they have well-defined spin quantum numbers, here we must construct irreducible flavor $SU(3)$ representations. These will have well-defined values of the quantum numbers we have chosen as the basis for our flavor $SU(3)$ states: isospin and strangeness.

So we require 27 linearly independent flavor states that form irreducible representations of flavor $SU(3)$. We have already seen in Section 4.3.2 that the relevant representations will arise from $3 \otimes 3 \otimes 3 = 10 \oplus 8 \oplus 8 \oplus 1$. The symmetric **10** consists of

$$|uuu\rangle, |ddd\rangle, |sss\rangle$$

$$\frac{1}{\sqrt{3}}(|uud\rangle + |udu\rangle + |duu\rangle), \quad \frac{1}{\sqrt{3}}(|ddu\rangle + |dud\rangle + |udd\rangle),$$

$$\frac{1}{\sqrt{3}}(|uus\rangle + |usu\rangle + |suu\rangle), \quad \frac{1}{\sqrt{3}}(|ssu\rangle + |sus\rangle + |uss\rangle),$$

$$\frac{1}{\sqrt{3}}(|dds\rangle + |dsd\rangle + |sdd\rangle), \quad \frac{1}{\sqrt{3}}(|ssd\rangle + |sds\rangle + |dss\rangle),$$

$$\frac{1}{\sqrt{6}}(|uds\rangle + |usd\rangle + |dus\rangle + |dsu\rangle + |sud\rangle + |sdu\rangle).$$

$$(6.15)$$

Any one of these may be combined with any of the previous spin-$\frac{3}{2}$ spin states, giving 10 different baryons, each with four possible S_z states. We have successfully constructed the spin-$\frac{3}{2}$ baryon decuplet.

There are eight states that are antisymmetric in quarks 1 and 2:

$$\frac{1}{\sqrt{2}}(|udu\rangle - |duu\rangle), \quad \frac{1}{\sqrt{2}}(|udd\rangle - |dud\rangle),$$

$$\frac{1}{\sqrt{2}}(|usu\rangle - |suu\rangle), \quad \frac{1}{\sqrt{2}}(|dsd\rangle - |sdd\rangle),$$

$$\frac{1}{\sqrt{2}} \left(|uss\rangle - |sus\rangle \right), \quad \frac{1}{\sqrt{2}} \left(|dss\rangle - |sds\rangle \right),$$

$$\frac{1}{\sqrt{4}} \left(|usd\rangle - |sud\rangle + |dsu\rangle - |sdu\rangle \right),$$

$$\frac{1}{\sqrt{12}} \left(|usd\rangle - |sud\rangle + |sdu\rangle - |dsu\rangle - 2\,|uds\rangle + 2\,|dus\rangle \right). \quad (6.16)$$

Notice that the last two of these both have quark content u, d, s. However, they are distinguished by their isospin: the first has $I = 1$, the second, $I = 0$, which can be seen by considering the symmetry properties under exchange of u and d only. This is exactly what we want for the spin-$\frac{1}{2}$ octet.

We can construct similar representations that are symmetric in quarks 1 and 3, and quarks 2 and 3. Finally, there is a fully antisymmetric state

$$\frac{1}{\sqrt{6}} \left(|uds\rangle - |dus\rangle + |dsu\rangle - |sdu\rangle + |sud\rangle - |usd\rangle \right). \quad (6.17)$$

At this point, we appear to have 10+8+8+8+1=35 flavor states. However, as with the spin states, not all of these are linearly independent. Either of the mixed-symmetry octets may be constructed from the other two, reducing the number of independent states to 27 as expected. So which of these octets is the correct one for describing the spin-$\frac{1}{2}$ baryons? The answer, as always, is a linear combination! We must construct a spin-flavor state that is fully symmetric, but we have only mixed-symmetry states available. We therefore combine spin and flavor states with the same symmetry and take the symmetric sum of all three to get the overall state. That is, if $|\psi_{ij}\rangle$ denotes a state antisymmetric in i and j, we take

$$\sqrt{\frac{4}{3}} \left(|\text{spin}_{12}\rangle\,|\text{flavor}_{12}\rangle + |\text{spin}_{13}\rangle\,|\text{flavor}_{13}\rangle + |\text{spin}_{23}\rangle\,|\text{flavor}_{23}\rangle \right).$$
$$(6.18)$$

Notice that the remaining antisymmetric flavor singlet, 1, has no spin state with which it can pair to produce a symmetric state. As such, the flavor singlet is unphysical. We have now reached a point where we can successfully explain the existence of the baryon octet and decuplet. However, we can go further and use the quark model to predict some of the properties of the baryons, including their mass, as we will see in Section 6.4.2. First, however, we should check that we can also explain the meson multiplets.

Building Mesons

The mesons are somewhat easier to construct than the baryons but were postponed until now because there is one aspect of their construction that is unusual, in that some of the physical flavor states will not be uniquely determined. The relative simplicity of meson states is due to the distinguishability of the constituent quark and antiquark. Since we can always tell the two apart, we need not consider symmetric and antisymmetric combinations. Instead, we can definitively assign a particular flavor to a particular constituent. Six of the available flavor states are given by

$$\left|u\overline{d}\right\rangle, \left|d\overline{u}\right\rangle, \left|u\overline{s}\right\rangle, \left|s\overline{u}\right\rangle, \left|d\overline{s}\right\rangle, \left|s\overline{d}\right\rangle, \tag{6.19}$$

directly corresponding to the mesons around the outside of the nonets. The remaining three possibilities would appear to be simply

$$\left|u\overline{u}\right\rangle, \left|d\overline{d}\right\rangle, \left|s\overline{s}\right\rangle. \tag{6.20}$$

However, since any one of these could potentially undergo a transformation into another through annihilation and pair-production, the physical particles may be linear combinations of these basis states. Exactly which combinations correspond to physical states, though, must be determined experimentally, by observing the relative frequency of decay modes. Isospin $SU(2)$ appears to be a good symmetry in both the spin-0 and spin-1 systems, with both the neutral pion and neutral ρ forming the $I_3 = 0$ component of an isospin triplet: $\frac{1}{\sqrt{2}}\left(\left|u\overline{u}\right\rangle - \left|d\overline{d}\right\rangle\right)$. In the case of the scalar mesons, experiment suggests that flavor $SU(3)$ is a reasonably good symmetry. The η' is, at least approximately, an $SU(3)$ flavor singlet $\frac{1}{\sqrt{3}}\left(\left|u\overline{u}\right\rangle + \left|d\overline{d}\right\rangle + \left|s\overline{s}\right\rangle\right)$ and the η the remaining linearly independent combination $\frac{1}{\sqrt{6}}\left(\left|u\overline{u}\right\rangle + \left|d\overline{d}\right\rangle - 2\left|s\overline{s}\right\rangle\right)$. In fact, since $SU(3)$ flavor is only an

FIGURE 6.6 The decay of a Δ^+ baryon at both hadron and quark levels.

approximation, the actual η and η' states are a mixture of the two idealized states given here, with a mixing angle of around 11.5°. On the other hand, in the case of the vector mesons, it appears that the strange quark does not play nicely with the other quarks, instead producing a physical state on its own with only minimal mixing. So the physical states in this case are (approximately) $\rho = \frac{1}{\sqrt{2}}\left(|u\bar{u}\rangle - |d\bar{d}\rangle\right)$, $\omega = \frac{1}{\sqrt{2}}\left(|u\bar{u}\rangle + |d\bar{d}\rangle\right)$ and $\phi = |s\bar{s}\rangle$. The values in these combinations tell us that, if we could repeatedly pull apart η mesons for example, then in six goes, we would find on average one $u\bar{u}$ pair, one $d\bar{d}$ pair, and four $s\bar{s}$ pairs. An important point about these particular combinations is that they are orthogonal: we can't make any one of them out of a linear combination of the other two. If we could, they would not be independent particles and would not have well-defined masses.

Quark Diagrams

With the quark model in place, we can begin to see that Feynman diagrams at the hadron level can be translated to diagrams at the quark level. For example, the decay of a Δ^+ baryon can be represented equally well at hadron and quark levels, as shown in Figure 6.6. Similarly, we also find that the weak interaction really works at the quark level, as shown in Figure 6.7.

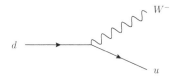

FIGURE 6.7 A weak interaction at the quark level.

6.4.2 Mass

A naive approach to predicting the hadron masses may be simply to add the masses of the constituent quarks. The problem with this approach is immediately obvious when we consider that the Λ and Σ^0 baryons have the same quark content, yet different masses. Clearly, there is something else contributing to the mass of the hadron. As we will see later in Section 10.5.3, an accurate approach to finding this something is a difficult problem that we will not be able to give a full solution to in this text. However, as a first approximation, we can account for the difference in mass by considering the spin coupling of the quarks. This approach is known as the static quark model. As discussed in Section 3.7, in the presence of a magnetic field, otherwise degenerate energy levels may split through the Zeeman effect. In atomic spectra, such splitting is caused by the magnetic field generated by the magnetic moment of the nucleus and gives rise to the fine structure in spectral lines. In the case of hadrons, the interaction of each pair of quarks will have an effect on the energy and thereby on the mass of the hadron. Since the energy of a magnetic dipole in an external magnetic field is given by $-\mu_i B_i$, the energy associated with a pair of quarks due to their mutual magnetic interactions should be of the form

$$E = A \frac{S_{1i} S_{2i}}{m_1 m_2}, \tag{6.21}$$

where S_1, S_2 are the quark spins, m_1, m_2 are the quark masses, and A is a constant to be determined from experimental data. This gives a formula for meson masses of the form

$$m_{\text{meson}} = m_1 + m_2 + A_{\text{meson}} \frac{S_{1i} S_{2i}}{m_1 m_2}, \tag{6.22}$$

and a similar expression for baryon masses:

$$m_{\text{baryon}} = m_1 + m_2 + m_3 + A_{\text{baryon}} \left(\frac{S_{1i} S_{2i}}{m_1 m_2} + \frac{S_{1i} S_{3i}}{m_1 m_3} + \frac{S_{2i} S_{3i}}{m_2 m_3} \right). \tag{6.23}$$

The values of the quark masses and constants A_{baryon}, A_{meson} may then be calculated by finding the best fit to the observed hadrons. This is left to the reader in Exercises 2 and 3.

This model is reasonably successful at predicting (or rather post-dicting) hadron masses, but it does seem to suggest that the quark masses are suspiciously high. In fact, this highlights a limitation of the model that we have employed, and we must be careful to make a distinction between two types of quark mass. The "effective mass" of a quark or "constituent quark mass" is the apparent mass according to this model, while the "current masses" (so called because of their role in the conserved currents that describe Standard Model interactions) are around 200 times smaller. The current masses, then, are not by themselves capable of explaining the much larger masses of hadrons. The additional mass in fact comes from the complicated internal dynamics of the hadron, with a sea of gluons and quark-antiquark pairs making up more than 98% of the total energy of the system. The real problem is that, in assuming that the quarks are static, we have implicitly assumed also that the system is non-relativistic. Since it is the combination of relativity with quantum mechanics that leads to the appearance of antiparticles, we have neglected all of this activity. The appearance of an effective quark mass in the static model is due to us carelessly attributing roughly one third of this additional binding energy to the mass-energy of each quark. We refer to the quark plus this lump of extra energy as a "constituent quark." Despite these limitations, however, there are at least two reasons why this is still a useful model. First, despite its flaws, the model still predicts hadron masses reasonably well. Second, in calculating the constituent quark masses, we now have an expression for the effective magnetic moment of each quark flavor, from which we can calculate the magnetic moments of hadrons. For example, with no internal orbital angular momentum, the magnetic moment of the proton is the sum of the magnetic moments of its constituent quarks. While this sounds simple, it must be borne in mind that the magnetic moment of each quark depends on that quark's relative spin orientation. As such, the symmetry properties of the proton's wavefunction, as discussed in previous sections, must be taken into account when performing the calculation. The result, however, is in reasonably good agreement with the experimentally determined value.

6.4.3 Angular Momentum, Parity, and Charge Parity

Angular Momentum

All of the mesons and baryons that we have considered so far have had a spin determined by the relative orientations of the spins of their constituents, since the orbital angular momentum has been assumed to be zero. This assumption only allows meson spins of 0 or 1, and baryon spins of $\frac{1}{2}$ or $\frac{3}{2}$, since these are the only possibilities when adding quark spins. In reality, many other hadron spins are realized, since the quarks may also have non-zero orbital momentum. Let's consider this possibility in the case of mesons. In a meson, with two independent spins, the only possibilities for the total spin are $S = 0$ and $S = 1$. For $L = 0$ (orbital angular momentum), these are also the values of J (total angular momentum), giving the scalar and vector mesons that we are already familiar with. For $L = 1$, we have the possibilities $J = 0, 1, 2$. However, we must be careful to distinguish two distinct states with $J = 1$. Specifically, there is the state in which $S = 0$ and $L = 1$, and there is the state in which $S = L = 1$ but in which the orientation of these angular momenta is such that their sum is $J = 1$. To put this another way, the $S = 0$ state is a singlet, while the $S = 1$ states form a triplet with J taking the values $J = 0, 1, 2$. This pattern then continues with higher values of L. In each case, there is a singlet with $S = 0$ and $J = L$, as well as a triplet with $S = 1$ and $J = L - 1, L, L + 1$. This information is most commonly summarized in spectroscopic notation

$$^{2S+1}L_J, \tag{6.24}$$

where $2S + 1$ is the multiplicity of the state, and L is conventionally represented by its spectroscopic letter rather than its numerical value ($S, P, D, F, G, \ldots$ standing for $L = 0, 1, 2, 3, 4, \ldots$). The meson states, then, are given by

$$^1S_0, {}^3S_1, {}^1P_1, {}^3P_{0,1,2}, {}^1D_1, {}^3D_{1,2,3}, \text{etc.} \tag{6.25}$$

In the case of baryons, there are now two independent contributions to the orbital angular momentum: one component, L_p, that measures the orbital momentum of a pair of quarks about their center of mass, and a second, L_r, that measures the orbital momentum of this pair's barycenter and the remaining quark about the overall center of mass. According to quantum mechanics, these angular momenta may combine in various ways. However, the possible values of L are still the integers. The possible values of S now are $\frac{1}{2}+\frac{1}{2}+\frac{1}{2} = \frac{3}{2}$ and $\frac{1}{2} + \frac{1}{2} - \frac{1}{2} = \frac{1}{2}$. Together, these give the following possibilities for baryons:

$$^2S_{\frac{1}{2}}, \, {}^4S_{\frac{3}{2}}, \, {}^2P_{\frac{1}{2},\frac{3}{2}}, \, {}^4P_{\frac{1}{2},\frac{3}{2},\frac{5}{2}}, \, {}^2D_{\frac{3}{2},\frac{5}{2}}, \, {}^4D_{\frac{1}{2},\frac{3}{2},\frac{5}{2},\frac{7}{2}}, \, {}^2F_{\frac{5}{2},\frac{7}{2}}, \, {}^4F_{\frac{3}{2},\frac{5}{2},\frac{7}{2},\frac{9}{2}} \, \cdots$$

$$(6.26)$$

Parity

The parity of all the lowest-mass mesons is -1. This simple statement of experimental fact is straightforward to deduce on theoretical grounds by looking at the constituent quarks, since we know that a quark-antiquark pair has a combined parity of -1. As stated in Section 4.2.2, the convention is to take the quark parity to be $+1$ and the antiquark parity to be -1. However, this is not necessary to determine the parity of the meson, since the alternative convention would also lead to the same result. Now consider, though, a meson with an orbital angular momentum ℓ. Here, the effect of a parity transformation on the entire system must be taken into account. It is not difficult to see that a parity transformation applied to two classical objects orbiting each other alters the system, with the transformed system appearing to rotate in the opposite direction from the original. This is found to be the case in the quantum system as well. While this argument will hopefully help to convince the reader that orbital momentum should play a role in determining parity, it is not intended to be taken too literally, since we must remember that it is the sign of the *wavefunction* under parity that matters. To proceed, then, we must consider the wavefunction of two quantum objects in mutual orbit. The solution of the Schrödinger equation in a spherically symmetric potential requires first the solution of Laplace's equation ($\nabla^2\Psi = 0$)

in spherical polar coordinates. The solutions may be expressed as the product of a radial part and an angular part. The radial part leads to the existence of a principal quantum number analogous to that in atomic orbitals, which labels the radial eigenfunction. We may neglect this part for our current purposes, however, since it is invariant under parity. The eigenfunctions of the angular part of the equation take the form of the spherical harmonics

$$Y_\ell^m(\theta, \phi) = K_\ell^m \times P_\ell^m(\cos\theta)e^{im\phi}, \tag{6.27}$$

where θ and ϕ are the polar and azimuthal angles and K_ℓ^m is a constant whose value is unimportant for the present discussion. P_ℓ^m is an associated Legendre polynomial, given by

$$P_\ell^m(x) = \frac{(-1)^m}{2^\ell \ell!}(1 - x^2)^{m/2}\frac{\mathrm{d}^{\ell+m}}{\mathrm{d}x^{\ell+m}}(x^2 - 1)^\ell, \tag{6.28}$$

where the labels ℓ and m are, respectively, the angular momentum and magnetic quantum numbers.

Consider, then, the effect of a parity inversion on the spherical harmonics. First, the spherical coordinates themselves transform as $\theta \mapsto \pi - \theta$ and $\phi \mapsto \phi + \pi$, as can be seen by considering the coordinates of an arbitrary point on the unit sphere. This implies, then, that $(\cos\theta)$ transforms to $(-\cos\theta)$ under parity. In turn, since the only "unsquared" appearance of $\cos\theta$ in the associated Legendre polynomial is in the derivative, we find that $P_\ell^m(\cos\theta) \mapsto (-1)^{\ell+m}P_\ell^m(\cos\theta)$. The exponential part of the spherical harmonic, on the other hand, clearly transforms according to

$$e^{im\phi} \mapsto e^{im(\phi+\pi)} = (-1)^m e^{im\phi}. \tag{6.29}$$

Overall, this gives the spherical harmonics a parity transformation of

$$Y_\ell^m(\theta, \phi) = (-1)^\ell Y_\ell^m(\theta, \phi). \tag{6.30}$$

From this, we can see that the behavior of the wavefunction under a parity transformation is determined by the magnitude of its orbital angular momentum (ℓ), but not by its orientation (m). So mesons with non-zero orbital angular momentum have an intrinsic parity of

$(-1)^{\ell+1}$, where the additional negative sign comes from the intrinsic parity of the quark and antiquark.

For baryons, we again have two angular momenta to consider, L_p and L_r. The quantum mechanical description of the system, then, will consist of a wavefunction that is formed from a product of two spherical harmonics, with independent values of ℓ. Under parity, each of these spherical harmonics will transform in the way discussed above, so we find that the parity of the baryon is given by

$$P_B = (-1)^{L_p+L_r}, \tag{6.31}$$

where there are three implicit factors of $+1$ to account for the intrinsic parities of the quarks. Unlike the case of mesons, in which the convention used to assign parities to fermions made no difference to the hadron parity, here we find that using the opposite convention would lead to $P_B = (-1)^3(-1)^{L_p+L_r} = -(-1)^{L_p+L_r}$. This is unsurprising, however, since baryons have distinct antiparticles, unlike mesons. So we would expect to have to impose our convention either at the quark level or the baryon level, and the change demonstrated here is simply a reflection of the consistency of the convention at both levels.

Charge Conjugation

We have already seen that the majority of particle species are not eigenstates of the charge conjugation operator. One group of particles that *are* eigenstates are those mesons that consist of a quark and its own antiquark. This is because the charge conjugation operator, $\mathbb{C}$, swaps each of the constituent particles for the other particle in the meson. To see what effect $\mathbb{C}$ has on a meson, there are three factors that we must consider. First, if the position of the antiquark relative to the quark is given by a position vector $\mathbf{r}$, then applying $\mathbb{C}$ effectively exchanges quark and antiquark, such that $\mathbf{r}$ is replaced by $-\mathbf{r}$. In other words, the effect that $\mathbb{C}$ has on the spatial part of the meson wavefunction is the same as the parity operator, giving a factor of $(-1)^L$, as before. Second, we have already seen that a symmetric spin state has $S = 1$, while an antisymmetric state has $S = 0$. So the spin part of the wavefunction will undergo a sign change if the sum of the quark spins is 0, but will be unchanged if the sum is 1. Finally, since the quark and antiquark are fermions, there is an overall negative sign that arises from their exchange. Putting this together, we find that the C-parity of a $\mathbb{C}$ eigenstate is given by $(-1)^{L+S}$. This expression also generalizes to the case of bound states of other particle-antiparticle pairs. In the case that the particles involved are bosons, the expression simplifies somewhat, since the sum of spins must necessarily be even in this case, and the overall negative for fermion exchange is not needed. In this case, then, the expression reduces to $(-1)^L$.

6.4.4 Larger Flavor Symmetries

The $SU(2)$ isospin symmetry, while not exact, is a good approximation. This is because the difference in masses between the u and d quarks is sufficiently small that the two look essentially identical as far as the strong force is concerned. It is true that they also differ in charge, but since the electromagnetic force is so weak compared with the strong force, this too has a negligible impact. Widening the group of "identical" quarks, we have extended the isospin symmetry to flavor $SU(3)$. This is not as good an approximation as isospin, as is evident in the larger differences in masses between the strange hadrons and those with zero strangeness. However, the symmetry is still a sufficiently good approximation that it is useful to arrange hadrons into $SU(3)$ multiplets. Again, the reliability of the approximation is due to the reasonably similar masses of the u, d, and s quarks.

Can we extend the flavor symmetry further, and include the charm quark as part of an $SU(4)$ symmetry? The answer is, unsatisfyingly, "yes and no." The charm quark mass is considerably higher than those of the three light quarks and, as such, the approximation begins to break down somewhat. However, at high energy, when the masses of all four of these quark flavors are negligible, the symmetry is restored. As for the three light quarks, we can plot the hadrons constructed from the four lightest quarks as representations of the relevant groups by choosing appropriate axes. The axes in this case are isospin and strangeness as before, with the additional quantum number, charm. The plots are given in Figure 6.8.

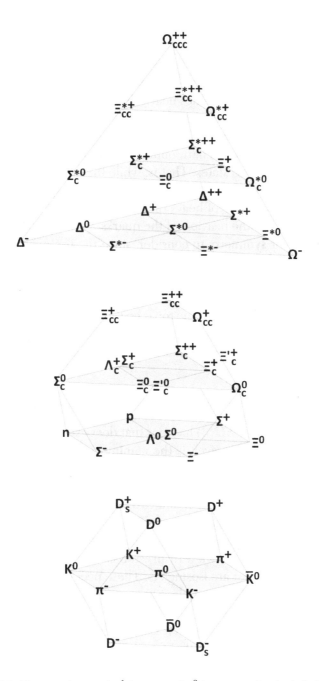

FIGURE 6.8 The ground-state spin-$\frac{1}{2}$ baryons, spin-$\frac{3}{2}$ baryons, and scalar (spin-0) mesons as $SU(4)$ representations. For clarity, only the π^0 is shown in place of four neutral mesons: $\pi^0, \eta, \eta', J/\Psi$.

Going even further, we could attempt to plot hadrons involving the bottom quark as $SU(5)$ representations (although the plots would become four dimensional!). However, by this point, the idea of a symmetry between the quarks is virtually meaningless, since the b quark mass is over 2,000 times heavier than the u! In fact, while the light quarks form mesons that are best described as superpositions of flavor states, the charm and bottom quarks form mesons of well-defined flavor known as charmonium and bottomonium (or collectively as quarkonium) states. That is, while the π^0, η, and η' mesons are composed of superpositions of the light quark flavors, the ground-state mesons consisting of heavy quarks are simply $|c\bar{c}\rangle$ and $|b\bar{b}\rangle$. By the time we reach the heaviest of the quarks—the top—there is certainly no flavor symmetry to consider, since the top quark does not even participate in hadron formation. The reason for this is the top's ludicrously large mass of 172 GeV, which is over 70,000 times the up quark mass, and almost 200 times heavier than even the *proton*. This is so high that the lifetime of the top is on the order of 10^{-25} s, around 100 times shorter than the typical time-scale of strong interactions: the top simply does not have time to form hadrons.

Naming Hadrons

One aspect of the $SU(3)$ hadrons that does extend to the hadrons containing c and b, however, is the naming conventions. The name of a baryon is determined by the number of u and d quarks it contains, with p and n the exceptions. Δ particles contain three u and d quarks, while Σ baryons contain two, Ξ contain one, and Ω none. The identity of the non-u or d quarks is assumed to be s unless shown otherwise with a subscript. Where spins of $\frac{1}{2}$ and $\frac{3}{2}$ are possible, the higher-spin baryon is shown with an asterisk (*). Finally, there are the spin-$\frac{1}{2}$ Λ baryons that are isosinglets. These necessarily contain u, d, and one other quark, whose identity is given as a subscript. Notice, however, that just as in the case of Σ^0 and Λ, with identical quark content but distinguished by their isospin multiplicity, there are similar doublings of states in systems such as usc. To distinguish these states, a prime is used. So for example, the Ξ_c and Ξ'_c both contain usc quarks, but the difference between them lies in exactly which parts of the wavefunction are symmetric or antisymmetric. These naming conventions are already in place despite the fact that several of these hadrons are predicted to be very massive and have not yet been observed in experiment. By itself, this system only allows us to name the ground state of a quark system: it does not allow us to distinguish between particles with the same quark content but different orbital momenta, for instance. For this reason, resonances are additionally labeled with their mass in brackets. For example, the full name of the negative Delta baryon is $\Delta^-(1232)$.

There is also a similar set of conventions used for the mesons. The class of meson is determined by the highest-mass quark present, with B mesons containing either b or $\bar{b}$, D containing c or $\bar{c}$, and K containing s or $\bar{s}$. If the remaining quark/antiquark has $I = 0$ (i.e., it is anything other than u, $\bar{u}$, d, or $\bar{d}$), it is labeled as a subscript. Finally, the charge is labeled as a superscript. If the meson has zero charge and contains one $I = \frac{1}{2}$-quark, then the particle and antiparticle are distinguished with a bar. Specifically, the meson with negative S, C, or $\tilde{B}$ is barred. The naming of neutral mesons, such as $s\bar{s}$, is less systematic. As with baryons, the spin-1 mesons are named exactly as their spin-0 equivalents, and distinguished with an asterisk (with the exception of the ρ mesons, which are essentially spin-1 pions).

6.4.5 Resonances

We have briefly mentioned those hadrons whose internal orbital angular momenta are non-zero, but we have not said a great deal about them. Any such hadron can be thought of as an excited state of one of the lighter hadrons. They have the same quark content, but differ in the spatial part of their wavefunction. As such, the energy (and hence mass) of such states tends to be considerably higher than the light hadrons. This places them, along with the spin-$\frac{3}{2}$ decuplet baryons and the vector (spin-1) meson nonet, in the class of *resonances*. As discussed in Section 5.3.2, a resonance is a particle whose mass is sufficiently high that its lifetime is too short for direct detection. In the case of hadronic resonances, the high transition rate for strong interactions puts the typical lifetime around 10^{-23} s. The existence of such particles is thus indirectly inferred through the appearance of resonance peaks in invariant-mass data.

Many hadronic resonances have been detected, and have even had their properties measured. The reader may find comprehensive lists of such resonances and their properties in the pages of the *Review of Particle Physics*.

EXERCISES

1. The χ_{b2} resonance is a bottomonium state with principal quantum number $n = 2$ and $^3P_2$. What are the spin, parity, and charge-parity of this state?

2. **(a)** Suppose we wish to find a value for the spin-interaction parameter, A_{meson} (Equation 6.22), making no assumptions about the constituent quark masses. If we are considering only those mesons in which the quarks have no orbital angular momentum, find a minimal set of mesons required for the calculation.
   (b) If we now assume that isospin $SU(2)$ symmetry is exact but make no further assumptions, show that the π^+ and ρ^+ mesons are sufficient to calculate A_{meson}.
   (c) By squaring the combination $\mathbf{S}_1 + \mathbf{S}_2$, show that $S_{1i}S_{2i}$ is $+1/4$ for the ρ^+ and $-3/4$ for the π^+.
   (d) Hence find an expression for the u and d masses, along with an expression for A_{meson}.
   (e) Find the numerical values of these expressions given that $m_\pi = 139.57$ MeV and $m_\rho = 775.4$ MeV.
   (f) Use your answers to the last part, along with the fact that $m_{\Delta^{++}} = 1232$ MeV, to find an estimate for A_{baryon} (Equation 6.23).

3. Use the constituent quark masses given in Appendix A to find a better estimate of A_{meson}.

4. **(a)** Find the spin, flavor, and color parts of the wavefunction for the Σ^+ baryon when its overall spin orientation is $+1/2$. Hence write down the combined wavefunction (ignoring the spatial part).

5. If flavor $SU(4)$ were an exact symmetry, outline how the group structure would work out to produce the observed $SU(4)$ multiplets given in Figure 6.8.

6. Color is an exact $SU(3)$ symmetry, with the three color of a quark transforming as the 3 of $SU(3)$. Represent this fact on an $SU(3)$ weight diagram and show graphically why a combination of all colors appears neutral. How can you also represent graphically that a quark-antiquark pair is color neutral?

7. Why isn't $e^+ + e^- \to e^+ + e^-$ used in the denominator of Equation 6.5?

8. Draw quark-level Feynman diagrams for the β-decay of a neutron and for the decay $\pi^0 \to \gamma + \gamma$.

RELATIVISTIC QUANTUM MECHANICS

As previously stated, particle physics is really the combination of quantum mechanics with relativity. As such, it is no longer appropriate to use the Schrödinger equation, since this is constructed from the non-relativistic energy momentum relation $E = p^2/2m$. Instead, we must look for a similar equation based on the relativistic relation $E^2 = p^2 + m^2$. As we will see, a direct attempt to do this will lead us to some problems that will require some restructuring of our previous understanding of quantum mechanics. In the process, it will lead us to the inevitability of antimatter in any relativistic quantum theory. We will also find that there is more than one approach to finding a relativistic quantum wave equation, and each approach describes particles of different spin. In this chapter, we will consider the equations for particles of spin 0 and spin 1.

7.1   THE KLEIN-GORDON EQUATION

The first attempt to construct a relativistic quantum theory was performed by Klein and Gordon in 1926. The equation that they arrived at, however, was not new, as it was originally written down by Schrödinger. Schrödinger discarded this equation, in favor of what is now the Schrödinger equation, partly because of the interpretational issues with the Klein-Gordon equation that we will consider shortly, and partly because it does not predict spin. Since Schrödinger was predominantly interested in describing spin-$\frac{1}{2}$ particles in an effort to explain emission spectra, he moved away from the relativistic equation.

7.1.1 A Relativistic Schrödinger Equation

Since the energy and momentum operators are given by $\widehat{E} = i\partial_0$ and $\widehat{\mathbf{p}} = -i\boldsymbol{\nabla}$, we can combine them into an energy-momentum four-vector operator $\widehat{p}_\mu = i\partial_\mu$, which we shall simply call the momentum operator from now on. In four-vector notation, the relativistic energy-momentum relation takes the simple form:

$$p^2 = m^2. \tag{7.1}$$

Following the same steps taken when deriving the non-relativistic Schrödinger equation, we promote the momentum in the relativistic energy-momentum relationship to the momentum operator, and introduce a wavefunction for the operators to act upon:

$$\widehat{p}_\mu \widehat{p}^\mu \phi = m^2 \phi$$
$$(i\partial_\mu)(i\partial^\mu)\phi = m^2 \phi \tag{7.2}$$
$$\left(\partial^2 + m^2\right)\phi = 0.$$

This is the Klein-Gordon equation. Notice that ϕ is a scalar, since it has no Lorentz index. This is consistent with it describing spin-0 particles, since we expect a spin-s particle to have $2s + 1$ degrees of freedom. Now the keen-eyed reader may at this point complain that the wavefunction was only a scalar by choice. It is true that we could equally have chosen a four-vector valued wavefunction ϕ^μ, in which case there would be four degrees of freedom. However, since there is no interaction between the degrees of freedom in this case, this would really describe four independent spin-0 particles, rather than one particle of higher spin. As we will see in Section 7.12, the situation is different when we write down a vector-valued wave equation in which different components are related.

7.1.2 Solutions of the Klein-Gordon Equation

To find solutions of the Klein-Gordon equation, it is enough to consider the same plane-wave solutions that were used earlier in deriving the Schrödinger equation:

$$\phi = \phi_0 e^{-ik \cdot x}. \tag{7.3}$$

Since this is exactly the same form we used for solutions to the Schrödinger equation, we know that it will be an eigenstate of the momentum operator with eigenvalue k^μ. As such, we are justified in calling the wave-vector p rather than k. The plane-wave is easily shown by direct substitution to be a solution to the equation as long as the particle has mass m:

$$\left(\partial_\mu \partial^\mu + m^2\right) \phi_0 e^{-ip_\nu x^\nu} = 0$$

$$\left((-ip_\mu)(-ip^\mu) + m^2\right) \phi_0 e^{-ip_\nu x^\nu} = 0 \tag{7.4}$$

$$\implies (p^2 - m^2)\phi = 0.$$

Since the Klein-Gordon equation is a linear PDE, any linear combination of solutions is itself a solution. Therefore, the plane waves form a basis for all solutions, since they are complete and orthogonal.

Looking at these solutions a little more carefully, however, a problem soon becomes apparent. Rather than a constraint on the mass of the particle, consider the necessary relation $p^2 - m^2 = 0$ as a condition on the energy. Then the plane wave is a solution if and only if $E = \pm\sqrt{\mathbf{p}^2 + m^2}$. This suggests that there are solutions to the Klein-Gordon equation with negative energy. Such solutions would appear to be nonsensical, but we cannot simply ignore them: as a general rule in quantum mechanics, the contribution of all possible solutions must be included in order to get the correct solutions to calculations. This means that the negative-energy solutions must be considered valid. We will come back to the correct interpretation of these solutions shortly. For now, we will persevere and attempt to find a probability density current for these particles.

7.1.3 Conserved Current

The conserved current for the Klein-Gordon equation can be found just as for the Schrödinger equation. The Klein-Gordon equation is multiplied by the complex conjugate of ϕ, while the conjugate equation is multiplied by ϕ. One is then subtracted from the other.

$$\phi^* \left(\partial^2 + m^2 \right) \phi = 0$$
$$\text{and} \quad \phi \left(\partial^2 + m^2 \right) \phi^* = 0$$
$$\text{subtracting:} \quad \phi^* \partial^2 \phi - \phi \partial^2 \phi^* = 0 \tag{7.5}$$
$$\implies \partial_\mu j^\mu = 0$$
$$\text{where} \quad j^\mu = i \left(\phi^* \partial^\mu \phi - \phi \partial^\mu \phi^* \right).$$

We have found a conserved quantity j^0, and an associated conserved current j^μ. In the non-relativistic case, the factor of i was included in j_i so that the quantity $|\psi|^2$ was real valued and so could be interpreted as a probability density. The factor of i here is likewise included to ensure that the conserved quantity j^0 is real, but the interpretation of j^0 as a probability is no longer viable. While the three space-like components of this new current have a very similar form to the non-relativistic case, the time-like component is very different. For the Schrödinger equation, it was given by $\phi^* \phi$, which is positive-definite. However, in the relativistic case, the time-like component is given by

$$j^0 = i \left(\phi^* \partial^0 \phi - \phi \partial^0 \phi^* \right). \tag{7.6}$$

The relative negative sign between the two terms in j^0 means that no choice of pre-factor can guarantee a positive value for all solutions ϕ. Clearly, a negative probability is meaningless, so we cannot interpret j^μ as a probability current. Things are looking bleak for the Klein-Gordon equation, but there is a way out of both of the apparent problems that have arisen. First, notice that the two problems are in fact related, since

$$j^\mu = i \left(\phi^* \partial^\mu \phi - \phi \partial^\mu \phi^* \right)$$

$$= i \left(A^* e^{i p_\nu x^\nu} \partial^\mu A e^{-i p_\sigma x^\sigma} - A e^{-i p_\sigma x^\sigma} \partial^\mu A^* e^{i p_\nu x^\nu} \right)$$

$$= i \left(|A|^2 e^{i p_\nu x^\nu} \left(-i p^\mu \right) e^{-i p^\sigma x^\sigma} - |A|^2 e^{-i p_\sigma x^\sigma} \left(i p^\mu \right) e^{i p_\nu x^\nu} \right)$$

$$= 2 |A|^2 p^\mu. \tag{7.7}$$

In particular, $\rho = j^0 = 2 |A|^2 E$, so it is the negative energy solutions that give rise to negative ρ. Notice that both these problems arise because of the second-order time derivative. In the Schrödinger equation, the time derivative (energy operator) is only first order. This means that there is no square root involved in finding the energy eigenvalues and the energy is always positive. This was Paul Dirac's motivation for attempting to derive a first-order relativistic analogue of the Schrödinger equation, and we will follow a similar path when we consider the Dirac equation in Chapter 8. However, while Dirac's equation was a truly remarkable discovery, his reason was actually quite unnecessary, since the Klein-Gordon equation *is* a valid description of relativistic particles, if looked at in the correct way.

Consider a negative-energy solution $\phi = A e^{-i(Et - p_x x - p_y y - p_z z)}$ (with $E < 0$). This describes a particle of mass m with energy E and momentum (p_x, p_y, p_z). Now consider the complex conjugate of this solution:

$$\phi^* = A^* e^{-i(-Et + p_x x + p_y y + p_z z)}. \tag{7.8}$$

This is still a solution of the Klein-Gordon equation with mass m, but it now has momentum $(-p_x, -p_y, -p_z)$, and more importantly has energy $-E$. So this conjugate solution describes some object with the same mass as the original particle but moving in the opposite direction, and in fact behaving in the opposite way to the original in every other respect. This solution describes an antiparticle! We have reinterpreted negative-energy particle solutions as positive-energy antiparticle solutions. Recall from Section 4.2.2 that crossing symmetry says an incoming particle of momentum p^μ behaves as an outgoing antiparticle of momentum $-p^\mu$. This demonstrates the necessity of introducing antiparticles in relativistic quantum mechanics.

Now to the other problem: that of negative probability density. The solution here is to reinterpret the current we derived as some other current that we know is conserved but which is allowed negative values. That is, the answer is to treat j^μ as a charge current. Suppose the particle described by the Klein-Gordon equation has an electric charge qe. Then redefining $j^\mu \mapsto qej^\mu$ gives us a conserved charge density current. A negative j^0 is now no problem. Also, notice that a positively charged particle moving with momentum p_i gives rise to the same current as its (negatively charged) antiparticle moving with momentum $-p_i$. In fact, this interpretation is more broadly applicable than to merely the electric charge: we can choose qe to be *any* quantum number associated with the particle. Since all quantities other than mass are reversed for an antiparticle, the same logic applies. So the correct interpretation of j^μ is as the flux of all the particle's quantum numbers, rather than its probability density. Notice that if ϕ is real-valued, then the expression for j^μ vanishes. Also, the state described by ϕ^* in this case is the same as that described by ϕ. So we see that a particle described by a real-valued wavefunction must be its own antiparticle and carry no non-zero quantum numbers.

The conserved current that we have derived is valid if there are no external electromagnetic fields. If we wish to construct a conserved current in the presence of external fields, we must make use, as we did in the non-relativistic case, of the substitution $p^\mu \mapsto p^\mu - qeA^\mu$. It can be shown in this case that the Klein-Gordon equation is modified to

$$\left((\partial^\mu + iqeA^\mu)\,(\partial_\mu + iqeA_\mu) + m^2 \right) \phi = 0$$
$$\implies \left(\partial^2 + m^2 \right) \phi = -iqe\,(2A^\mu \partial_\mu \phi + \phi \partial_\mu A^\mu) + q^2 e^2 A_\mu A^\mu \phi, \tag{7.9}$$

while the conserved current becomes

$$qej^\mu = iqe\,(\phi^* \partial^\mu \phi - \phi \partial^\mu \phi^*) - 2q^2 e^2 \phi^* \phi A^\mu. \tag{7.10}$$

The reader is invited to demonstrate this in Exercise 5.

7.2 THE MAXWELL AND PROCA EQUATIONS

7.2.1 Derivation of the Maxwell Equation

Given that the photon is known to be a spin-1 particle, a good place to start in finding a relativistic spin-1 wave equation is the Maxwell equation. In Lorentz-covariant form, Maxwell's equations can be written as

$$\partial_\mu F^{\mu\nu} = j^\nu,$$
$$F_{\mu\nu} = \partial_\mu A_\nu - \partial_\nu A_\mu, \qquad (7.11)$$

where j^ν is the electric current density and A_μ is the electromagnetic potential. Combining these equations, we find Maxwell's equation in terms of the potential:

$$\partial_\mu \left(\partial^\mu A^\nu - \partial^\nu A^\mu \right) = j^\nu$$
$$\partial^2 A^\nu - \partial^\nu \partial \cdot A = j^\nu. \qquad (7.12)$$

Given that photons are quanta of the electromagnetic field, we expect that the equation of the photon wavefunction should behave as the classical electromagnetic field. Therefore, the equation for the photon, and so for any other spin-1 particle, is simply a reinterpretation of Equation 7.12, in which A^μ is now the vector-valued wavefunction.

7.2.2 Solutions of the Maxwell Equation

In free space, the current density j^μ is set to zero. In this case, it is straightforward to check that the plane waves $A^\mu = A_0 \varepsilon^\mu e^{-ip \cdot x}$, where A_0 is a normalization constant and ε^μ is a constant unit vector, are solutions of the Maxwell equation as long as the four-momentum, p_ν, is such that $p^2 = 0$. That is, the solutions to the free-space Maxwell equation are massless, as we would expect for photons. The vector ε^μ defines the polarization of the photon, and can be any constant unit vector. For concreteness, we choose a basis for the polarization vector that depends on the momentum of the photon. That is, if the photon is traveling in a direction determined by its three-momentum **p**, then we select a basis vector in the direction of travel $\varepsilon_3^\mu = (0, \mathbf{k}/|\mathbf{k}|)$, and one in the time-like direction $\varepsilon_0^\mu = (1, 0, 0, 0)$. For the remaining two basis vectors, we choose an *arbitrary* unit vector **n** perpendicular to **k**. The final two basis vectors are then $\varepsilon_1^\mu = (0, \mathbf{n})$ and $\varepsilon_2^\mu = (0, \mathbf{n} \times \mathbf{k}/|\mathbf{k}|)$. The specific polarization vector for a photon may then be constructed from this basis, as

$$\varepsilon^\mu = \sum_{A=0}^{3} \varepsilon_A^\mu, \tag{7.13}$$

where μ is a Lorentz index and A labels the basis vectors.

We appear to have four distinct degrees of freedom for the photon, since its polarization can be any linear combination of these four polarization basis vectors. However, we know that the physical electromagnetic field only has two possible polarizations. The discrepancy arises from the fact that the Maxwell equation contains redundancy in the form of gauge freedom. Even classically, we know that the electromagnetic potentials may be redefined without altering the physical electric and magnetic fields. In Lorentz-covariant form, this gauge invariance appears as an ability to shift the four-potential according to

$$A^\mu(x) \mapsto A^\mu(x) + \partial^\mu \chi(x), \tag{7.14}$$

where $\chi(x)$ is any scalar function of x^μ. It is easy to see by substitution that this shift has no effect on the physical fields $F^{\mu\nu}$. In order to specify the potential uniquely, it is necessary to "fix the gauge"—that is, to choose an additional condition that must be satisfied by the potential. Since this choice is arbitrary, it will have no bearing on the physical properties of the system. The same procedure is necessary in the quantum case, and it is this gauge-fixing that will reduce the apparent degrees of freedom of the system from four to the physical two.

A common condition to impose is the Lorenz condition, $\partial \cdot A = 0$, since this simplifies the Maxwell equation considerably.[1] Specifically, the second term of the Maxwell equation clearly vanishes. In free space, this reduces the Maxwell equation to

$$\partial^2 A^\mu = 0. \tag{7.15}$$

This is just shorthand for four copies of the massless Klein-Gordon equation. In Section 7.1.1, we said that such a system would behave as four independent spin-0 particles. However, there is an important distinction to be made this time, in that this Klein-Gordon equation has only arisen because of the imposition of a further constraint, $\partial \cdot A = 0$. This constraint has already linked the components of A^μ so that they are necessarily interdependent.

Consider a gauge transformation applied *after* the Lorenz condition has been imposed, of the form $A^\mu \mapsto A^\mu + \partial^\mu \chi_h$, where χ_h is a harmonic function (satisfying $\partial^2 \chi = 0$). Then A^μ clearly still obeys the Maxwell equation, as we have seen that this is true for *any* gauge transformation. Crucially, however, this type of transformation also leaves the Lorenz condition invariant. This demonstrates that the Lorenz condition is in fact only a *partial* gauge fixing condition, and leaves a residual gauge freedom. In fact, the Lorenz condition reduces the degrees of freedom only to three, not the physical two. To specify a unique solution, we must impose a further condition. Although there are many gauges available, the simplest example is the Coulomb gauge $\nabla \cdot A = 0$. In fact, it can be shown that physically meaningful solutions that obey the Coulomb gauge condition must also obey the Lorenz condition, so it could be said that the former implies the latter. However, we will treat them as independent conditions. Applying both conditions to the plane-wave solution, we find the following conditions on the polarization vector

$$p \cdot \varepsilon = 0 \quad \text{and} \quad \mathbf{p} \cdot \varepsilon = 0. \tag{7.16}$$

---

[1] The reader will often see this misspelled as the "Lorentz condition." The Lorenz condition is named after Ludvig Lorenz, unlike Lorentz invariance and the Lorentz group, both of which are named after Hendrik Lorentz.

Considering the polarization basis vectors, we see that the second condition rules out any contribution from ε_3^μ, and, in turn, the first rules out ε_0^μ. Thus, we are left with a polarization vector composed of a linear combination of only two of the basis vectors, ε_1^μ and ε_2^μ.

The physical degrees of freedom of the photon have thus been successfully reduced to two. However, since the choice of n from which the polarization basis was constructed was itself arbitrary, we are apparently still free to decide what the two physical polarization states should be. An obvious choice may seem to be simply ε_1^μ and ε_2^μ, corresponding to linear polarization, in which the polarizations are aligned with either the "horizontal" or "vertical" axes. Equally, we could choose

$$\varepsilon_R^\mu = \frac{1}{\sqrt{2}} \left(\varepsilon_1^\mu + i\varepsilon_2^\mu \right), \quad \varepsilon_L^\mu = \frac{1}{\sqrt{2}} \left(\varepsilon_1^\mu - i\varepsilon_2^\mu \right), \tag{7.17}$$

corresponding to circular polarization. The advantage of this scheme, as we will see in Section 7.2.3, is that these polarization vectors are eigenstates of the helicity operator, which measures the orientation of the photon's spin with respect to the direction of motion.

Whatever basis we choose, it should obey an orthonormality condition:

$$\varepsilon_A^{*\mu}(p)\varepsilon_B^\nu(p)g_{\mu\nu} = g_{AB}, \tag{7.18}$$

and a completeness relation

$$\varepsilon_A^{*\mu}(p)\varepsilon_B^\nu(p)g^{AB} = g^{\mu\nu}. \tag{7.19}$$

7.2.3 Including Mass: The Proca Equation

The solutions to any relativistic wave equation must also obey the Klein-Gordon equation, since this is simply a statement of the relativistic energy-momentum relation. We have already seen this principle in the Maxwell equation: the Maxwell equation is equivalent to four copies of the massless Klein-Gordon equation with additional constraints linking those copies together. It is not too difficult, therefore, to see how to include a mass into the equation. We simply replace the massless Klein-Gordon operator ∂^2 with the massive operator $\partial^2 + m^2$, to arrive at the Proca equation:

$$\partial^2 A^\mu - \partial^\mu \partial \cdot A + m^2 A^\mu = j^\mu. \tag{7.20}$$

It is important to appreciate that gauge symmetry has been lost with the introduction of this mass term. Indeed, a transformation of the form $A^\mu \mapsto A^\mu + \partial \chi$ now leads to a transformed equation that differs from the original by $m^2 \partial^\mu \chi$. From this, we see that gauge invariance is restored if and only if $m = 0$. However, in the massive case, the Lorenz condition is necessarily satisfied by the solutions. To see this, we act on the Proca equation with the differential operator ∂_μ, giving

$$m^2 \partial \cdot A = 0. \tag{7.21}$$

Hence, in the case that $m \neq 0$, the Lorenz condition must hold. It is accurate to say, then, that the Proca equation is equivalent to four massive Klein-Gordon equations for the components of A^μ together with the Lorenz condition. This reduces the degrees of freedom to three, as we would expect for a spin-1 particle. This does not, of course, prove that the spin is 1, but it does show at least that the equation is consistent with a spin-1 particle.

7.2.4   Spin of Vector Particles

In order to demonstrate conclusively that the spin of a vector particle is 1, it is necessary to analyze the behavior of solutions under Lorentz transformations. In particular, it can be shown that the spin operators for a solution to the Proca equation are given by

$$\Sigma_1 = \begin{pmatrix} 0 & 0 & 0 & 0 \\ 0 & 0 & 0 & 0 \\ 0 & 0 & 0 & -i \\ 0 & 0 & i & 0 \end{pmatrix}, \Sigma_2 = \begin{pmatrix} 0 & 0 & 0 & 0 \\ 0 & 0 & 0 & i \\ 0 & 0 & 0 & 0 \\ 0 & -i & 0 & 0 \end{pmatrix},$$

$$\Sigma_3 = \begin{pmatrix} 0 & 0 & 0 & 0 \\ 0 & 0 & -i & 0 \\ 0 & i & 0 & 0 \\ 0 & 0 & 0 & 0 \end{pmatrix}. \tag{7.22}$$

The demonstration of this is left as an exercise for the reader but is deferred until the next chapter when we will see a similar derivation for the (slightly) simpler case of a spin-1/2 particle (Exercise 10). In turn, these give a total spin operator of

$$\Sigma^2 = \Sigma_1^2 + \Sigma_2^2 + \Sigma_3^2 = 2 \begin{pmatrix} 0 & 0 & 0 & 0 \\ 0 & 1 & 0 & 0 \\ 0 & 0 & 1 & 0 \\ 0 & 0 & 0 & 1 \end{pmatrix}, \tag{7.23}$$

from which we can see that $s(s+1) = 2$, implying a spin of 1. Notice that a time-like polarization vector would be canceled out by this operator, since it contains only zeroes in the left-most column. This would imply that a time-like polarized vector particle would have spin 0. This is not a problem, however, since such polarizations are unphysical.

In order to characterize the state of a particle, it is necessary of course to provide the spin orientation quantum number. We can choose to measure this orientation along any axis, but any such choice is arbitrary. A more natural approach is to allow the state itself to determine a spin polarization axis for us. We define the helicity operator as the component of spin along the axis determined by the particle's momentum. That is:

$$h(p) = \frac{\Sigma \cdot \mathbf{p}}{|\mathbf{p}|} = \frac{1}{|\mathbf{p}|} \begin{pmatrix} 0 & 0 & 0 & 0 \\ 0 & 0 & -ip_z & ip_y \\ 0 & ip_z & 0 & -ip_x \\ 0 & -ip_y & ip_x & 0 \end{pmatrix}. \tag{7.24}$$

Using the helicity operator, we find that the circularly polarized solutions form a complete set of spin states, since ε_L^μ, ε_3^μ, and ε_R^μ have helicities of $-1, 0,$ and $+1$ respectively.

In the massless case, there is no option to choose an arbitrary spin quantization axis. Such a choice strictly takes place in the particle's rest frame, a concept which does not apply to massless particles. Hence, in the massless case, we have no choice but to characterize particles by their helicity. If the momentum of the massless particle is $\mathbf{p}$, then the four momentum is of the form

$$p^\mu = (|\mathbf{p}|, \mathbf{p}), \tag{7.25}$$

while the longitudinal polarization vector is

$$\varepsilon_3^\mu = \left(0, \frac{\mathbf{p}}{|\mathbf{p}|}\right). \tag{7.26}$$

If we write the polarization vector of an arbitrary state in terms of the basis states, $\varepsilon^\mu = \alpha^A \varepsilon_A^\mu$, and impose the Lorenz condition, we find

$$p_\mu \varepsilon^\mu = 0$$

$$\sum_{A=0}^{3} \alpha^A (|\mathbf{p}|, \mathbf{p}) \cdot \varepsilon_A = 0. \tag{7.27}$$

This condition is clearly respected by the ε_1^μ and ε_2^μ (or ε_L^μ, ε_R^μ) basis vectors, placing no constraints on the values of α_1 and α_2. However, since the condition is not obeyed by ε_0^μ and ε_3^μ individually, it imposes a constraint on the corresponding coefficients α^0, α^3:

$$\alpha^0 \left[(|\mathbf{p}|, \mathbf{p}) \cdot (1, 0)\right] + \alpha^3 \left[(|\mathbf{p}|, \mathbf{p}) \cdot \left(0, \frac{\mathbf{p}}{|\mathbf{p}|}\right)\right] = |\mathbf{p}| \alpha^0 - |\mathbf{p}| \alpha^3 = 0, \tag{7.28}$$

or $\alpha^0 = \alpha^3$. This suggests that the full set of physical polarization vectors for massless particles is given by ε_L^μ, ε_R^μ, and $\varepsilon_0^\mu + \varepsilon_3^\mu$. However, the latter basis vector corresponds to a plane-wave of the form

$$A^\mu = A_0 \left(1, \frac{\mathbf{p}}{|\mathbf{p}|}\right) e^{-i(|\mathbf{p}|, \mathbf{p}) \cdot x}, \tag{7.29}$$

which can be written as

$$A^\mu = \partial^\mu \left(A_0 i |\mathbf{p}| e^{-i(|\mathbf{p}|, \mathbf{p}) \cdot x}\right). \tag{7.30}$$

Since this solution is expressible as a total derivative, it is gauge equivalent to the solution $A^\mu = 0$, and therefore has no physical significance. We can deduce, then, that the longitudinal polarization available to massive particles is not a viable solution for massless particles.

7.3 COMBINING EQUATIONS: HOW DO PARTICLES INTERACT?

Just as we introduced the effect of an electromagnetic field into the Klein-Gordon equation in Section 7.1.3, we can also capture the effect of a charged particle on the behavior of a photon by including terms in the Maxwell equation. In particular, we know that the term on the right of the Maxwell equation is the electromagnetic current. But we already have an expression for the current carried by a scalar particle: we simply use the electric charge, qe, as the relevant quantum number in the conserved current formula. We therefore find a pair of coupled equations for the photon and the scalar particle:

$$\left(\partial^2 + m^2\right)\phi = -iqe\left(2A^\mu\partial_\mu\phi + \phi\partial_\mu A^\mu\right) + q^2e^2 A_\mu A^\mu\phi,$$
$$\partial^2 A^\mu - \partial^\mu\partial\cdot A = iqe\left(\phi^*\partial^\mu\phi - \phi\partial^\mu\phi^*\right) - 2q^2e^2\phi^*\phi A^\mu.$$
$$(7.31)$$

How should we interpret these equations though? Recall that the right side of the Maxwell equation is the source term for the electromagnetic field. That is, a non-zero value on the right is capable of *producing* an electromagnetic field. We can say that the current acts as a *source term* for photons, capable of producing photons from the movement or merely the existence of the charged scalar particle. So we see from the second equation that a pair of scalars can produce a photon via the terms with derivatives, and a combination of two scalars and a photon can produce another photon via the term $2q^2\phi^*\phi A^\mu$. Similarly, in the first equation, a scalar can arise from an existing scalar and either one or two photons.

This all points toward the idea of using Feynman diagrams to calculate physical quantities. A Feynman diagram is a stylized representation of what is happening at the microscopic level, with virtual particles being created and exchanged. Each one of these events, though, corresponds to some term in the relevant equations of motion for the system. In particular, by introducing a set of *Feynman rules*, derived from the equations of motion, we can translate directly from the Feynman diagram to an algebraic expression for the transition amplitude for the process. To state this idea more clearly, we must take a slight detour into the world of quantum field theory. This is a huge topic in its own right and the following is by no means intended as an in-depth introduction. It will, however, serve to introduce some of the concepts.

7.3.1   Quantum Field Theory without the Math

One of the counterintuitive aspects of quantum mechanics (there are several!) is the concept of wave-particle duality, in which objects exhibit behavior characteristic of both particles and waves depending on the measurements being taken. How are we to make sense of this fact? The answer, or at least, *an* answer comes in the form of quantum field theory. A full introduction to the quantum behavior of fields is far beyond the scope of this book, but some of the ideas of the subject may be discussed without delving into the full mathematical treatment. The most familiar example of a quantum field is the electromagnetic field. Classically, electromagnetism is described in terms of a field, with waves propagating through it transferring energy. The quantum picture of electromagnetism is very different, of course, with photons mediating the electromagnetic force between charged particles. The continuous nature of the electromagnetic field gives way to a description of the same phenomena built from discrete chunks of energy with well-defined properties of momentum and spin. The means to reconcile these two pictures is the quantized electromagnetic field. The field is quantized in the same way that a particulate system is quantized, with the promotion of observables to operators, and the imposition of certain commutation relations between these operators. The difference is in the dynamic variables used to describe the system. For a simple quantum mechanical system, the dynamic variables are the position and momentum of the relevant particles. For a quantum field, the dynamic variables are the value and rate of change of the field at each point in space-time. The result is that the wave-modes of the field are then quantized: that is, discretized. The lowest-energy state of the field is then identified as the vacuum, while higher-energy excitations are identified as particles. A wave propagating through space can have an amount of energy and momentum that allows for its identification as a photon, while higher states may be identifiable as collections of two, three, or more photons. That is, it is the excitations of the field above the ground state that we perceive as particles. Quantum field theory extends this idea to other particles. Where the photon is the quantized excitation of the electromagnetic field, the electron is an excitation in the electron field, and so on. The standard model, therefore, consists of a set of 37 such fields: one for each particle species in the theory. Each of these fields pervades all of space-time, so it should come as no surprise that one field should be capable of transferring its energy and momentum to another field. However, only some of these fields are capable of interacting, leading to the allowed and forbidden interactions of the standard model.

A rigorous derivation of the invariant amplitudes, $\mathcal{M}$, of scattering and decay processes requires a full field-theoretical treatment, so we shall merely state the result. If particles are really the quantized excitations of a set of fields, then a scattering process is really the conversion of some initial field configuration to a final configuration. In any theory in which the fields interact, even single-particle states are found to be complicated solutions to the field equations, in that they are not analytical. However, they may be approximated by the corresponding free-theory states found in the limit that the coupling constants of the theory vanish. Any scattering process and its associated amplitude is similarly tricky to calculate, but can be approximated as a power series in the interactions of free-theory states. In particular, the solution is written as a power series in the relevant coupling. The process is similar to non-relativistic perturbation theory, but allows for an interesting interpretation of the underlying dynamics. We can calculate a scattering process by writing down all of the Feynman diagrams that *could* contribute to the process and summing their contributions, found via the Feynman rules. That is, the amplitude for the overall process can be found by adding together all of the ways in which it *could have happened*.

There is a second approach to quantization, which leads to precisely the same quantum theory as the canonical approach of promoting observables to operators. This is the path integral method, and can be thought of as an extension of the famous double-slit experiment. In this experiment, it was shown that if a particle travels through a screen with a double slit, but no measurement is taken of which slit the particle traverses, then the probability of finding the particle at a particular point on the far side of the screen is given by an interference pattern. This suggests that the particle has, in a sense, made both possible journeys and interfered with itself on the far side of the screen. If the same experiment were conducted with a greater number of screens, each with a large number of slits, then all possible paths through the screens would have to be summed in order to find the probability density distribution for the particle's position. In the limit that the number of screens and number of slits per screen tends to infinity, this system is indistinguishable from empty space. To find the likelihood of a particle reaching a particular point in free space, then, it is necessary to sum over *all possible* paths that the particle could have taken to get to that point. More precisely, this "sum over histories" has each path weighted by a factor $e^{iS[\text{path}]}$ where the quantity $S = \int_{\text{path}} \mathcal{L}(x)\mathrm{d}x$ is the action for that path. Such sums oscillate wildly for small variations in S when we are far from the classical solution, where S is stationary. This causes destructive interference between paths far from the classical path, such that the greatest contributions to the quantum behavior of a system come from those paths closest to the classical path. This leads to the remarkable result that this sum over histories is in fact responsible for the classical principle of least action. In the case of an interacting field theory, the path integral is further complicated by the fact that viable paths involve the particle undergoing various interactions *en route*. For example, in the case of an electron traveling through free-space, the electron could simply move from one position to another. On the other hand, it could emit one or more virtual photons, which it then re-absorbs. Or, it could emit a photon that itself splits into an electron-positron pair before re-combining to be re-absorbed by the original electron.

All of these paths are valid, and so all must be counted toward the probability amplitude for the process being considered. In fact, there are infinitely many such contributions, but as long as the coupling is small, higher-order interactions have a less significant contribution and the sum converges. Generalizing this idea, since we cannot know the exact nature of the virtual particles exchanged in an interaction, we must "average over our ignorance." Both quantization schemes thus arrive at the conclusion that we must sum the contributions of Feynman diagrams. Due to the perturbative nature of these calculations, in practice, only Feynman diagrams with up to a specific number of vertices are considered for any given calculation, with more vertices giving a more precise answer.

Increasing the number of vertices in a diagram typically introduces closed internal loops, which we will consider in Section 9.6. The simplest diagrams for a process have no internal loop and are referred to as tree-level diagrams. Even at tree level, however, there may be more than one contribution to a given process. For example, the interaction of an electron and a positron via photon exchange can occur in two different ways:

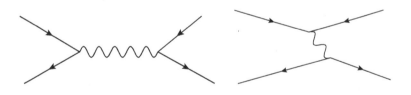

7.3.2 Feynman Rules

Before we write down the Feynman rules and calculate an amplitude, it is worth pointing out that Equation 7.31, along with the complex conjugate of the scalar equation, can be derived from a Lagrangian of the form

$$\mathcal{L} = \partial_\mu \phi^* \partial^\mu \phi - \frac{1}{4} F_{\mu\nu} F^{\mu\nu} - iqe A^\mu \left(\phi^* \partial_\mu \phi - \phi \partial_\mu \phi^* \right), \quad (7.32)$$

via the Euler-Lagrange equations

$$\partial_\mu \left(\frac{\partial \mathcal{L}}{\partial (\partial_\mu \phi)} \right) = \frac{\partial \mathcal{L}}{\partial \phi},$$

$$\partial_\mu \left(\frac{\partial \mathcal{L}}{\partial (\partial_\mu \phi^*)} \right) = \frac{\partial \mathcal{L}}{\partial \phi^*}, \tag{7.33}$$

$$\partial_\mu \left(\frac{\partial \mathcal{L}}{\partial (\partial_\mu A^\nu)} \right) = \frac{\partial \mathcal{L}}{\partial A^\nu},$$

and that the allowed vertices in Feynman diagrams can also be deduced directly from this Lagrangian. In fact, this is the more typical approach to determining Feynman rules.

To calculate an invariant amplitude, we use the following procedure. First, label all the internal and external particles with a four-momentum, such that momentum is conserved at each vertex. Also, at each point that a photon joins a vertex, label it with a Lorentz index (for the photon's polarization vector). Next, write down the relevant factor for each feature of the diagram, including internal and external particles, and each vertex. The relevant factors are as follows:

- For each scalar-scalar-photon vertex with Lorentz index μ, include a factor of $iqe(p_1 + p_2)^\mu$ where p_1, p_2 are the momenta of the scalars, following the direction of the arrows on the diagram (see the example in Equation 7.40).

- For each two-photon–two-scalar vertex with Lorentz indices μ, ν, include a factor of $-2i(qe)^2 g^{\mu\nu}$.

- For each incoming external photon with momentum p connected to a vertex with Lorentz index μ, include its polarization vector $\varepsilon_\mu(p)$.

- For each outgoing external photon with momentum p connected to a vertex with Lorentz index μ, include a conjugate polarization vector $\varepsilon_\mu^*(p)$.

- For each internal (virtual) scalar particle with momentum p, include a *propagator*, $\frac{i}{p^2 - m^2}$, where m is the scalar's mass.

- For each internal photon with momentum p, include a photon propagator, $\frac{-ig_{\mu\nu}}{p^2}$. (This is modified to $\frac{-ig_{\mu\nu}}{p^2 - m^2}$ in the case of a vector particle of mass m.)

- If an internal momentum q is unconstrained, we must sum over all possible values: $\int d^4 q / (2\pi)^4$.

These rules, along with the other Feynman rules covered in this text, are summarized in Appendix B. The product of all the factors arising from a diagram gives a contribution to $-i\mathcal{M}$, where $\mathcal{M}$ is the transition amplitude. Notice that I have not mentioned external scalar particles in the previous list. This is because the external scalar contribution is trivial, giving a factor of 1, and so we can simply ignore it.

It is worth taking a further moment to appreciate the origin of these rules. The momentum in the vertex factor comes from the derivative in the interaction term of the equation of motion. In the case of the vertex with two photons and the photon propagator, the two Lorentz indices must be the same, hence the metrics. A polarization vector for external states is reasonably intuitive. The least clear of these rules are the two propagators. To derive these, we use a Fourier transform to write the non-interacting part of the Lagrangian in momentum-space. The propagator is then the inverse of the operator sandwiched between the two momentum-space wavefunctions. Let us take the scalar propagator as an example to see why this works. The wavefunction for a non-interacting scalar of course obeys the Klein-Gordon equation. Suppose now that we wish to produce a scalar particle in an otherwise empty system by introducing some sort of disturbance or source. We do this at a particular place at a particular time, so the source can be approximated as a delta function. Taking this disturbance to be at the origin, we have

$$\left(\partial^2 + m^2\right)\phi(x) = \alpha\delta^4(x), \tag{7.34}$$

where α is some proportionality constant. Writing $\phi(x)$ in terms of its Fourier transform

$$\phi(x) = \int \frac{d^4 k}{(2\pi)^4} \, \widetilde{\phi}(k) e^{-ik\cdot x}, \tag{7.35}$$

and also writing the delta function in terms of a k integral, we find

$$(\partial^2 + m^2) \int \frac{d^4k}{(2\pi)^4} \, \tilde{\phi}(k) e^{-ik\cdot x} = \alpha \int \frac{d^4k}{(2\pi)^4} \, e^{-ik\cdot x}$$

$$\int \frac{d^4k}{(2\pi)^4} \left[\left(-k^2 + m^2 \right) \tilde{\phi}(k) - \alpha \right] e^{-ik\cdot x} = 1$$

$$\implies \tilde{\phi}(k) = \frac{-\alpha}{k^2 - m^2}.$$
$$(7.36)$$

Substituting this back into Equation 7.35, we find that at a point x, where the wavefunction would otherwise be trivially zero (for an empty system), the behavior of the wavefunction is modified by the presence of the source at the origin according to

$$\phi(x) = \int \frac{d^4k}{(2\pi)^4} \frac{(-\alpha)e^{-ik\cdot x}}{k^2 - m^2}. \qquad (7.37)$$

A more rigorous derivation would show that we require $\alpha = -i$ so

$$\phi(x) = \int \frac{d^4k}{(2\pi)^4} \frac{ie^{-ik\cdot x}}{k^2 - m^2}, \qquad (7.38)$$

or, in momentum space,

$$\phi(k) = \frac{i}{k^2 - m^2}. \qquad (7.39)$$

Looking at the denominator of this propagator, it is clear that a virtual particle that is close to being on shell carries more weight than one that is far from the mass-shell. In this way, the propagator ensures that the most likely processes are those in which the exchange particle is "almost real."

Writing down the relevant factors for a Feynman diagram and multiplying all of these factors together gives a contribution to $-i\mathcal{M}$, where $\mathcal{M}$ is the invariant amplitude for the diagram. The full amplitude is found by summing all similar terms from relevant Feynman diagrams. This may then be squared and substituted into

Equation 5.41 to find the differential cross-section for the process. As an example, consider the diagram

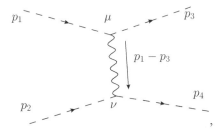

in which two particles of mass m_1 and m_2 and charges q_1 and q_2 interact via photon exchange. The arrows on the scalar lines, incidentally, show the flow of quantum numbers rather than momentum, so an antiparticle line would appear to point backward. Following the Feynman rules, we can write the amplitude for this diagram as

$$-i\mathcal{M} = iq_1 e(p_1 + p_3)^\mu \, iq_2 e(p_2 + p_4)^\nu \left(\frac{-ig_{\mu\nu}}{(p_1 - p_3)^2} \right)$$

$$\implies \mathcal{M} = -\frac{q_1 q_2 e^2 (p_1 + p_3) \cdot (p_2 + p_4)}{(p_1 - p_3)^2}.$$

(7.40)

This amplitude can now be substituted into Equation 5.41 to find the cross-section for this interaction.

EXERCISES

1. **(a)** Given that $\phi(t, \mathbf{x})$ is a plane-wave solution of the Klein-Gordon equation, show that $\Psi(t, \mathbf{x}) = \phi(t, \mathbf{x})e^{imt}$ is an energy eigenstate whose eigenvalue is equal to the kinetic energy of ϕ.

   (b) Show that, in the non-relativistic limit $(E_k << m)$, the Klein-Gordon equation for ϕ reduces to the Schrödinger equation.

2. Show that the plane wave $A^\mu = A_0 \varepsilon^\mu \exp(-ip \cdot x)$ is a solution of the Maxwell equation if $p^2 = 0$.

3. Show that the Proca equation is not invariant under a gauge transformation such that $A^\mu \mapsto A^\mu + \partial^\mu \chi$.

4. Show that the helicity basis states for vector particles, together with the time-like polarization vector, form a complete orthonormal set.

5. Making use of the substitution $p^\mu \mapsto p^\mu - qeA^\mu$, derive the form of the Klein-Gordon equation and the conserved current in the case of a charged particle in an electromagnetic field (Equations 7.9, 7.10).

6. Draw a Feynman diagram to depict a charged scalar particle and antiparticle annihilating to produce a photon, which then pair-produces another scalar particle and antiparticle. Label the diagram appropriately including particle momenta and use the Feynman rules to calculate the amplitude for this process. How would this amplitude differ if the vector were massive?

7. Derive the equations of motion for the scalar, its conjugate, and the photon from the Lagrangian in Equation 7.32 via the Euler-Lagrange equations.

CHAPTER 8

THE DIRAC EQUATION

8.1   A LINEAR RELATIVISTIC EQUATION

As we saw in Chapter 7, the simplest relativistic wave equation leads to some difficulties involving negative-energy solutions and negative charge densities. We have also seen how to overcome these difficulties with a reinterpretation of the equation. Initially, however, when first discovered, these issues appeared to be a very serious stumbling block for a relativistic quantum theory, suggesting that the Klein-Gordon equation was simply wrong. The non-relativistic Schrödinger equation is free from these problems because it is a first order differential equation with respect to time. Since the time derivative is proportional to the energy operator, this is just another way of saying that the Schrödinger equation is linear with respect to energy. The quadratic nature of the relativistic energy-momentum relation, on the other hand, leads to an equation that is second-order in time, which in turn means that negative-energy solutions cannot be ruled out. This led Paul Dirac in 1928 to search for a different equation, with two important properties. First, its solutions should obey the correct relativistic energy-momentum relation. Second, in order to retain the interpretation of the non-relativistic equation, it should be first order in time. In particular, it should in fact *be* a Schrödinger equation $i\partial_t \psi = \widehat{H}\psi$, with a Hamiltonian suitable for a relativistic theory. Lorentz invariance requires that the treatment of time and space coordinates should be the same, and so it is immediately clear that the Hamiltonian can be only first order in spatial coordinates as well. Another term that is compatible with these first derivatives in terms of units is the mass. Dirac thus postulated a Hamiltonian of the form $\widehat{H} = \alpha \cdot \widehat{\mathbf{p}} + \beta m$, where α and β are constants, and m is the mass of the particle. The equation can be put in a more obviously Lorentz-covariant form if multiplied through by β^{-1}:

$$i\beta^{-1}\partial_t \psi = \beta^{-1}\alpha \cdot \widehat{\mathbf{p}} + m\psi$$
$$\gamma^\mu \widehat{p}_\mu \psi = m\psi, \tag{8.1}$$

where we have defined γ^μ by:

$$\gamma^0 = \beta^{-1} \quad \text{and} \quad \gamma^i = \beta^{-1}\alpha^i. \tag{8.2}$$

Another way to view this equation is as an eigenvalue equation: the solution ψ is an eigenfunction of the operator $i\gamma^\mu \partial_\mu$, with eigenvalue m. As such, we can operate on both sides of the equation a second time with this operator, to arrive at

$$(\gamma^\mu \widehat{p}_\mu \gamma^\nu \widehat{p}_\nu)\psi = m^2\psi. \tag{8.3}$$

Since the two momentum operators in this expression clearly commute, the operator $\widehat{p}_\mu \widehat{p}_\nu$ is symmetric in μ and ν. Using this fact and relabeling the indices gives

$$\begin{aligned} m^2\psi &= (\gamma^\mu \gamma^\nu \widehat{p}_\mu \widehat{p}_\nu)\psi \\ &= (\gamma^\mu \gamma^\nu \widehat{p}_\nu \widehat{p}_\mu)\psi \\ &= (\gamma^\nu \gamma^\mu \widehat{p}_\mu \widehat{p}_\nu)\psi. \end{aligned} \tag{8.4}$$

We can, therefore, write $m^2\psi$ as the average of these two expressions, leading to

$$m^2\psi = \frac{1}{2}(\gamma^\mu \gamma^\nu + \gamma^\nu \gamma^\mu)\widehat{p}_\mu \widehat{p}_\nu \psi. \tag{8.5}$$

Since the relativistic energy-momentum relation can be written $g^{\mu\nu}p_\mu p_\nu = m^2$, this places a constraint on the γ coefficients:

$$\{\gamma^\mu, \gamma^\nu\} = 2g^{\mu\nu}, \tag{8.6}$$

where $\{A, B\} = AB + BA$ denotes the anticommutator of two objects. When a set of objects obeys a relation of the previous kind, they are said to form a Clifford algebra, and this is the defining property of the objects γ^μ.

A moment's thought will show that the Clifford algebra cannot be satisfied by ordinary numbers, since we have the requirement that, for example, $\left(\gamma^0\right)^2 = 1$ and $\left(\gamma^1\right)^2 = -1$, and yet $\gamma^0\gamma^1 = -\gamma^1\gamma^0$. These objects are instead something new, and potentially abstract. In fact, the nature of these objects is not particularly important, since everything physical that they represent can be deduced directly from the Clifford algebra relation (Equation 8.6). As with group theory, however, we can represent these objects with a set of matrices, and so this is how they are most commonly thought of. In fact, they are generally referred to simply as the gamma matrices. The smallest matrices that can satisfy the defining relation are 4×4 matrices. Put another way, the smallest representation of the algebra defined by Equation 8.6 has dimension 4. In turn, this means that the wavefunction itself is also not a simple number, but is an object with (at least) four components. The obvious candidate is of course a four-vector, but an analysis of the transformation properties of the Dirac equation shows that the wavefunction cannot be a vector. It is a new type of object, known as a Dirac spinor. We will explore this further in Section 8.3.

8.2 REPRESENTATIONS OF THE GAMMA MATRICES

There are a continuous infinity of possible representations of the gamma matrices, but in practice only a few are ever used. Each of the commonly used representations has its own useful properties, but it is worth re-emphasizing that the physical properties of a system cannot depend on the representation that is chosen. All physically significant results must be true in any representation and can, therefore, be derived directly from the defining property of the gamma matrices. Having said this, the properties of a particular representation may lend themselves to solution of a particular problem. We will consider two representations in this text, which will be explored in the following sections. There are some properties of the gamma matrices that are worth deriving in a representation-free manner, as this demonstrates that the useful properties derived in this way are not an artifact of the specific representation used. In particular, it is worth noting that, since the Hamiltonian must be Hermitian, we must have $\alpha^{i\dagger} = \alpha^i$ and $\beta^\dagger = \beta$, which in turn give $\gamma^{0\dagger} = \gamma^0$ and $\gamma^{i\dagger} = -\gamma^i$. These relations can be summarized neatly by

$$\gamma^{\mu\dagger} = \gamma^0 \gamma^\mu \gamma^0. \tag{8.7}$$

8.2.1 The Dirac Representation

Arguably the "standard" representation for massive particles, the Dirac representation is given by

$$\gamma^0 = \begin{pmatrix} \mathbb{1}_2 & 0 \\ 0 & -\mathbb{1}_2 \end{pmatrix}, \quad \gamma^i = \begin{pmatrix} 0 & \sigma_i \\ -\sigma_i & 0 \end{pmatrix}, \tag{8.8}$$

written in block notation, in which each entry is itself a 2×2 matrix, with $\mathbb{1}$ the 2×2 identity and σ_i the i-th Pauli matrix.

8.2.2 The Weyl Representation

While the Dirac representation is the most common representation for massive particles, it is not particularly well suited for describing massless particles. The reasons for this will become clear when we discuss chirality later in this chapter. For now, it is sufficient to say that another representation more suitable for massless particles is the Weyl representation. This representation shares the same γ^i as the Dirac representation but differs in γ^0. In block form, it is given by

$$\gamma^0 = \begin{pmatrix} 0 & \mathbb{1}_2 \\ \mathbb{1}_2 & 0 \end{pmatrix}, \quad \gamma^i = \begin{pmatrix} 0 & \sigma_i \\ -\sigma_i & 0 \end{pmatrix}. \tag{8.9}$$

Notice that all four matrices are characterized by their vanishing block diagonal elements. This is the reason that the representation is suited to massless states, since it allows the spinor to be split into two 2-component spinors, as we will see in Section 8.4.4.

8.3   SPINORS AND LORENTZ TRANSFORMATIONS

We can derive the transformation properties of a Dirac spinor by considering the Dirac equation in two different frames. In particular, consider a second reference frame related to the first by a Lorentz transformation $\Lambda^\mu{}_\nu$. Since we do not yet know the corresponding transformation law for the spinor, let us write it as some matrix S. Since the gamma matrices are constants, they require no transformation between frames. The derivative on the other hand will transform as a covariant four-vector. In the new frame, then, the Dirac equation may be written

$$\left(i\gamma^\mu \left(\Lambda^{-1}\right)^\nu{}_\mu \partial_\nu - m \right) S\psi = 0. \tag{8.10}$$

Multiplying from the left by the inverse transformation, S^{-1}, gives

$$\left(iS^{-1}\gamma^\mu \left(\Lambda^{-1}\right)^\nu{}_\mu S\partial_\nu - m \right) \psi = 0. \tag{8.11}$$

Notice that the derivative does not act on S since S is a transformation *between* frames whereas the derivative measures the rate of change with respect to *one* frame. Since the $\left(\Lambda^{-1}\right)^\nu{}_\mu$ is just a collection of numbers, it must commute with S and S^{-1}. Therefore, we have

$$\left(i \left(\Lambda^{-1}\right)^\nu{}_\mu S^{-1}\gamma^\mu S\partial_\nu - m \right) \psi = 0. \tag{8.12}$$

If this equation is to remain invariant under Lorentz transformations, the above form must be identical to the original. We can, therefore, identify $\left(\Lambda^{-1}\right)^{\nu}{}_{\mu} S^{-1} \gamma^{\mu} S$ with γ^{ν}. Since we already know that $\left(\Lambda^{-1}\right)^{\nu}{}_{\mu} \Lambda^{\mu}{}_{\rho} = \delta^{\nu}_{\rho}$, we can see that

$$S^{-1} \gamma^{\mu} S = \Lambda^{\mu}{}_{\nu} \gamma^{\nu}. \tag{8.13}$$

It can be shown (as the reader is invited to verify in Exercise 6) that the appropriate transformation matrix is given by

$$S(\omega_{\mu\nu}) = e^{-\frac{i}{4}\omega_{\mu\nu}\sigma^{\mu\nu}} \tag{8.14}$$

where $\omega_{\mu\nu}$ is an antisymmetric set of rotation angles and boost velocities defining the Lorentz transformation (as in Exercise 3), and $\sigma^{\mu\nu} = \frac{i}{2}[\gamma^{\mu}, \gamma^{\nu}]$.

Let's take a spinor and rotate it about the z axis by some angle θ.[1] This corresponds to $\omega_{12} = -\omega_{21} = \theta$ and $\omega_{\mu\nu} = 0$ otherwise. When we do this, we find that

$$
\begin{aligned}
S(\Lambda) &= \exp\left(-\frac{1}{4}\left(\theta\sigma^{12} - \theta\sigma^{21}\right)\right) \\
&= \exp\left(-\frac{2\theta}{4}\sigma^{12}\right) \\
&= \exp\left(-\frac{\theta}{2}\left[\gamma^{1}, \gamma^{2}\right]\right) \\
&= \exp\left(-\frac{\theta}{2}\begin{pmatrix} i & 0 & 0 & 0 \\ 0 & -i & 0 & 0 \\ 0 & 0 & i & 0 \\ 0 & 0 & 0 & -i \end{pmatrix}\right) \\
&= \begin{pmatrix} e^{\frac{i\theta}{2}} & 0 & 0 & 0 \\ 0 & e^{-\frac{i\theta}{2}} & 0 & 0 \\ 0 & 0 & e^{\frac{i\theta}{2}} & 0 \\ 0 & 0 & 0 & e^{-\frac{i\theta}{2}} \end{pmatrix},
\end{aligned} \tag{8.15}
$$

[1] More accurately, we will rotate our reference frame through an angle θ around the spinor.

where the last step only follows because the matrix is diagonal. For a full rotation of 2π, this gives

$$S(2\pi \text{ rotation}) = \begin{pmatrix} -1 & 0 & 0 & 0 \\ 0 & -1 & 0 & 0 \\ 0 & 0 & -1 & 0 \\ 0 & 0 & 0 & -1 \end{pmatrix}. \tag{8.16}$$

Rotating a spinor one full turn gives us back the negative of what we started with! We must rotate through two full turns to get back to where we started:

$$S(4\pi \text{ rotation}) = \begin{pmatrix} 1 & 0 & 0 & 0 \\ 0 & 1 & 0 & 0 \\ 0 & 0 & 1 & 0 \\ 0 & 0 & 0 & 1 \end{pmatrix}. \tag{8.17}$$

A spinor, then, is an object that in a sense "rotates more slowly" than a vector, and is therefore able to resolve the structure of the Lorentz group more finely than a vector. There are some very neat demonstrations (usually involving ribbons) of the fact that the group of rotations really does contain additional structure that vectors miss. This is not a rotation in the usual sense, however, since the components of the spinor do not correspond to directions in space. It's more a case that, as we rotate a system in space, we must make a simultaneous "rotation" of the components in a related spinor space.

With the spinor, we have found a representation of the Lorentz group that cannot be constructed as a tensor from the fundamental vector representation. Since the spinor resolves the structure of the Lorentz group more finely than a vector, it appears to be the more fundamental object. A question that presents itself, then, is whether we can express scalars, vectors, and higher-order tensors in terms of spinors. We can certainly construct a single real number out of a spinor by using the Hermitian conjugate: $\psi^\dagger \psi$. However, since $\psi^\dagger$ transforms under Lorentz transformations according to $\psi^{\dagger\prime} = \psi^\dagger S^\dagger$, and S is not unitary, this combination is *not* a scalar. In fact, it can be shown that $S^\dagger = \gamma^0 S^{-1} \gamma^0$, leading to the conclusion that the appropriate scalar quantity is the combination $\psi^\dagger \gamma^0 \psi$. For this reason, we find it useful to define the *adjoint* of a spinor as $\overline{\psi} = \psi^\dagger \gamma^0$. With this definition, it can also be shown that $\overline{\psi}\gamma^\mu\psi$ transforms as a vector. In fact, following a similar procedure to that used in the cases of the spin-0 and spin-1 equations, it can also be shown that $\overline{\psi}\gamma^\mu\psi$ is the conserved current for the Dirac equation. This requires the introduction of the adjoint Dirac equation. Recall that for the Klein-Gordon equation, both ϕ and ϕ^* were solutions. This is because the Klein-Gordon operator is real, so taking the conjugate of the equation is equivalent to taking the conjugate of the wavefunction. In the case of the Dirac equation, the operator is not real and is also matrix-valued, so taking the Hermitian conjugate gives a slightly altered version of the Dirac equation that is satisfied by the adjoint spinor:

$$[(i\gamma^\mu\partial_\mu - m)\,\psi]^\dagger = 0$$
$$\psi^\dagger\left(-i\gamma^{\mu\dagger}\overleftarrow{\partial}_\mu - m\right) = 0 \qquad (8.18)$$
$$\overline{\psi}\left(i\gamma^\mu\overleftarrow{\partial}_\mu + m\right) = 0,$$

where $\overleftarrow{\partial}_\mu$ denotes a derivative that acts on everything to the *left*, not to the right. The last line follows from the identity $\gamma^{\mu\dagger} = \gamma^0\gamma^\mu\gamma^0$, after canceling an overall factor of γ^0. From this and the original Dirac equation, we can show that the conserved current for a spinor is given by $i\overline{\psi}\gamma^\mu\psi$, as the reader is invited to show in Exercise 9.

8.4 SOLUTIONS OF THE DIRAC EQUATION

As with previous relativistic equations, we will assume a plane-wave solution to the Dirac equation of the form

$$\psi = u(p)e^{-ip\cdot x}, \qquad (8.19)$$

where $u(p)$ is a normalized constant spinor that depends on the momentum of the particular plane wave in question. Acting with the momentum operator $\widehat{p}^\mu$ shows that this is a momentum eigenstate with eigenvalue p^μ, and substituting the plane wave into the Dirac equation gives

$$(\gamma^\mu p_\mu - m)\,\psi = 0. \tag{8.20}$$

This is the momentum-space Dirac equation, and in fact holds for any solution of the real-space equation, as can be seen by taking the Fourier transform of the original equation. Notice that the gamma matrices are contracted with the momentum. Such contractions occur so regularly when working with the Dirac equation that it is convenient to introduce a shorthand notation. When a quantity is written with a "slash" through it, this represents contraction with the gamma matrices. Hence $\slashed{p} = \gamma^\mu p_\mu$ and $\slashed{\partial} = \gamma^\mu \partial_\mu$. In the case of the plane-wave solutions, then, we have

$$\left(\slashed{p} - m\right) u(p) e^{-ip \cdot x} = 0. \tag{8.21}$$

Since the exponential factor cannot take a value of 0, this places a necessary constraint on the spinorial factor of plane-wave solutions:

$$\left(\slashed{p} - m\right) u(p) = 0. \tag{8.22}$$

Multiplying from the left by $\left(\slashed{p} + m\right)$ gives $\left(\slashed{p} + m\right)\left(\slashed{p} - m\right) u(p) = 0$, which Equation 8.6 reduces to $\left(p^2 - m^2\right) u(p) = 0$. Since the bracket is now a scalar, it must be the case that $p^2 = m^2$, so the plane wave describes a particle of mass m as expected. However, it also means that the plane wave is a solution as long as $E = \pm\sqrt{\mathbf{p}^2 + m^2}$: the negative-energy solutions that were the main motivation for Dirac's equation are still present! Luckily we, unlike Dirac, are already in the fortunate position of having a handy interpretation for such solutions, so they need not worry us. In fact, it is Dirac whom we must thank for that interpretation, as it was he who realized that the negative-energy solutions arising from his equation were describing a type of matter that had not previously been detected or predicted.

Multiplying Equation 8.22 out to form a scalar in a sense "averaged out" the constraints on the individual components of $u(p)$ to show that our solution describes a relativistic particle. If we keep the components separate, Equation 8.22 contains additional information. Writing the spinor as

$$u(p) = N \begin{pmatrix} u_A(p) \\ u_B(p) \end{pmatrix}, \tag{8.23}$$

where $u_A(p)$ and $u_B(p)$ are two-component spinors, we can rewrite the constraint in block form in the Dirac representation as:

$$\left[\begin{pmatrix} \mathbb{1} & 0 \\ 0 & \mathbb{1} \end{pmatrix} E + \begin{pmatrix} 0 & \sigma_i \\ -\sigma_i & 0 \end{pmatrix} p_i - m \begin{pmatrix} \mathbb{1} & 0 \\ 0 & \mathbb{1} \end{pmatrix} \right] \begin{pmatrix} u_A(p) \\ u_B(p) \end{pmatrix} = 0, \tag{8.24}$$

or

$$\begin{pmatrix} E - m & -\sigma \cdot \mathbf{p} \\ \sigma \cdot \mathbf{p} & -E - m \end{pmatrix} \begin{pmatrix} u_A(p) \\ u_B(p) \end{pmatrix} = 0. \tag{8.25}$$

From this, we find that we can express the two-component spinors in terms of each other, as

$$\begin{aligned} u_A(p) &= \frac{\sigma \cdot \mathbf{p}}{E - m} u_B(p) \\ u_B(p) &= \frac{\sigma \cdot \mathbf{p}}{E + m} u_A(p). \end{aligned} \tag{8.26}$$

Therefore, only one of these objects is independent: once it is chosen, the other is fixed. There is freedom, however, to choose the first of these objects as we please. If we let $u_A = \begin{pmatrix} 1 \\ 0 \end{pmatrix}$, then

$$\begin{aligned} u_B &= \frac{\sigma \cdot \mathbf{p}}{E + m} u_A \\ &= \frac{1}{E + m} \left(\begin{pmatrix} 0 & 1 \\ 1 & 0 \end{pmatrix} p_x + \begin{pmatrix} 0 & -i \\ i & 0 \end{pmatrix} p_y \right. \\ &\quad \left. + \begin{pmatrix} 1 & 0 \\ 0 & -1 \end{pmatrix} p_z \right) \begin{pmatrix} 1 \\ 0 \end{pmatrix} \\ &= \frac{1}{E + m} \begin{pmatrix} p_z \\ p_x + ip_y \end{pmatrix}. \end{aligned} \tag{8.27}$$

So one possible solution to the Dirac equation (in the Dirac representation) is given by

$$u_1(p) = N \begin{pmatrix} 1 \\ 0 \\ \frac{p_z}{E+m} \\ \frac{p_x+ip_y}{E+m} \end{pmatrix}. \tag{8.28}$$

Notice that, as $\mathbf{p} \to 0$,

$$
\begin{aligned}
E + m &= m \pm \sqrt{\mathbf{p}^2 + m^2} \\
&= m \left(1 \pm \sqrt{1 + \frac{\mathbf{p}^2}{m^2}} \right) \\
&\to m \left(1 \pm \left(1 + \frac{1}{2} \frac{\mathbf{p}^2}{m^2} \right) \right) \\
&= m \pm m \pm \frac{\mathbf{p}^2}{2m},
\end{aligned} \tag{8.29}
$$

where the $\pm$ sign depends on whether we are considering positive- or negative-energy solutions.

If $E < 0$, then $E+m = -\frac{\mathbf{p}^2}{2m}$ so $u_B \to \infty$ in the limit of vanishing momentum, which is clearly nonsensical. On the other hand, if $E > 0$, then $E + m = 2m$, leading to a well-defined $u_B(p)$. Since we can always transform to a frame in which the three-momentum is zero, for consistency we must identify the u_1 solution as a positive-energy solution.

We could have chosen the $u_A(p)$ differently as $u_A(p) = \begin{pmatrix} 0 \\ 1 \end{pmatrix}$, in which case we would have found a different $u_B(p)$. Analysis similar to the previous one gives two independent positive-energy solutions:

$$u_1 = N \begin{pmatrix} 1 \\ 0 \\ \frac{p_z}{E+m} \\ \frac{p_x+ip_y}{E+m} \end{pmatrix} \quad u_2 = N \begin{pmatrix} 0 \\ 1 \\ \frac{p_x-ip_y}{E+m} \\ \frac{-p_z}{E+m} \end{pmatrix} \tag{8.30}$$

and two independent negative-energy solutions:

$$u_3 = N \begin{pmatrix} \frac{p_z}{E-m} \\ \frac{p_x+ip_y}{E-m} \\ 1 \\ 0 \end{pmatrix} \quad u_4 = N \begin{pmatrix} \frac{p_x-ip_y}{E-m} \\ \frac{-p_z}{E-m} \\ 0 \\ 1 \end{pmatrix}. \tag{8.31}$$

The final thing to consider in constructing the basis spinors is the value of the normalization constant, N. Since we are describing relativistic particles, we use the relativistic normalization, in which there are $2E$ particles in a unit volume. This requires $\psi^\dagger \psi = u^\dagger u = 2E$. Since we can show that $u^\dagger u = 2EN^2/(E+m)$, this requires

$$N = \sqrt{E+m}. \tag{8.32}$$

8.4.1 Spin

In order to find the spin of the particles described by the Dirac equation, it is necessary to find the representation of the Lorentz group that is appropriate for spinors. To do this, we look at the Dirac equation in two infinitesimally close frames. If the spinor as a function of space-time is given by $\psi(x)$ in one frame, then there is a corresponding function ψ' that takes the transformed coordinate $\overline{x}^\mu = \Lambda^\mu{}_\nu x^\nu$ as its argument. Equivalently, we can express x in terms of $\overline{x}$, as $x^\nu = \left(\Lambda^{-1}\right)^\nu{}_\mu \overline{x}^\mu$. To move between these descriptions, we have the transformation law

$$\psi'(\overline{x}) = S\psi(x). \tag{8.33}$$

In the case of an infinitesimal transformation, we can expand both S and x to first order to give

$$\psi'(\overline{x}) = \left(\mathbb{1} - \frac{i}{4}\delta\omega_{\mu\nu}\,\sigma^{\mu\nu}\right)\psi\left(\overline{x}^\mu - g^{\mu\nu}\delta\omega_{\nu\rho}\overline{x}^\rho\right), \tag{8.34}$$

or, as a Taylor expansion of ψ,

$$\psi'(\overline{x}) = \left(\mathbb{1} - \frac{i}{4}\delta\omega_{\mu\nu}\,\sigma^{\mu\nu}\right)\left(1 - g^{\mu\nu}\delta\omega_{\nu\rho}\,\overline{x}^\rho\frac{\partial}{\partial\overline{x}^\mu}\right)\psi(\overline{x}). \tag{8.35}$$

We now write this in terms of the Lorentz generators $J^{\mu\nu}$, in the form

$$\psi'(\overline{x}) = \left(1 - \frac{i}{2}\delta\omega_{\mu\nu}\, J^{\mu\nu}\right)\psi(\overline{x}), \qquad (8.36)$$

where the factor of $1/2$ accounts for the over-counting due to $\omega_{\mu\nu}$, including each parameter twice. Relabeling dummy indices so that we may factorize out the $\delta\omega$ and ψ, and neglecting the second-order term, we have

$$-\frac{i}{2}\delta\omega_{\mu\nu}\, J^{\mu\nu}\psi = \delta\omega_{\mu\nu}\left(-\frac{i}{4}\sigma^{\mu\nu} - g^{\rho\mu}x^\nu\partial_\rho\right)\psi. \qquad (8.37)$$

From this, we find that we can identify the Lorentz generators for spinors as

$$J^{\mu\nu} = \frac{1}{2}\sigma^{\mu\nu} + x^\mu p^\nu - x^\nu p^\mu, \qquad (8.38)$$

where we have made use of the identities $p^\mu = i\partial^\mu$ and $\omega_{\mu\nu} = \frac{1}{2}(\omega_{\mu\nu} - \omega_{\nu\mu})$. In particular, looking just at the spatial parts:

$$J^{ij} = \frac{1}{2}\sigma^{ij} + x^i p^j - x^j p_i. \qquad (8.39)$$

Here, we recognize the second and third terms as the orbital angular momentum operator, as in Equation 4.25, and so we identify the first term as the spin operators, in tensor form. Converting this to three-vector form (see Exercise 5), the spin operators can then be identified as

$$\Sigma_i = \frac{1}{2}\begin{pmatrix} \sigma_i & 0 \\ 0 & \sigma_i \end{pmatrix}, \qquad (8.40)$$

and the total spin as

$$\Sigma^2 = \frac{3}{4}\mathbb{1}_4. \qquad (8.41)$$

Since the eigenvalue of the total spin is $\sqrt{s(s+1)}$, we see that the spinor describes a spin-$\frac{1}{2}$ particle as expected.

To fully appreciate a key moment in the history of particle physics, it is important that the reader be comfortable with the fact that spinors really do describe spin-$\frac{1}{2}$ particles. So if further evidence is needed, consider also the fact that the Dirac equation is a relativistic Schrödinger equation with Hamiltonian $\widehat{H} = \alpha \cdot \widehat{p} + \beta m$. Any conserved quantity should, therefore, commute with this Hamiltonian. It is reasonably straightforward to verify that the orbital angular momentum operator $\widehat{L}$ does not commute with $\widehat{H}$, but that the total angular momentum operator $\widehat{J} = \Sigma + \widehat{L}$ does commute. Hence, if we are to retain the conservation of angular momentum (which is a reasonable assumption!), then the spin must be given by the above spin operators, leading again to a spin-$\frac{1}{2}$ particle.

8.4.2 Antiparticles

The reason for emphasizing the spin-$\frac{1}{2}$ nature of spinors in Section 8.4.1 is that when Dirac first constructed his equation, he too derived the spin of the objects he had described, and so found that the equation was appropriate for describing electrons. However, with a spin of $\frac{1}{2}$, he would have expected as we should that there would be only two degrees of freedom. There appear, therefore, to be two additional degrees of freedom. Furthermore, these additional states have negative energy. It was for this reason that Dirac originally postulated the concept of antimatter. In fact, as we saw in Chapter 1, Dirac's concept of antimatter was slightly different from the modern understanding, and was based on the idea of a sea of negative-energy states held up by the exclusion principle. While this hole theory of antimatter is very elegant, it is unfortunately flawed, since we now know that bosons can also have antiparticles. In this case, the hole theory does not work since bosons are not subject to the exclusion principle and so would all decay away to infinitely negative energy states. For a full understanding of the nature of antiparticles, it is necessary to introduce the quantum theory of fields, which we will discuss only briefly in Chapter 9, since it is somewhat beyond the scope of this text. The outcome, however, is that negative-energy particle solutions can be reinterpreted as positive-energy antiparticle solutions and vice versa, regardless of the spin of the particles in question. In the case of a scalar particle, we moved between these equivalent descriptions using the complex conjugate. In the case of spinors, things are not quite so straightforward, and we must use a "charge conjugate" operation to switch between descriptions. This will be introduced in Section 8.6.

For now, let us find a set of basis spinors more suited to describing the negative-energy states, to complement those derived in Section 8.4. Starting again from the assumption of a plane-wave solution $\psi \propto e^{-ip\cdot x} \propto e^{-i(Et-\mathbf{p}\cdot\mathbf{x})}$, but insisting now that E is to be defined as the positive root $(E = +\sqrt{|\mathbf{p}|^2 + m^2})$, then in order to write the negative-energy solutions, it is necessary to change the sign in the exponent:

$$\psi = v(p)e^{ip\cdot x} = v(p)e^{-i(-Et+\mathbf{p}\cdot\mathbf{x})}. \tag{8.42}$$

Notice that we now use $v(p)$ for the basis spinor to differentiate the negative energy-type solutions. Substituting this into the Dirac equation gives

$$(i\gamma^\mu \partial_\mu - m)\, v e^{ip\cdot x} = 0$$
$$(-\gamma^\mu p_\mu - m)\, v = 0 \tag{8.43}$$
$$(\not{p} + m)\, v = 0.$$

Recall that the corresponding constraint for u-type solutions is given by $(\not{p} - m)\, u(p) = 0$, so we have different constraints for the two types of solution. Writing the basis spinor as $v = N \begin{pmatrix} v_A \\ v_B \end{pmatrix}$ and following the same steps as for the u-type spinors leads to the constraints

$$v_A(p) = \frac{\sigma \cdot \mathbf{p}}{E + m} v_B(p)$$
$$v_B(p) = \frac{\sigma \cdot \mathbf{p}}{E - m} v_A(p), \tag{8.44}$$

and ultimately to the basis spinors

$$v_1 = N \begin{pmatrix} \frac{p_x - ip_y}{E+m} \\ \frac{-p_z}{E+m} \\ 0 \\ 1 \end{pmatrix}, \quad v_2 = N \begin{pmatrix} \frac{p_z}{E+m} \\ \frac{p_x + ip_y}{E+m} \\ 1 \\ 0 \end{pmatrix},$$

$$v_3 = N \begin{pmatrix} 1 \\ 0 \\ \frac{p_z}{E-m} \\ \frac{p_x+ip_y}{E-m} \end{pmatrix}, \quad v_4 = N \begin{pmatrix} 0 \\ 1 \\ \frac{p_x-ip_y}{E-m} \\ \frac{p_z}{E-m} \end{pmatrix}. \tag{8.45}$$

As with the u-type solutions, consistency demands that two of these spinors have negative energy and two have positive energy. However, it is important to note that, in this case, we have defined E to be the negative of the energy, since the v-type solutions have $\psi = v(p)e^{ip\cdot x}$. So v_1 and v_2 have positive E and negative energy, while v_3 and v_4 have negative E and positive energy.

We now have eight basis spinors: four u-type and four v-type. However, only four of these are linearly independent. Therefore, the standard choice of basis for spinors in the Dirac representation is to use u_1, u_2, v_1 and v_2 since all of these have positive E. The u-type solutions are then used to describe particles while the v-type solutions are used for antiparticles. The standard basis is thus:

$$u_1 = N \begin{pmatrix} 1 \\ 0 \\ \frac{p_z}{E+m} \\ \frac{p_x+ip_y}{E+m} \end{pmatrix}, \quad u_2 = N \begin{pmatrix} 0 \\ 1 \\ \frac{p_x-ip_y}{E+m} \\ \frac{-p_z}{E+m} \end{pmatrix},$$

$$v_1 = N \begin{pmatrix} \frac{p_x-ip_y}{E+m} \\ \frac{-p_z}{E+m} \\ 0 \\ 1 \end{pmatrix}, \quad v_2 = N \begin{pmatrix} \frac{p_z}{E+m} \\ \frac{p_x+ip_y}{E+m} \\ 1 \\ 0 \end{pmatrix}. \tag{8.46}$$

These form a suitable basis, since they are complete, in the sense that

$$\sum_{A=1}^{2} u_A(p) \cdot \overline{u}_A(p) = \not{p} + m, \quad \sum_{A=1}^{2} v_A(p) \cdot \overline{v}_A(p) = \not{p} - m, \tag{8.47}$$

and orthonormal, in that

$$\overline{u}_A(p) \cdot u_B(p) = 2m\delta_{AB}, \quad \overline{v}_A(p) \cdot v_B(p) = -2m\delta_{AB}, \tag{8.48}$$

as the reader is invited to verify in Exercise 12.

When analyzing antiparticle solutions, it is necessary to use slightly modified operators for observables. This is most obvious in the case of the energy, since acting on $\psi = v_1 e^{ip \cdot x}$ with $i\frac{\partial}{\partial t}$ gives $-E$, where we know that the value of E is positive. That is, although the basis spinors have been defined in such a way as to make E positive, the v-type solutions still have negative energy. To understand why, it is important to realize that all of the solutions we have found are really *particle* solutions. We are still describing antiparticles in terms of particles via crossing symmetry. The physical energy of the antiparticle is then the negative of that found by acting with the energy operator on the negative-energy particle solution.

With this point in mind, it is now easy to see that other operators should also be modified in order to find the physical values of observables for antiparticles. For example, the physical spin of an antiparticle is given by $-\frac{1}{2}\Sigma$, since a spin-up particle is equivalent to a spin-down antiparticle.

8.4.3 Helicity

The helicity of a spin-$\frac{1}{2}$ particle is defined as for a spin-1 particle, as the component of spin along the axis defined by the particle's momentum. Since the spin operators are already defined for the spinor, we can define the helicity operator as

$$
\begin{aligned}
h(p) &= \frac{\Sigma \cdot \mathbf{p}}{|\mathbf{p}|} \\
&= \frac{1}{2} \begin{pmatrix} \frac{\sigma \cdot \mathbf{p}}{|\mathbf{p}|} & 0 \\ 0 & \frac{\sigma \cdot \mathbf{p}}{|\mathbf{p}|} \end{pmatrix}.
\end{aligned}
\tag{8.49}
$$

The standard massive basis states introduced in Section 8.4.2 are not helicity eigenstates. In fact, they are not, in general, eigenstates of any of the spin operators. The only exception is that a state with momentum in the z direction has a well-defined z-component of spin. It may be useful, therefore, to find a general set of helicity eigenstates. It is possible to construct a basis of helicity states, though exploration of this set will be left to the reader in Exercise 11.

A spinor can be either right-helical or left-helical, with helicity eigenvalues of $+1/2$ and $-1/2$ respectively. The reader may see these states referred to as left- and right-handed states, but this terminology is avoided here since it can sometimes also refer to the chirality of a state, leading to confusion. It is worth noting that the helicity of a state is a conserved quantity: that is, the helicity operator commutes with the Hamiltonian. This seems intuitive since conservation of angular momentum would suggest that a particle that is "spinning" one way cannot spontaneously transform into a particle spinning the other way. There is, however, one way in which we can apparently change the helicity. Consider a right-helical particle with a velocity v in some reference frame. Now boost into a reference frame with a velocity in the same direction as the particle but with a larger magnitude. In other words, consider overtaking the particle. From the point of view of the observer in this boosted frame, the particle will now have a relative velocity in the opposite direction from that in the original frame. However, the boost has no effect on the apparent spin state of the particle. The particle's spin was aligned with its velocity in the original frame, but that spin is now *anti-parallel* to the new velocity. The particle which was right-helical is now left-helical. That the helicity is not Lorentz-invariant should not really come as a surprise, since the form of the operator depends only on the three-momentum, rather than on the full momentum four-vector.

8.4.4 Chirality

The chirality of a state also comes in left- and right-handed varieties, but is a much more abstract concept than the helicity. Whereas the helicity has a direct physical interpretation as the alignment of the particle's spin, the chirality has a less clear physical interpretation, relating to the transformation properties of a spinor. Specifically, it is due to the fact that the Dirac spinor is a reducible representation of the Lorentz group, which can be broken down into two irreducible representations. It is these irreducible representations that are referred to as the left- and right-chiral components of the spinor. To see how the concept arises, we first introduce a fifth gamma matrix $\gamma^5 = i\gamma^0\gamma^1\gamma^2\gamma^3$. This is traditionally referred to as γ^5, since its introduction dates back to a time when use of the indices $1, 2, 3, 4$ for space and time was more common than $0, 1, 2, 3$. We say "fifth gamma matrix" because it can be shown to obey the same Clifford algebra as the other matrices:

$$\{\gamma^\mu, \gamma^5\} = 0, \quad \{\gamma^5, \gamma^5\} = 2. \tag{8.50}$$

In fact, if we were to construct spinors in a five-dimensional space-time, γ^5 would play the role of the extra gamma matrix. Notice that the above relations are true in any representation, and can be derived directly from the definition of γ^5 and the anticommutation relations for the other gamma matrices. In a particular representation, of course, γ^5 can be written explicitly. We could, if we wished, write the chirality operator in the Dirac representation. For what follows, however, we will find it most useful to work in the Weyl representation, since this is specifically designed for working with chiral states. Indeed, it is sometimes referred to as the chiral representation. In this representation, γ^5 is given by

$$\gamma^5 = \begin{pmatrix} -\mathbb{1}_2 & 0 \\ 0 & \mathbb{1}_2 \end{pmatrix}. \tag{8.51}$$

Next we introduce two operators $\widehat{L} = \frac{1}{2}\left(\mathbb{1}_4 - \gamma^5\right)$ and $\widehat{R} = \frac{1}{2}\left(\mathbb{1}_4 + \gamma^5\right)$, known as the left- and right-projection operators. Between them, these operators have three important properties. First, their sum is the identity operator, so that $\widehat{L} + \widehat{R}$ acting on any state gives back the same state. Second, they are both *idempotent*, which is to say that they each square to themselves: $\widehat{L}\widehat{L} = \widehat{L}$ and $\widehat{R}\widehat{R} = \widehat{R}$. The third property is that the product of the two operators vanishes: $\widehat{L}\widehat{R} = \widehat{R}\widehat{L} = 0$. These properties are what characterize the operators as projection operators. To see why, define the states ψ_L and ψ_R by $\psi_L = \widehat{L}\psi$ and $\psi_R = \widehat{R}\psi$ for some arbitrary spinor ψ. We will call these the left-chiral and right-chiral projections of ψ. The first property guarantees that $\psi_L + \psi_R = \psi$, so the two projected components sum to give the original state. In the Weyl representation, these projection operators take the simple form

$$\widehat{L} = \begin{pmatrix} \mathbb{1}_2 & 0 \\ 0 & 0 \end{pmatrix}, \quad \widehat{R} = \begin{pmatrix} 0 & 0 \\ 0 & \mathbb{1}_2 \end{pmatrix}, \tag{8.52}$$

and it is easy to see that they have the effect of projecting out the top two and bottom two components of the spinor respectively. For this reason, in the Weyl representation we introduce 2 two-component spinors χ_L and χ_R such that the Dirac spinor is constructed out of these as

$$\psi = \begin{pmatrix} \chi_L \\ \chi_R \end{pmatrix}. \tag{8.53}$$

A common abuse of notation is to use ψ_L to refer to both a left-chiral Dirac spinor *and* the two-component spinor forming the top two components in the Weyl representation. To avoid any confusion, we will make a clear distinction here, using ψ for four-component objects and χ for two-component objects. Notice that a left-chiral spinor can now be written $\psi_L = (\ \chi_L,\ 0\ )^{\mathrm{T}}$. Similarly, $\psi_R = (\ 0,\ \chi_R\ )^{\mathrm{T}}$.

Recall that the infinitesimal Lorentz transformation of a spinor is given by

$$S\psi = \left(\mathbb{1} - \frac{i}{4}\omega_{\mu\nu}\sigma^{\mu\nu} \right)\psi. \tag{8.54}$$

Let us now consider this expression in terms of its components. First, recall that the antisymmetric parameter matrix $\omega_{\mu\nu}$ consists of six independent parameters: three angles $\theta_{1,2,3}$ which parametrize rotations about the three spatial axes, and three rapidities $\xi_{1,2,3}$ that parametrize boosts. These parameters can be extracted from the parameter matrix according to

$$\xi_i = \omega_{i0} = \frac{1}{2}\left(\omega_{i0} - \omega_{0i} \right)$$
$$\text{and} \quad \theta_i = \frac{1}{2}\varepsilon_{ijk}\omega_{jk}. \tag{8.55}$$

Furthermore, in the Weyl representation, the generator matrices $\sigma^{\mu\nu} = \frac{i}{2}\left[\gamma^\mu, \gamma^\nu \right]$ can be split into

$$\sigma^{i0} = i\begin{pmatrix} \sigma^i & 0 \\ 0 & -\sigma^i \end{pmatrix}, \quad \sigma^{ij} = \varepsilon^{ijk}\sigma_k. \tag{8.56}$$

This allows us to write the previous infinitesimal transformation in the form

$$S\psi = \begin{pmatrix} \mathbb{1}_2 - \frac{i}{2}\theta_k\sigma_k - \frac{1}{2}\xi_k\sigma_k & 0 \\ 0 & \mathbb{1}_2 - \frac{i}{2}\theta_k\sigma_k + \frac{1}{2}\xi_k\sigma_k \end{pmatrix} \begin{pmatrix} \chi_L \\ \chi_R \end{pmatrix}.$$
(8.57)

So we see that the transformation laws for left- and right-chiral spinors are given by

$$\chi_L \mapsto \left(\mathbb{1}_2 - \frac{i}{2}\theta_k\sigma_k - \frac{1}{2}\xi_k\sigma_k \right) \chi_L \quad \text{and}$$

$$\chi_R \mapsto \left(\mathbb{1}_2 - \frac{i}{2}\theta_k\sigma_k + \frac{1}{2}\xi_k\sigma_k \right) \chi_R.$$
(8.58)

In particular, the two irreducible representations transform identically under rotations but have opposite behavior under boosts.

In the case of massive particles, the helicity and the chirality are essentially completely unrelated concepts, but they do have somewhat complementary properties. Namely, while the helicity is conserved but not Lorentz-invariant, the chirality is Lorentz-invariant but not conserved. Indeed, as we will see in Section 8.5, the mass term in the Dirac equation mixes the chiral components of a spinor, so it does not even make sense to talk of the chirality of a massive particle. However, since the chiral components of a spinor transform independently under Lorentz transformations, the chiral components are clearly individually Lorentz-covariant.

8.5 MASSLESS PARTICLES

Working again in a representation-free scheme, since γ^5 anticommutes with the gamma matrices, we have $i\displaystyle{\not\partial}\gamma^5 = -i\gamma^5\not\partial$. Since the identity obviously commutes with this differential operator, it is easy to see that $i\not\partial\widehat{L} = \widehat{R}i\not\partial$ and $i\not\partial\widehat{R} = \widehat{L}i\not\partial$. Clearly, then, acting on either chiral component of a spinor with the differential operator found in the Dirac equation reverses the chirality of that component. Therefore, when written in chiral components, the Dirac equation becomes a set of coupled equations:

$$i\partial\!\!\!/\psi_L = m\psi_R, \quad i\partial\!\!\!/\psi_R = m\psi_L. \tag{8.59}$$

Notice that these equations are only coupled in the case of a massive particle. If m is set to zero, the equations decouple to

$$i\partial\!\!\!/\psi_L = 0, \quad i\partial\!\!\!/\psi_R = 0, \tag{8.60}$$

which, in the Weyl representation, simplifies to a pair of massless Pauli equations

$$i\sigma_i\partial_i\chi_L = 0, \quad i\sigma_i\partial_i\chi_R = 0. \tag{8.61}$$

We see that the left- and right-chiral components of a massless spinor behave completely independently of each other, so it is, in fact, more appropriate to consider each as a separate particle, each with a fixed chirality.

The standard basis spinors derived in Section 8.4 are not appropriate for massless particles, since they are ultimately based on the spin of the particle in its rest frame, and a massless particle *has no* rest frame. The appropriate basis for massless particles is, therefore, the helicity basis (see Exercise 11). Notice that the argument given in Section 8.4.3 based on overtaking the particle to change its helicity no longer applies in the case of massless particles: since massless particles must travel at the speed of light, we *cannot* overtake them. So in the massless case, helicity is both Lorentz-invariant *and* conserved. Similarly, since chirality of a massless particle is fixed, the chirality is also both Lorentz-invariant and conserved. In fact, it can be shown that the two concepts become identical for massless particles. In particular, the left-chiral state has a negative helicity, while the right-chiral state has positive helicity. It is for this reason that chiral states are labeled as left and right, but only in the massless case do these labels mean anything!

8.6   CHARGE CONJUGATION

For scalar particles, the complex conjugate was sufficient to define the charge conjugate of a particle. In the case of a spinor, since it is a multi-component object, the complex conjugate will no longer necessarily do the job. Instead, we assume that the charge conjugate of a spinor is given by

$$\psi^c = C\psi^*, \tag{8.62}$$

where C is some matrix. In particular, we assume that for a basis spinor u, there is a charge conjugate spinor $v = Cu^*$. Recall that antiparticle solutions must satisfy the constraint

$$\left(\slashed{p} + m\right) v(p) = 0. \tag{8.63}$$

Also, if v is to be an antiparticle solution of the Dirac equation, then its charge conjugate, u, should be a particle solution. Therefore, taking the complex conjugate of the Dirac equation:

$$\left(-i\gamma^{\mu*}\partial_\mu - m\right) u^* e^{ip\cdot x} = 0$$
$$\left(\gamma^{\mu*}p_\mu - m\right) u^* = 0 \tag{8.64}$$
$$\left(\gamma^{\mu*}p_\mu C^{-1}\right) v = 0.$$

The final expressions in Equations 8.43 and 8.64 must be equivalent, so multiplying the latter from the left by C, we find

$$\left(C\gamma^{\mu*}C^{-1}p_\mu - m\right) v = 0, \tag{8.65}$$

so equivalence requires

$$C\gamma^{\mu*}C^{-1} = -\gamma^\mu. \tag{8.66}$$

From this, we can verify that $C = i\gamma^2$ is an appropriate charge conjugation matrix. Incidentally, there is nothing significant about the fact that C is related to γ^2. This does not imply anything special about γ^2

or the y direction. The fact is that this C is representation-specific, and the relation between C and γ^2 is an artifact of the Dirac and Weyl representations. Using this C, we can also verify that, of the standard Dirac-representation basis spinors given in Section 8.4, we have $v_1 = u_1^c$ and $v_2 = u_2^c$. Note, this is the reason for the overall negative sign in v_2.

An important property of the charge conjugation operation, which we will make use of when formalizing the Standard Model, is that it inter-converts left-chiral and right-chiral spinors.

8.7   DIRAC, WEYL, AND MAJORANA SPINORS

Spinors come in several varieties in particle physics, though they are all related. To understand these relations, it is necessary to delve a little into the mathematics of the special orthogonal groups. Specifically, we have already seen in Section 4.3.2 that the rotation group has a double cover, in the form of $SU(2)$. This means that, while all $SO(3)$ representations are also representations of $SU(2)$, the converse is not true. $SU(2)$ has extra representations in addition to those of the rotation group, but which must still be considered in quantum mechanics, since only squared amplitudes have physical meaning. This idea generalizes to larger rotation groups, though the double cover in most cases does not coincide with one of the simple Lie groups. Instead, the double cover of $SO(n)$ is referred to as $Spin(n)$, and the representations of $Spin(n)$ that are not present in $SO(n)$ are the spinor representations. For odd dimensions, n, there is only one fundamental spinor representation, but when n is even, there are two. In the even case, the two fundamental representations are chiral, with one left-chiral and one right-chiral, whereas there is no concept of chirality in odd-dimensional spaces. The chiral spinors are irreducible representations and are known as Weyl spinors. They are complex-valued representations with $2^{\left(\frac{n}{2}-1\right)}$ components. In the case of the Lorentz group, therefore, the Weyl spinors are the two-component chiral spinors appearing as the upper and lower blocks of the Dirac spinor in Weyl representation. The Dirac spinors, on the other hand, are reducible, and can be composed from two Weyl spinors, as we have already seen. The Dirac spinor, then, has $2^{\left(\frac{n}{2}\right)}$ complex components, giving a four-component spinor for the Lorentz group.

Both Weyl and Dirac spinors are complex representations of their algebras, but there is a third type of spinor representation. We can impose a reality condition on a Dirac spinor, $\psi^* = \psi$, reducing the number of components by half. In Minkowski space, then, we can define a real-valued spinor with two components. A spinor defined in this way is called a Majorana spinor, and a fermion described by such a spinor is a Majorana fermion, but what would be its properties? First, it is important to realize that, if a Majorana spinor is to obey a Dirac equation, then since the mass term is clearly real valued, we must have

$$(i\gamma^\mu)^* = i\gamma^\mu. \tag{8.67}$$

That is, all non-zero entries in the gamma-matrices must be purely imaginary, and the representations we have used so far do not satisfy this requirement: the Majorana representation is used instead.

If we were to attempt to apply a reality condition to a Weyl spinor, we would find the attempt unsuccessful. There simply is not enough freedom in a Weyl spinor to restrict the components further. But the notion of chirality is related to which Weyl spinor representation a fermion belongs to. With this in mind, then, a Majorana fermion is non-chiral. It does, however, have a helicity. Recall that a real scalar must describe a particle that is its own antiparticle. In a similar vein, we find that the antiparticle of a Majorana fermion is the same fermion with opposite helicity! This in turn would imply that the Majorana fermion can carry no fermion number, since the left- and right-handed particles would necessarily have to carry both the same and the opposite number. There is still some debate over the possibility of Majorana neutrinos, since all neutrinos are believed to be left-handed, while all antineutrinos appear to be right-handed. A smoking gun for the Majorana nature of neutrinos would be the observation of neutrinoless double-beta decay, since this would require an interaction of the form

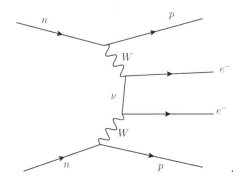

Notice the lack of arrows on the internal neutrino line, due to the lack of a fermion number. Any attempt to direct these lines with arrows results in two lines pointing toward the same vertex, demonstrating that this process cannot occur if neutrinos are not Majorana particles. To be clear, there is currently no evidence for this process, but the possibility is an intriguing one.

8.8   BILINEAR COVARIANTS

Returning to the slightly more familiar territory of Dirac spinors, we know that the number of components is four in 4D space. Taking a product of two Dirac spinors, then, should give us an object with 16 components. As with the combination of spins under $SU(2)$, however, the object formed in this way is not irreducible, but it should be expressible as a sum of irreducible representations of the Lorentz group. These are the bilinear covariants for a Dirac spinor, and they are characterized by their behavior under proper Lorentz transformations and parity.

Bilinear form	type	transformation	parity	components
$\bar{\psi}\psi$	scalar	1	$+1$	1
$\bar{\psi}\gamma^{\mu}\psi$	vector	$\Lambda^{\mu}{}_{\nu}$	-1	4
$\bar{\psi}\sigma^{\mu\nu}\psi$	antisymmetric tensor	$\Lambda^{\mu}{}_{\rho}\Lambda^{\nu}{}_{\tau}$	$+1$	6
$\bar{\psi}\gamma^{\mu}\gamma^{5}\psi$	pseudo-vector	$\Lambda^{\mu}{}_{\nu}$	$+1$	4
$\bar{\psi}\gamma^{5}\psi$	pseudo-scalar	1	-1	1

Summing the right column, we see that this accounts for all 16 degrees of freedom, so no other irreducible Lorentz-group representations may be constructed from two spinors. This allows us to determine the types of currents that we can build into a theory. In particular, we see that $\overline{\psi}\gamma^\mu\psi$ gives a vector current that we may couple to, for example, the electromagnetic field. This allows us to construct a theory of a spin-$\frac{1}{2}$ particle in an electromagnetic field:

$$
\begin{aligned}
\partial_\mu F^{\mu\nu} &= qe\overline{\psi}\gamma^\nu\psi \\
(i\gamma^\mu\partial_\mu - m)\,\psi &= qe\gamma^\mu\psi A_\mu.
\end{aligned}
\tag{8.68}
$$

Both of these equations may be derived from a Lagrangian of the form

$$
\mathcal{L} = \overline{\psi}\left(i\gamma^\mu\partial_\mu - m\right)\psi - qe\overline{\psi}\gamma^\mu\psi A_\mu - \frac{1}{4}F_{\mu\nu}F^{\mu\nu}.
\tag{8.69}
$$

This is the Lagrangian for quantum electrodynamics, of which much more will be said in Chapter 9.

Note, from the bilinear covariants, that we can also construct an axial current (pseudo-vector current), of the form $\overline{\psi}\gamma^\mu\gamma^5\psi$. This will prove useful when we wish to formulate a theory of the weak interactions. In particular, a vector boson may couple to a current of the form $\overline{\psi}\gamma^\mu\psi$ while a pseudo-vector may couple through $\overline{\psi}\gamma^\mu\gamma^5\psi$. Both of these are parity-preserving interactions, since vectors and pseudo-vectors both have a well-defined and consistent behavior under parity transformations. However, if a boson were to couple to a combination of these, such as $\overline{\psi}\left(\alpha\gamma^\mu + \beta\gamma^\mu\gamma^5\right)\psi$, then the resulting theory would violate parity. As we will see in Chapter 11, the weak interactions are found to be of this form. In fact, because the weak current is an *equal* combination of the two currents, we say that they violate parity *maximally*.

Also from the bilinear covariants, we see that the term $\overline{\psi}\psi$ is a scalar, to which we can couple a scalar field. Adding the relevant source terms to the Dirac and Klein-Gordon equations gives

$$
\begin{aligned}
\left(i\partial\!\!\!/ - m\right)\psi &= y\phi\psi, \\
\left(\partial^2 + m^2\right)\phi &= -y\overline{\psi}\psi,
\end{aligned}
\tag{8.70}
$$

where y is the coupling. This is equivalent to adding a term $-y\phi\bar{\psi}\psi$ to the Lagrangian and gives a new Feynman rule: each fermion-fermion-scalar vertex gives a factor of $-y$. These latter interactions are referred to as Yukawa interactions, since they have the same Lorentz structure as Yukawa's original formulation of the nuclear interaction.

EXERCISES

1. Working in the Dirac representation, show that the momentum-space Dirac equation can be written in the form

$$
\begin{pmatrix}
E - m & 0 & -p_z & -p_x + ip_y \\
0 & E - m & -p_x - ip_y & p_z \\
p_z & p_x - ip_y & -E - m & 0 \\
p_x + ip_y & -p_z & 0 & -E - m
\end{pmatrix} u(p) = 0.
$$

2. Write the helicity operator for a Dirac spinor with momentum only in the z direction. Hence show that a solution of the form $\psi = u_1(p)e^{-i(Et - p_z z)}$ is a helicity eigenstate and find its helicity.

3. (a) Use the defining Clifford algebra relations for the γ-matrices to show that $\left(\gamma^5\right)^2 = \mathbb{1}$ regardless of representation.

(b) Hence show that the projection operators $\widehat{L} = \frac{1}{2}\left(\mathbb{1} - \gamma^5\right)$ and $\widehat{R} = \frac{1}{2}\left(\mathbb{1} + \gamma^5\right)$ are idempotent (square to themselves).

4. Show explicitly in the Weyl representation that the left- and right-chiral parts of a spinor evolve independently of each other in the case of a massless fermion.

5. From Equation 8.39, show that the three-vector form of the spin operators is as given in Equation 8.40. (Hint: Consider that angular momentum in three-space is defined by a cross-product, $\ell = \mathbf{r} \times \mathbf{p}$. How does this definition extend to more than three dimensions and how can you write it in index notation?)

6. **(a)** Use the γ-matrix Clifford algebra relation to show that

$$[\sigma^{\mu\nu}, \gamma^\rho] = 2i\left(g^{\nu\rho}\gamma^\mu - g^{\mu\rho}\gamma^\nu\right),$$

where $\sigma^{\mu\nu} = \frac{i}{2}[\gamma^\mu, \gamma^\nu]$.

   (b) From the spinor transformation law (Equation 8.14), write down an appropriate form of S and S^{-1} for an infinitesimal transformation. Hence show that this transformation obeys the condition given in Equation 8.13. (Hint: Recall the infinitesimal form of Lorentz transformations given in Exercise 3.)

7. Working in the Dirac representation, show that the plane wave

$$\psi = \begin{pmatrix} 1 \\ 0 \\ 0 \\ \frac{k}{E+m} \end{pmatrix} e^{-i(Et-kx)}$$

   is an eigenstate of the operator $i\gamma^\mu \partial_m u$ and find its eigenvalue. Interpret this state physically.

8. Verify that $v_1 = u_1^c$ and $v_2 = u_2^c$.

9. Starting from the Dirac equation and its adjoint, and following a similar procedure as for the scalar particle, show that $j^\mu = i\bar{\psi}\gamma^\mu\psi$ is a conserved current for the Dirac spinor.

10. The Lorentz generators for a vector particle take the form $(J^{\mu\nu})^\alpha{}_\beta$, where the indices μ, ν play a different role from α and β. Compare this with the corresponding expression for the spinor Lorentz generators.

(a) What are the roles of the indices?

(b) Following a similar procedure to that used for spin-$\frac{1}{2}$ particles in Section 8.4.1, show that the Lorentz generators $(J^{\mu\nu})^\alpha{}_\beta$ obey

$$\left(\delta^\alpha_\beta - \frac{i}{2}\,(\omega_{\mu\nu} J^{\mu\nu})^\alpha{}_\beta\,\omega_{\mu\nu}\right) A^\beta =$$
$$A^\alpha + g^{\alpha\rho}\omega_{\rho\beta} A^\beta - g^{\mu\rho}\omega_{\rho\tau} x^\tau \partial_\mu A^\alpha.$$

(c) Relabel indices as appropriate to show that the generators are given by

$$(J^{\mu\nu})^\alpha{}_\beta = i\left(g^{\alpha\mu}\delta^\nu_\beta - g^{\alpha\nu}\delta^\mu_\beta\right) + (x^\mu p^\nu - x^\nu p^\mu)\,\delta^\alpha_\beta.$$

(Hint: You will also need to insert a couple of Kronecker deltas to make the indices work out right.)

(d) Hence show that the spin operators for a vector particle are as given in Equation 7.22.

11. (a) Write down the energy-momentum four-vector for a Dirac solution of the form

$$\psi = \left(\begin{array}{c} \chi \\ \frac{k}{E+m}\chi \end{array}\right) e^{-i(Et - kx\sin\theta\cos\phi - ky\sin\theta\sin\phi - kz\cos\theta)}.$$

(b) For a solution of this form, show that

$$\frac{\sigma \cdot \mathbf{p}}{|\mathbf{p}|} = \left(\begin{array}{cc} \cos\theta & \sin\theta e^{-i\phi} \\ \sin\theta e^{i\phi} & -\cos\theta \end{array}\right).$$

(c) Show that χ is an eigenstate of $\frac{\sigma \cdot \mathbf{p}}{|\mathbf{p}|}$ for arbitrary θ and ϕ.

(d) Hence show that ψ is a helicity eigenstate and find its eigenvalue.

(e) Find a similar helicity eigenstate with the opposite helicity.

12. Show that

$$\sum_{A=1}^{2} u_A(p) \cdot \overline{u}_A(p) = \not{p} + m, \quad \sum_{A=1}^{2} v_A(p) \cdot \overline{v}_A(p) = \not{p} - m$$

and

$$\overline{u}_A(p) \cdot u_B(p) = 2m\delta_{AB}, \quad \overline{v}_A(p) \cdot v_B(p) = -2m\delta_{AB}.$$

13. (a) Starting from Equation 8.26, show that the positive-energy solutions to the Dirac equation approximately satisfy

$$u_A(p) = \frac{\sigma \cdot \mathbf{p}}{E_k} \quad \text{and } u_B(p) = 0$$

in the non-relativistic limit, where E_k is the kinetic energy.

(b) Hence show that the two-component spinor $\psi = \frac{1}{2m}u_A(p)e^{imt}$ obeys the Pauli equation in the absence of electromagnetic fields (Equation 3.55).

14. Construct a set of four 4×4 matrices with no real entries that obey the Clifford algebra and so could function as the γ-matrices in the Majorana representation.

CHAPTER 9

QUANTUM ELECTRODYNAMICS

9.1   *U*(1) SYMMETRY IN WAVE EQUATIONS

In non-relativistic quantum mechanics, wavefunctions must be normalized in a consistent way, but this choice is arbitrary, since it is only distinct *directions* in the Hilbert space of state vectors that correspond to physically distinct states. It may seem intuitive, then, that changing the phase of a wavefunction may lead to a distinct state. However, this is not the case, as can be seen by considering the correspondence between the wave and particle descriptions of a system. The energy and momentum of a particle are related to the frequency and wavelength of the corresponding wave respectively. The overall phase of the wave, though, has no bearing on the nature of the corresponding particle. So a phase shift in the wavefunction is a symmetry of the system. This can be seen directly from the Schrödinger equation by adding a constant phase to the exponent for a plane wave or, more generally, multiplying the wavefunction by a phase $e^{i\alpha}$. The derivatives in the Schrödinger equation clearly have no effect on a constant multiplicative factor, so we find that the phase cancels throughout the equation:

$$i\frac{\partial}{\partial t}e^{i\alpha}\psi = -\frac{1}{2m}\partial_i\partial_i e^{i\alpha}\psi + Ve^{i\alpha}\psi$$

$$\implies i\frac{\partial}{\partial t}\psi = -\frac{1}{2m}\partial_i\partial_i\psi + V\psi. \tag{9.1}$$

That is, multiplying ψ by $e^{i\alpha}$ has no effect on the Schrödinger equation. It is easy to see that the same logic applies equally to all of the relativistic wave equations we have considered so far as well. Such a symmetry in the wavefunction is known as a global $U(1)$ symmetry: global because the same transformation is applied at all space-time points, and $U(1)$ since the set of such transformations forms a $U(1)$ symmetry group (notice that the single generator in this case is the identity).

This continuous symmetry must come with a conserved quantity and associated current, and indeed it does. In fact we have already met this conserved current, but it takes a slightly different form for each of the different wave equations. For the Schrödinger equation it is the probability density and probability density current. For the Klein-Gordon equation it is given by $j^\mu = i\left(\phi^*\partial^\mu\phi - \phi\partial^\mu\phi^*\right)$, while for the Dirac equation it is given by $j^\mu = \overline{\psi}\gamma^\mu\psi$, and so on. It is the charge current that we have derived independently for each wave equation, and we see now that such a current was guaranteed by Noether's theorem.

Notice that this $U(1)$ symmetry is not local, however. That is, we cannot make a different transformation at each point in space-time independently. The intuitive argument for this statement is that making a different transformation at two points in space would lead to the wave-form expanding in some places and contracting in others. This change in the wavelength of the wavefunction *would* have an effect on the corresponding particle, altering its momentum. Such a transformation, then, cannot be a symmetry of the system. Equivalently, it is also straightforward to see that multiplication by a position-dependent phase leads to new terms in the wave equation. Taking the Dirac equation as an example, since it will be the most important in what follows, we find that a transformation of the form $\psi \mapsto e^{i\alpha(x)}\psi$ leads to

$$(i\gamma^\mu \partial_\mu - m) \, e^{i\alpha(x)}\psi = ie^{i\alpha(x)}\gamma^\mu \partial_\mu \psi$$
$$- i\gamma^\mu \left(\partial_\mu e^{i\alpha(x)} \right) \psi - me^{i\alpha(x)}\psi$$
$$= e^{i\alpha(x)} \left(i\gamma^\mu \partial_\mu - m \right) \psi - e^{i\alpha(x)}\gamma^\mu \left(\partial_\mu \alpha(x) \right) \psi.$$
$$(9.2)$$

If ψ is itself a solution of the Dirac equation, then the first term in the last line vanishes, leaving only the second term. Since this is not necessarily zero, the transformed spinor is not a solution of the Dirac equation.

9.2 LOCALIZING THE $U(1)$ SYMMETRY

Consider now an electrically charged spin-$\frac{1}{2}$ with charge qe, where e is the fundamental electromagnetic charge (the charge on an electron) and q is the relative charge. That is $q = -1$ for the electron, $+\frac{2}{3}$ for the up quark, and so forth. The current produced by this fermion, then, is $j^\mu = qe\overline{\psi}\gamma^\mu\psi$. This gives an inhomogeneous Maxwell equation $\partial^2 A^\mu - \partial^\mu \partial \cdot A = qe\overline{\psi}\gamma^\mu\psi$. Similarly, the modified Dirac equation in the presence of an electromagnetic field is $(i\gamma^\mu \partial_\mu - qe\gamma^\mu A_\mu - m)\,\psi = 0$. Put another way, this system can be summarized by the Lagrangian

$$\mathcal{L} = \overline{\psi} \left(i\gamma^\mu \partial_\mu - qe\gamma^\mu A_\mu - m \right) \psi - \frac{1}{4}F_{\mu\nu}F^{\mu\nu}. \qquad (9.3)$$

This is the Lagrangian for quantum electrodynamics (QED), a theory of charged spin-$\frac{1}{2}$ particles interacting with the electromagnetic field. The $U(1)$ phase symmetry on the spinor in this case is more general than in the previous section, since we now have the freedom to change the phase independently at each point in spacetime. This additional freedom is granted by the gauge symmetry of the electromagnetic field. Making a local phase transformation as in

Equation 9.2 leads to

$$\left(i\gamma^\mu\partial_\mu - qe\gamma^\mu A_\mu - m\right)e^{i\alpha(x)}\psi$$
$$= e^{i\alpha(x)}\left(i\gamma^\mu\partial_\mu - qe\gamma^\mu A_\mu - \gamma^\mu\left(\partial_\mu\alpha(x)\right) - m\right)\psi. \tag{9.4}$$

Since the photon has a gauge symmetry, we are free to redefine A^μ according to $A^\mu \mapsto A^\mu + \frac{1}{qe}\partial^\mu\alpha(x)$. In this way, we reduce Equation 9.4 to the same modified Dirac equation obeyed by the initial spinor ψ. By coupling to the electromagnetic field, we have localized the $U(1)$ symmetry that was previously only global. We can now flip this entire argument on its head, and ask in what situation we can localize the global phase symmetry in the Dirac equation.

Looking again at the demonstration from the previous section that the symmetry is not local in the general case, clearly the problem lies in the derivative, as this is what generates the additional term. Specifically, while the mass term transforms the same way as the spinor itself under phase transformations, the problem is that the derivative term transforms differently. So what is needed is a modified derivative D_μ with the property that $D_\mu\psi \mapsto e^{i\alpha(x)}D_\mu\psi$. Such a derivative is known as a gauge-covariant derivative. We assume that this derivative takes the form of a simple partial derivative with an additional term, which for consistency must be a four-vector field: $D_\mu = \partial_\mu + iqeA_\mu$, where we have taken out a factor of iqe in anticipation of what is to follow. The required transformation law for the covariant derivative is

$$D_\mu\left(e^{i\alpha(x)}\psi\right) = e^{i\alpha(x)}D_\mu\psi, \tag{9.5}$$

but substituting in the assumed form of D_μ gives

$$\left(\partial_\mu + iqeA_\mu\right)e^{i\alpha(x)}\psi = e^{i\alpha(x)}\left(\partial_\mu + iqeA_\mu\right)\psi + ie^{i\alpha(x)}\left(\partial_\mu\alpha(x)\right)\psi. \tag{9.6}$$

In order to remove the offending final term, we need the freedom to redefine the vector field A_μ according to $A_\mu \mapsto A_\mu + \frac{1}{qe}\partial_\mu\alpha(x)$. That is, we *require* gauge freedom in A_μ in order to localize the phase symmetry in ψ. This leads to a Lagrangian of the form

$$\mathcal{L} = \overline{\psi}\left(i\not{D} - m\right)\psi, \tag{9.7}$$

with local $U(1)$ phase symmetry and gauge symmetry. Recall that a massive vector particle has no gauge invariance, so if A_μ is to describe a particle, it must be massless. We now ask the question "are there any other terms that could be added to this Lagrangian without destroying these symmetries?" To see that the answer is yes, notice that the second gauge-covariant derivative of a spinor also transforms as a spinor: $D_\mu D_\nu e^{i\alpha(x)}\psi = e^{i\alpha(x)}D_\mu D_\nu \psi$. So a second derivative is also invariant under local $U(1)$. The commutator of two such derivatives gives

$$\begin{aligned} [D_\mu, D_\nu] &= [\partial_\mu + iqeA_\mu, \partial_\nu + iqeA_\nu] \\ &= iqe\left(\partial_\mu A_\nu - \partial_\nu A_\mu\right) \\ &= iqeF_{\mu\nu}. \end{aligned} \tag{9.8}$$

In deriving this result, we must be careful with the meaning of the partial derivatives. In the first line, the partial derivatives act on everything to their right, including an implicit wavefunction. In the second line, the additional terms generated by the derivatives' effect on the implicit wavefunction have vanished, and the derivatives now act only on A. While this object is gauge-invariant, with two Lorentz indices it is not Lorentz-invariant. A suitable additional term for the Lagrangian must be a scalar, so a term proportional to $F_{\mu\nu}F^{\mu\nu}$ will fit the bill.

Simply by imposing a local $U(1)$ phase symmetry, we have arrived at a theory of a fermion interacting with a massless spin-1 particle with gauge invariance. Local $U(1)$ has led directly to a theory of a charged fermion interacting with the photon, or at least something that behaves very much like a photon. This approach to deriving quantum electrodynamics, and its generalization to larger symmetry groups, will prove useful in constructing the remainder of the Standard Model of particle physics, as in the following chapters. Theories built on local symmetries in this way are known as gauge theories.

9.3   THE LINK WITH CLASSICAL PHYSICS

If quantum electrodynamics is to describe the behavior of charged particles in the presence of an electromagnetic field, then in particular it should agree with the classical theory in the appropriate limit. In order to check that this is the case, it is sufficient to compute the interaction energy between two static charged particles. It is instructive to do this first for the case of a massive scalar exchange particle: we will return shortly to the case of a massless vector exchange particle like the photon. Since we are considering a static situation, the Klein-Gordon equation for the exchange particles reduces to

$$\left(-\partial_i \partial_i + m^2\right) \phi(\mathbf{x}) = \rho(\mathbf{x}), \tag{9.9}$$

where $\rho(\mathbf{x})$ is a source of the exchange particles. In the presence of a single static source of charge $Q_1 = q_1 e$ at position $\mathbf{x}_1$, then, this becomes

$$\left(-\partial_i \partial_i + m^2\right) \phi(\mathbf{x}) = Q_1 \delta^3(\mathbf{x} - \mathbf{x}_1). \tag{9.10}$$

Other than working in only three dimensions, the procedure is now identical to the derivation of the propagator as in Section 7.3, and we find

$$\phi(\mathbf{x}) = Q_1 \int \frac{d^3 k}{(2\pi)^3} \frac{e^{i\mathbf{k}\cdot(\mathbf{x}-\mathbf{x}_1)}}{\mathbf{k}^2 + m^2}. \tag{9.11}$$

This then describes, in a sense, the distribution of exchange particles around the source in the static approximation. If there is a second source of charge $Q_2 = q_2 e$ at position $\mathbf{x}_2$, then the energy of interaction between this second source and the exchange particles due to the first source is given by

$$
\begin{aligned}
U &= -\int d^3 x \, \phi(\mathbf{x}) Q_2 \delta^3(\mathbf{x} - \mathbf{x}_2) \\
&= -Q_1 Q_2 \int d^3 x \int \frac{d^3 k}{(2\pi)^3} \frac{e^{i\mathbf{k}\cdot(\mathbf{x}-\mathbf{x}_1)}}{\mathbf{k}^2 + m^2} \delta^3(\mathbf{x} - \mathbf{x}_2) \\
&= -\frac{Q_1 Q_2}{(2\pi)^3} \int d^3 k \, \frac{e^{i\mathbf{k}\cdot(\mathbf{x}_2-\mathbf{x}_1)}}{\mathbf{k}^2 + m^2}
\end{aligned} \tag{9.12}
$$

If we choose our coordinate system such that the direction of $(\mathbf{x}_2 - \mathbf{x}_1)$ is aligned with the z axis, and then convert to spherical polar coordinates, (r, θ, ζ), we find

$$\mathbf{k} \cdot (\mathbf{x}_2 - \mathbf{x}_1) = \ell r \cos \theta, \tag{9.13}$$

where $\ell \equiv |\mathbf{k}|$ and $r \equiv |\mathbf{x}_2 - \mathbf{x}_1|$. So the previous expression for U becomes

$$
\begin{aligned}
U &= -\frac{Q_1 Q_2}{(2\pi)^3} \int_0^\infty d\ell \int_0^{2\pi} d\zeta \int_0^\pi d\theta \, \ell^2 \sin \theta \, \frac{e^{i\ell r \cos \theta}}{\ell^2 + m^2} \\
&= -\frac{Q_1 Q_2}{(2\pi)^2} \int_0^\infty d\ell \int_0^\pi d\theta \, \ell^2 \sin \theta \, \frac{e^{i\ell r \cos \theta}}{\ell^2 + m^2} \\
&= \left[\frac{Q_1 Q_2}{(2\pi)^2} \int_0^\infty d\ell \, \frac{\ell^2}{i\ell r} \frac{e^{i\ell r \cos \theta}}{\ell^2 + m^2} \right]_{\theta=0}^\pi \\
&= -\frac{Q_1 Q_2}{(2\pi)^2} \int_0^\infty d\ell \, \frac{\ell}{ir} \frac{e^{i\ell r} - e^{-i\ell r}}{\ell^2 + m^2} \\
&= -\frac{Q_1 Q_2}{(2\pi)^2 \, ir} \int_{-\infty}^\infty d\ell \, \frac{\ell e^{i\ell r}}{\ell^2 + m^2},
\end{aligned}
\tag{9.14}
$$

where this last step follows from a change of integration variable, $\ell \mapsto -\ell$, in the second term.

To complete this calculation, it is necessary to use the tools of complex analysis. Specifically, we wish to compute this integral along the entire real line. However, since r is necessarily positive, for large imaginary values of ℓ, we have $e^{i\ell r} \to 0$ and the integrand vanishes. We can, therefore, compute the integral over a closed loop that runs the length of the real axis and then closes in the positive imaginary half-plane. Since the semicircular part of this contour has an integrand of zero, the value of the whole contour integral is equal to the value of the integral along just the real line. This allows us to apply Cauchy's integral formula: for an integrand of the form

$$\frac{f(z)}{z - z_0}, \tag{9.15}$$

where $f(z)$ is any analytic function, the integral over a closed contour C, performed counterclockwise, is given by

$$\oint_C \frac{f(z)}{z - z_0} = \begin{cases} 2\pi i \times f(z_0) & \text{if } z_0 \text{ is contained in } C \\ 0 \end{cases} . \quad (9.16)$$

Writing the integrand in Equation 9.14 as

$$\frac{\ell e^{i\ell r}}{\ell^2 + m^2} = \frac{\ell e^{i\ell r}}{(\ell + im)(\ell - im)}, \quad (9.17)$$

we see that it is of the correct form to use Cauchy's formula, with

$$f(\ell) = \frac{\ell e^{i\ell r}}{\ell + im}. \quad (9.18)$$

Hence, we arrive at an interaction energy between our two sources of

$$\begin{aligned} U &= -\frac{Q_1 Q_2}{(2\pi)^2 \, ir} \, 2\pi i \, \frac{im \, e^{-mr}}{2im} \\ &= -\frac{Q_1 Q_2 e^{-mr}}{4\pi r} \end{aligned} \quad (9.19)$$

This is the classical potential for a force with a range of order $1/m$. If the exchange particle is massless, this reduces to the potential for a force with infinite range:

$$U = -\frac{Q_1 Q_2}{4\pi r}. \quad (9.20)$$

Notice that this force is attractive ($-dU/dr < 0$) for like charges and repulsive ($-dU/dr > 0$) for opposite charges. This may seem counterintuitive but in fact is the correct result when we consider a force mediated by scalar exchange particles. As an example, consider the residual nuclear force between nucleons: this is mutually attractive since it is mediated (to a good approximation) by spin-0 pions. The more familiar situation of a repulsive force between like charges arises when we consider a spin-1 mediator. In this case, we must also consider the polarization state of the virtual exchange particle. In the low-energy semi-classical limit in which we are working, it is the time-like component A^0 of the photon that contributes to the potential. This is because A^μ couples to the current produced by the particles acting as sources, and in the static limit, the only non-zero part of this current is the time-like component. Since the photon propagator takes the form $-ig_{\mu\nu}/k^2$, considering only the A^0 component, we see that it behaves as a scalar with propagator $-i/k^2$, whereas a true massless scalar would have propagator i/k^2. This relative negative sign gives rise to the familiar Coulomb potential as a result of photon exchange:

$$U = \frac{Q_1 Q_2}{4\pi r}. \tag{9.21}$$

9.4   A WELL-TESTED THEORY

Quantum electrodynamics has the impressive distinction of being the most accurate scientific theory ever devised. To clarify this statement, some of the predictions of quantum electrodynamics match with experimentally determined values to the highest precision of any such quantities in a physical theory. Among these is the magnetic moment of the electron. As we saw in Section 3.8, the spin g-factor of the electron is predicted to be 2 by the Pauli equation, and also therefore by the Dirac equation. This is, however, not in exact agreement with the experimentally determined value. There is a small deviation of g from 2, usually quantified as $\frac{g-2}{2}$, and known as the anomalous magnetic moment. The experimental value of this anomalous moment is $(1.15965218073(28)) \times 10^{-3}$, and quantum electrodynamics is able to match this to the tenth significant figure with a calculated value of $(1.15965218178(77)) \times 10^{-3}$. There are other precision measurements that have been performed to test QED and all find that the theory is in excellent agreement with experiment. Such precision arises from calculating higher-order Feynman diagrams, so let's explore this procedure.

9.5   CALCULATIONS IN QED

We have already seen, in Section 7.3, how to calculate a simple Feynman diagram. In QED, the calculations tend to get somewhat trickier. This is partly because there are often multiple diagrams to consider for each process, but the main complication is from the Feynman rules. The Feynman rules for QED are such that even an amplitude with a single contribution can be difficult to compute, since it typically contains a product of multiple γ-matrices.

Before performing a full calculation, it is worth noting that the Feynman diagram formalism also allows for a very simple estimation procedure. Specifically, if we write down the simplest diagrams contributing to a particular process and then simply count the number of vertices, this gives an order-of-magnitude estimate for the strength of a process. For example, consider the processes of Møller scattering, $e^- + e^- \to e^- + e^-$, and Delbrück scattering, $\gamma + \gamma \to \gamma + \gamma$. The lowest-order contributions to each of these processes are

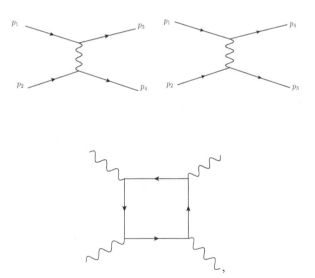

and

respectively, so we can see, just from the number of vertices, that Delbrück scattering is typically an order of $\alpha = \frac{e^2}{4\pi} \approx \frac{1}{137}$ weaker than Møller scattering.

9.5.1 Feynman Rules for QED

The Feynman rules for QED are as follows. First, label each internal and external particle with a momentum consistent with momentum conservation. Additionally, label all external fermions and photons with a polarization state index. Internal and external photons contribute the same factors as in Section 7.3. The additional fermion-specific rules are as follows, where all particles are taken to have momentum p and spin-polarization A:

- An incoming fermion contributes a spinor $u_A(p)$.

- An outgoing fermion contributes a spinor $\bar{u}_A(p)$.

- An incoming *anti*-fermion contributes a spinor $\bar{v}_A(p)$.

- An outgoing *anti*-fermion contributes a spinor $v_A(p)$.

- An internal fermion contributes a propagator:

$$\frac{i(\not{p}+m)}{p^2 - m^2}\left(= i\left(\not{p}-m\right)^{-1}\right).$$

- A fermion-fermion-photon vertex contributes a factor $-iqe\gamma^\mu$, where μ is the Lorentz index for the photon connecting to the vertex.

When applying these rules, there are three more points we must be aware of. First, spinors must be written in the correct order. To achieve this, we follow a fermion line backward, against the flow of the fermion number denoted by the arrows on the lines. If the fermion forms a closed internal loop, we take the trace of the resulting string of gamma matrices. Second, we include an additional factor of -1 for any closed fermion loop. And third, if any two diagrams are related by the relabeling of external fermion lines, we give one of them an additional overall factor of -1. A full list of these rules is given in Appendix B.

9.5.2 Calculating Amplitudes

Distinguishable Particles

Let's see how a full calculation goes through. The most obvious amplitude we could try to calculate, and what may intuitively feel as though it should be the simplest, is the elastic scattering of two electrons. However, as we can see from the above diagrams for Møller scattering, there are two contributions to this process at the lowest order, and this complicates matters. The reason we have two diagrams is due to the fact that electrons are indistinguishable, so we must account for the possibility that the outgoing electron with momentum p_3 originally had momentum p_1, but also the possibility that it originally had momentum p_2. This issue can be avoided, therefore, if we calculate the scattering amplitude for two *different* particles. With this in mind, let's calculate the scattering of a particle of mass m_1 and charge q_1 with a particle of mass m_2 and charge q_2. To lowest order, this amplitude is given by the diagram:

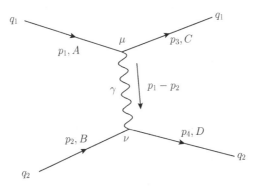

where A, B, C, D are the spin states of the particles. Following the QED Feynman rules, we can write the amplitude as

$$-i\mathcal{M} = \overline{u}_C(p_3)\left(-iq_1 e\gamma^\mu\right)u_A(p_1)$$
$$\times \left(-\frac{ig_{\mu\nu}}{(p_1 - p_3)^2}\right)\overline{u}_D(p_4)\left(-iq_2 e\gamma^\nu\right)u_B(p_2)$$

$$\text{or } \mathcal{M} = -\frac{q_1 q_2 e^2 g_{\mu\nu}}{(p_1 - p_3)^2}\overline{u}_C(p_3)\gamma^\mu u_A(p_1) \cdot \overline{u}_D(p_4)\gamma^\nu u_B(p_2). \quad (9.22)$$

We should stress at this point that the previous expression is the full amplitude for this scattering process. All that is left to do is to work through the process of simplifying this expression to something that can be plugged into the differential cross-section formula, and we will have calculated our first physically measurable quantity. The *concept* of using Feynman rules to calculate an amplitude, then, is very straightforward. We shouldn't lose sight of that in what follows, because the algebra can get a little tricky in this next part.

If the spins are specified—that is, if we know the electron and the muon to have particular helicities—then we could at this point substitute in particular values for each spinor, choosing from an appropriate basis. We would then have to perform matrix multiplications to find the final amplitude. This can be time-consuming, but it is reasonably straightforward. More commonly, however, we wish to know the spin-averaged amplitude, since we will wish to compare with experiments that typically do not measure spins. Rather than finding the amplitude for each spin state individually and then averaging (which would be a tedious process), we use a shortcut. First, recall that the physically meaningful quantity is not $\mathcal{M}$ itself but $|\mathcal{M}|^2$. It turns out that spin averaging *this* quantity is much simpler than averaging before squaring. So we find

$$
\begin{aligned}
|\mathcal{M}|^2_{\substack{\text{spin} \\ \text{average}}} &= \frac{1}{4} \sum_{A=1}^{2} \sum_{B=1}^{2} \sum_{C=1}^{2} \sum_{D=1}^{2} \mathcal{M}\mathcal{M}^\dagger \\
&= -\frac{q_1^2 q_2^2 e^4 g_{\mu\nu} g_{\rho\sigma}}{4(p_1 - p_3)^4} \sum_{A,B,C,D} (\overline{u}_C(p_3)\gamma^\mu u_A(p_1)) \qquad (9.23) \\
&\quad \times (\overline{u}_D(p_4)\gamma^\nu u_B(p_2)) \\
&\quad \times (\overline{u}_C(p_3)\gamma^\rho u_A(p_1))^\dagger \times (\overline{u}_D(p_4)\gamma^\sigma u_B(p_2))^\dagger .
\end{aligned}
$$

The bracketed factors in the last expression commute, since each is just a four-vector. We wish to combine the first and third, but to do so, we first rewrite the third factor:

$$
\begin{aligned}
(\overline{u}_C(p_3)\gamma^\rho u_A(p_1))^\dagger &= \left(u_C^\dagger(p_3)\gamma^0\gamma^\rho u_A(p_1)\right)^\dagger \\
&= u_A^\dagger(p_1)\gamma^{\rho\dagger}\gamma^0 u_C(p_3) \\
&= u_A^\dagger(p_1)\gamma^0\gamma^\rho\gamma^0\gamma^0 u_C(p_3) \\
&= \overline{u}_A(p_1)\gamma^\rho u_C(p_3),
\end{aligned}
\tag{9.24}
$$

which follows from Equation 8.7. Taking the first and third factors now, together with the sums over A and C, gives

$$
\begin{aligned}
\sum_{A,C} &(\overline{u}_C(p_3)\gamma^\mu u_A(p_1)) \cdot (\overline{u}_C(p_3)\gamma^\rho u_A(p_1))^\dagger \\
&= \sum_{A,C} (\overline{u}_C(p_3)\gamma^\mu u_A(p_1)) \, (\overline{u}_A(p_1)\gamma^\rho u_C(p_3)) \, .
\end{aligned}
\tag{9.25}
$$

Using the completeness relation (Equation 8.47), the sum over A gives

$$
\sum_A \overline{u}_C(p_3)\gamma^\mu \left(\slashed{p}_1 + m_1\right) \gamma^\rho u_C(p_3).
\tag{9.26}
$$

For the next step, we first take a slight detour. Consider a 1×2 matrix (row vector), $\mathbf{a}$, and a 2×1 matrix (column vector), $\mathbf{b}$. Multiplying these in the order $\mathbf{ab}$ gives a scalar:

$$
\mathbf{ab} = \left(\begin{array}{cc} a_1 & a_2 \end{array} \right) \left(\begin{array}{c} b_1 \\ b_2 \end{array} \right) = a_1 b_1 + a_2 b_2.
\tag{9.27}
$$

On the other hand, multiplying them in the reverse order gives a 2×2 matrix

$$
\mathbf{ba} = \left(\begin{array}{c} b_1 \\ b_2 \end{array} \right) \left(\begin{array}{cc} a_1 & a_2 \end{array} \right) = \left(\begin{array}{cc} a_1 b_1 & a_2 b_1 \\ a_1 b_2 & a_2 b_2 \end{array} \right).
\tag{9.28}
$$

Notice that taking the trace of this matrix gives back the same scalar as the first multiplication:

$$
\mathrm{tr}\,(\mathbf{ba}) = a_1 b_1 + a_2 b_2 = \mathbf{ab}.
\tag{9.29}
$$

This relation generalizes and we can say that, for arbitrary matrices,

$$A_{1 \times n} M_{n \times n} B_{n \times 1} = \operatorname{tr}(M_{n \times n} B_{n \times 1} A_{1 \times n}). \qquad (9.30)$$

This is, incidentally, also the origin of the trace rule for closed fermion loops in Feynman diagrams.

Returning to Equation 9.26 and applying Equation 9.30, we find

$$\sum_A \overline{u}_C(p_3) \gamma^\mu \left(\not{p}_1 + m_1 \right) \gamma^\rho u_C(p_3)$$

$$= \sum_A \operatorname{tr} \left(\gamma^\mu \left(\not{p}_1 + m_1 \right) \gamma^\rho u_C(p_3) \overline{u}_C(p_3) \right) \qquad (9.31)$$

$$= \operatorname{tr} \left(\gamma^\mu \left(\not{p}_1 + m_1 \right) \gamma^\rho \left(\not{p}_3 + m_1 \right) \right).$$

This trace can now be evaluated using the standard trace relations for the γ matrices given in Appendix C:

$$\operatorname{tr} \left(\gamma^\mu \left(\not{p}_1 + m_1 \right) \gamma^\rho \left(\not{p}_3 + m_1 \right) \right)$$

$$= \operatorname{tr} \left(\gamma^\mu \gamma^\alpha \gamma^\rho \gamma^\beta \right) (p_1)_\alpha (p_3)_\beta + m_1^2 \operatorname{tr} \left(\gamma^\mu \gamma^\rho \right) + \operatorname{tr} \left(\text{odd } \# \text{ of } \gamma\text{'s} \right)$$

$$= 4 \left(g^{\mu\alpha} g^{\rho\beta} + g^{\mu\beta} g^{\rho\alpha} - g^{\mu\rho} g^{\alpha\beta} \right) (p_1)_\alpha (p_3)_\beta + 4 m_1^2 g^{\mu\rho}$$

$$= 4 \left(p_1^\mu p_3^\rho + p_3^\mu p_1^\rho - g^{\mu\rho} \left(p_1 \cdot p_3 - m_1^2 \right) \right).$$

$$(9.32)$$

This is the result for the first and third factors of 9.23, and we find a similar result for the second and fourth factors. Putting everything together, then, we have

$$|\mathcal{M}|^2_{\substack{\text{spin} \\ \text{average}}} = -\frac{q_1^2 q_2^2 e^4 g_{\mu\nu} g_{\rho\sigma}}{4(p_1 - p_3)^4}$$

$$\times 4 \left(p_1^\mu p_3^\rho + p_3^\mu p_1^\rho - g^{\mu\rho} p_1 \cdot p_3 + m_1^2 g^{\mu\rho} \right) \qquad (9.33)$$

$$\times 4 \left(p_2^\nu p_4^\sigma + p_4^\nu p_2^\sigma - g^{\nu\sigma} p_2 \cdot p_4 + m_2^2 g^{\nu\sigma} \right).$$

All that remains is to contract Lorentz indices. This gives 16 terms in the first instance, but many of these cancel or combine. After the dust settles, we are left with

$$|\mathcal{M}|^2_{\substack{\text{spin} \\ \text{average}}} = -\frac{8q_1^2 q_2^2 e^4}{(p_1 - p_3)^4} \Big((p_1 \cdot p_2)(p_3 \cdot p_4) + (p_1 \cdot p_4)(p_2 \cdot p_3)$$

$$- m_1^2 (p_2 \cdot p_4) - m_2^2 (p_1 \cdot p_3) + 2m_1^2 m_2^2 \Big)$$

(9.34)

as the invariant amplitude for elastic scattering of distinguishable charged particles.

Indistinguishable Particles

We will not perform the full calculation of the electron-electron (Møller) scattering amplitude, since the reader can hopefully now see how to do it for themselves. We will, however, address the additional complications arising from the fact that the particles involved in this process are indistinguishable. Specifically, we now have two contributing diagrams at lowest order:

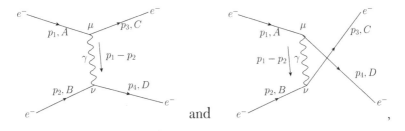

with amplitudes

$$\mathcal{M}_1 = -\frac{e^2 g_{\mu\nu}}{(p_1 - p_3)^2} \overline{u}_C(p_3)\gamma^\mu u_A(p_1) \cdot \overline{u}_D(p_4)\gamma^\nu u_B(p_2)$$

$$\text{and } \mathcal{M}_2 = -\frac{e^2 g_{\mu\nu}}{(p_1 - p_4)^2} \overline{u}_D(p_4)\gamma^\mu u_A(p_1) \cdot \overline{u}_C(p_3)\gamma^\nu u_B(p_2),$$

(9.35)

respectively, where we have made use of the fact that here $q_1 = q_2 = -1$. The overall amplitude, then, is the *difference* between these individual amplitudes, $\mathcal{M}_1 - \mathcal{M}_2$, since the diagrams are related by a relabeling of fermions, and fermions anticommute. The reader may be curious as to which of the amplitudes should be negative in this difference. Fortunately, the answer is "it doesn't matter," since it is only the square of the amplitude that has physical meaning. The important point to remember is that the two terms have a *relative* minus sign, since it is this that will give the correct cross-terms when squaring the amplitude. Incidentally, one view of quantum physics is that it is this squaring of the amplitude that leads to quantum behavior. For instance, the interference patterns in an electron double-slit experiment arise from the cross-terms in the squared amplitude. A classical approach to the same situation would sum without squaring, leading to no interference. The overall spin-averaged amplitude for $e^- \text{-} e^-$ scattering, then, is given by

$$
\begin{aligned}
|\mathcal{M}|^2_{\substack{\text{spin} \\ \text{average}}} &= \frac{1}{4} \sum_{A,B,C,D} |\mathcal{M}_1 - \mathcal{M}_2|^2 \\
&= \frac{1}{4} \sum_{A,B,C,D} \left(|\mathcal{M}_1|^2 + |\mathcal{M}_2|^2 - \mathcal{M}_1 \mathcal{M}_2^\dagger - \mathcal{M}_1^\dagger \mathcal{M}_2 \right).
\end{aligned}
$$

$$(9.36)$$

The first two terms in this expression are identical to the previous example for distinguishable particles, and the calculation of these terms goes through in exactly the same way. The third and fourth terms are different, however, since they have a different trace structure. To see what we mean by this, consider the third term in the previous equation:

$$
\begin{aligned}
\mathcal{M}_1 \mathcal{M}_2^\dagger &= \frac{e^4 g_{\mu\nu} g_{\rho\sigma}}{(p_1 - p_3)^2 (p_1 - p_4)^2} \bar{u}_C(p_3)\gamma^\mu u_A(p_1) \cdot \bar{u}_D(p_4)\gamma^\nu u_B(p_2) \\
&\quad \cdot (\bar{u}_D(p_4)\gamma^\rho u_A(p_1))^\dagger \cdot (\bar{u}_C(p_3)\gamma^\sigma u_B(p_2))^\dagger \\
&= \frac{e^4 g_{\mu\nu} g_{\rho\sigma}}{(p_1 - p_3)^2 (p_1 - p_4)^2} \bar{u}_C(p_3)\gamma^\mu u_A(p_1) \cdot \bar{u}_D(p_4)\gamma^\nu u_B(p_2) \\
&\quad \cdot \bar{u}_A(p_1)\gamma^\rho u_D(p_4) \cdot \bar{u}_B(p_2)\gamma^\sigma u_C(p_3).
\end{aligned}
$$

$$(9.37)$$

Taking the spin average, then, gives one trace of the form

$$\text{tr} \left(\gamma^{\mu} (\not{p}_1 + m) \gamma^{\rho} (\not{p}_4 + m) \gamma^{\nu} (\not{p}_2 + m) \gamma^{\sigma} (\not{p}_3 + m) \right), \qquad (9.38)$$

which when multiplied out would require the evaluation of several four-matrix and six-matrix traces, as well as a trace of eight gamma matrices. The eight-matrix trace alone consists of 105 terms, so the reader will appreciate how complicated these amplitude calculations can become. Of course, most of these terms either cancel or combine with one another, and symmetry arguments applied as early as possible in the calculation can simplify the problem.

Other Amplitudes

There are other amplitudes that we can calculate with these methods. In particular, in the first example of the scattering of distinguishable particles, we can replace one of the particles with its antiparticle with only a minimal effect on the subsequent calculation. The amplitude in this case contains antiparticle spinors, and the ordering is slightly different:

$$-i\mathcal{M} = \overline{u}_C(p_3) \left(-iq_1 e \gamma^{\mu} \right) u_A(p_1) \left(\frac{-ig_{\mu\nu}}{(p_1 - p_3)^2} \right)$$
$$\times \overline{v}_B(p_2) \left(-iq_2 e \gamma^{\nu} \right) v_D(p_4) \qquad (9.39)$$
$$\mathcal{M} = -\frac{q_1 q_2 e^2 g_{\mu\nu}}{(p_1 - p_3)^2} \overline{u}_C(p_3) \gamma^{\mu} u_A(p_1) \cdot \overline{v}_B(p_2) \gamma^{\nu} v_D(p_4),$$

leading to a reordering of the trace:

$$|\mathcal{M}|^2 = \frac{q_1^2 q_2^2 e^4 g_{\mu\nu} g_{\rho\sigma}}{(p_1 - p_3)^4} \text{tr} \left(\gamma^{\mu} \left(\not{p}_1 + m_1 \right) \gamma^{\rho} \left(\not{p}_3 + m_1 \right) \right) \times$$
$$\times \text{tr} \left(\gamma^{\nu} \left(\not{p}_4 - m_2 \right) \gamma^{\sigma} \left(\not{p}_2 - m_2 \right) \right), \qquad (9.40)$$

as well as the appearance of some $\left(\not{p} - m \right)$ factors, but the methods are the same.

Similarly, the completeness relation for polarization vectors (Equation 7.19), means that we can use the same trace method for photons. For example, the amplitude for Compton (electron-photon) scattering is given by

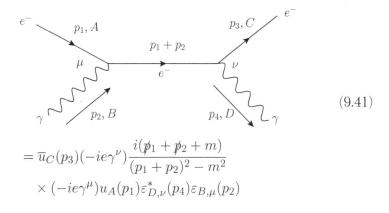

$$= \overline{u}_C(p_3)(-ie\gamma^\nu)\frac{i(\not{p}_1 + \not{p}_2 + m)}{(p_1 + p_2)^2 - m^2}$$
$$\times (-ie\gamma^\mu)u_A(p_1)\varepsilon^*_{D,\nu}(p_4)\varepsilon_{B,\mu}(p_2)$$

(9.41)

so we find

$$\mathcal{M} = \frac{e^2}{(p_1 + p_2)^2 - m^2}$$
$$\times \overline{u}_C(p_3)\gamma^\nu(\not{p}_1 + \not{p}_2 + m)\gamma^\mu u_A(p_1)\varepsilon^*_{D,\nu}(p_4)\varepsilon_{B,\mu}(p_2)$$
$$\mathcal{M}^\dagger = \frac{e^2}{(p_1 + p_2)^2 - m^2}\overline{u}_A(p_1)\gamma^\rho(\not{p}_1 + \not{p}_2 + m)\gamma^\sigma u_C(p_3)\varepsilon_{D,\sigma}\varepsilon^*_{B,\rho}$$

(9.42)

and, therefore

$$|\mathcal{M}|^2_{\substack{\text{spin}\\\text{average}}} = \frac{e^4}{(p_1 + p_2)^4}\text{tr}\left(\gamma^\nu(\not{p}_1 + \not{p}_2 + m)\gamma^\mu(\not{p}_1 + m)\gamma^\rho\right.$$
$$\left.(\not{p}_1 + \not{p}_2 + m)\gamma^\sigma(\not{p}_3 + m)\right)g_{\mu\rho}g_{\nu\sigma},$$

(9.43)

where the metrics arise from the completeness relations on the polarization vectors.

Mandelstam Variables

Notice that the tree-level Feynman diagrams for $2 \to 2$ scattering processes come in three distinct shapes. First, there are those in which the total initial momentum, $p_1 + p_2$ is carried by the virtual particle. Then there are two diagram topologies in which a virtual particle is exchanged between two quite distinct particles. In particular, the momentum carried by the exchange particle may be $p_1 - p_3$ or $p_1 - p_4$ depending on the diagram's topology: this is especially noticeable in the case of indistinguishable particles, where both of these latter possibilities occur. Since the internal propagator always carries one of three values of momentum, then, it is useful to introduce a shorthand for these allowed values. These are the Mandelstam variables, defined by $s = (p_1 + p_2)^2$, $t = (p_1 - p_3)^2$ and $u = (p_1 - p_4)^2$.

We find some useful relationships between these variables that reflect the symmetries hidden in Feynman diagrams. As a simple example, if we already have an expression for the amplitude in the "t-channel" for Møller scattering, written in terms of s, t, and u, then we may write down the amplitude for the "u-channel" without any further calculation, simply by making the replacement $t \leftrightarrow u$.

9.5.3 Calculating the Differential Cross-Section

Now that we have a set of transition amplitudes, we can find the differential cross-section for these interactions. Since all of the processes we have considered so far are $2 \to 2$ processes, with two initial-state particles and two final-state particles, the procedure is the same in all cases. First, let's make the general cross-section formula (Equation 5.38) specific to a $2 \to 2$ process. Collecting numerical factors, we have:

$$\frac{d\sigma}{d^3 p_1 d^3 p_2} = \frac{S |\mathcal{M}|^2}{64\pi^2 \sqrt{(p_A \cdot p_B)^2 - (m_A m_B)^2}} \tag{9.44}$$
$$\times \frac{1}{E_1 E_2} \delta^4 (p_A + p_B - p_1 - p_2).$$

If this process occurs in a typical symmetric collider experiment, then the initial-state particles have equal and opposite momentum. Even if this is not the case, we can always choose to work in the center-of-mass reference frame ($\mathbf{p}_A + \mathbf{p}_B = 0$) and then transform the final result back to the laboratory frame afterward. As such, we have

$$p_A^\mu = (E_A, \mathbf{k}), \; p_B^\mu = (E_B, -\mathbf{k}), \; p_1^\mu = (E_1, \mathbf{p}_1), \; p_2^\mu = (E_2, \mathbf{p}_2), \tag{9.45}$$

where $\mathbf{k}$ is the initial three-momentum. Notice that we make no assumption about the final-state three-momenta. We know in advance of course that these must also be equal and opposite thanks to conservation of momentum. However, we do not feed this into the calculation, as the delta function will automatically ensure it for us. In fact, because of the delta function, to make this assertion in advance would lead to a divergent integral during the subsequent calculation. By substituting these four-vectors into the previous cross-section, we can rewrite the factor in the square root in the form $(E_A + E_B) |\mathbf{k}|$. Separating the delta function into its time-like and space-like parts, we find

$$\frac{\mathrm{d}\sigma}{\mathrm{d}^3 p_1 \mathrm{d}^3 p_2} = \frac{S |\mathcal{M}|^2}{64\pi^2 (E_A + E_B) |\mathbf{k}|} \\ \times \frac{1}{E_1 E_2} \delta (E_1 + E_2 - E_A - E_B) \, \delta^3 (\mathbf{p}_1 + \mathbf{p}_2), \tag{9.46}$$

where we have chosen to multiply the arguments of both delta functions by -1 to avoid unnecessary negative signs later. This is a legitimate step, since the delta function is an even function. The three-dimensional delta function now sets $\mathbf{p}_1 = -\mathbf{p}_2$ when we perform the $\mathrm{d}^3 p_2$ integral:

$$\frac{\mathrm{d}\sigma}{\mathrm{d}^3 p_1} = \frac{S |\mathcal{M}|^2}{64\pi^2 (E_A + E_B) |\mathbf{k}|} \frac{1}{E_1 E_2} \delta (E_1 + E_2 - E_A - E_B). \tag{9.47}$$

Since the remaining integral that we need to compute is with respect to the momentum of one of the particles, we rewrite the energies in the delta function in terms of momentum:

$$\frac{d\sigma}{d^3 p_1} = \frac{S|\mathcal{M}|^2}{64\pi^2(E_A + A_B)|\mathbf{k}|} \frac{1}{E_1 E_2} \delta$$
$$\times \left(\sqrt{|\mathbf{p}_1|^2 + m_1^2} + \sqrt{|\mathbf{p}_1|^2 + m_2^2} - E_A - E_B \right). \tag{9.48}$$

Notice that both final-state particles now have the same momentum, $\mathbf{p}_1$. Since the remaining delta is one-dimensional and depends only on the magnitude of the momentum $\mathbf{p}_1$, we rewrite the integration variable in polar form as $d^3 p_1 = |\mathbf{p}_1|^2 d|\mathbf{p}_1| d\Omega$, where Ω is the solid angle into which the particle with momentum p_1 is scattered. Also, notice that the delta function does not take the magnitude of the momentum directly as its argument, instead taking as argument a *function* of the magnitude. Therefore, in order to perform the $|\mathbf{p}_1|$ integral, we use the following identity for the delta function:

$$\delta(f(x) - f_0) = \sum_i \frac{\delta(x - x_i)}{|f'(x_i)|}, \tag{9.49}$$

where $\{x_i\}$ is the set of all values of x such that $f(x_i) = f_0$, and where $f'(x_i)$ is the derivative of the function f evaluated at x_i. In the case of Equation 9.48, the function in question is

$$f(|\mathbf{p}_1|) = \sqrt{|\mathbf{p}_1|^2 + m_1^2} + \sqrt{|\mathbf{p}_1|^2 + m_2^2} \tag{9.50}$$

so the delta function becomes

$$\frac{\delta(|\mathbf{p}_1| - |\mathbf{p}|)}{|\mathbf{p}| \left(\left(|\mathbf{p}|^2 + m_1^2\right)^{-1/2} + \left(|\mathbf{p}|^2 + m_2^2\right)^{-1/2} \right)} = \frac{\delta(|\mathbf{p}_1| - |\mathbf{p}|)}{|\mathbf{p}| \left(E_1^{-1} + E_2^{-1} \right)}, \tag{9.51}$$

where $|\mathbf{p}|$ is defined by

$$f(|\mathbf{p}|) = \sqrt{|\mathbf{p}| + m_1^2} + \sqrt{|\mathbf{p}|^2 + m_2^2} = E_A + E_B. \tag{9.52}$$

That is, $|\mathbf{p}|$ is that particular value for the magnitude of the final momentum that ensures energy conservation.

Putting everything together now, we find

$$
\frac{1}{|\mathbf{p}_1|^2} \frac{d\sigma}{d|\mathbf{p}_1| \, d\Omega} = \frac{S \, |\mathcal{M}|^2}{64\pi^2 (E_A + E_B) \, |\mathbf{k}|} \frac{1}{E_1 E_2}
$$

$$
\times \frac{1}{|\mathbf{p}_1| \left(E_1^{-1} + E_2^{-1} \right)} \, \delta \left(|\mathbf{p}_1| - |\mathbf{p}| \right)
$$

$$
\implies \frac{d\sigma}{d\Omega} = \int \frac{S \, |\mathcal{M}|^2}{64\pi^2 (E_A + E_B) \, |\mathbf{k}|} \frac{|\mathbf{p}_1|}{E_1 + E_2} \delta \left(|\mathbf{p}_1| - |\mathbf{p}| \right) d|\mathbf{p}_1|
$$

$$
= \frac{S \, |\mathcal{M}|^2 \, |\mathbf{p}|}{64\pi^2 (E_A + E_B)^2 \, |\mathbf{k}|} ,
$$

$$
(9.53)
$$

where this last step follows from the energy conservation imposed by the final delta function.

To proceed further, we must now substitute in a particular invariant amplitude, $\mathcal{M}$. Since there are no more delta functions to simplify the computation of integrals, to integrate with respect to the angular variables in Ω, we must now have an explicit dependence on the initial and final momenta. Since the computations typically become numerical at this stage, we will not proceed any further with exact calculations. However, with a suitable approximation, we can find the differential cross-section for high-energy scattering processes, as the reader is invited to explore in Exercise 4. Notice also that we were only able to proceed as far as we did without choosing a specific $\mathcal{M}$ because of the restrictions on a two-particle final state. The three-momentum delta function directly relates the two final momenta, while the energy delta function uniquely determines their magnitude. If we had attempted a similar analysis with a three-particle (or higher) final state, the particles would not be so restricted. Conservation of energy and momentum in this case may be respected in infinitely many different ways, by shifting momentum between the final-state particles. As such, in order to calculate the cross-section for processes with more than two particles in the final state, we must include an explicit form for $\mathcal{M}$ from the outset. Again, this then typically requires a numerical computation for the cross-section.

9.6 BEYOND LEADING ORDER: RENORMALIZATION

So far, all of the amplitudes that we have considered have been at the tree level. How do these methods generalize to higher-order diagrams? The answer is, unsurprisingly, "with additional complications." In particular, as soon as we move beyond the leading order and introduce loops into the calculations, we find we must consider the renormalization process. There will not be a full analysis of this procedure here, but we will demonstrate exactly where the issue arises and show how it may be addressed in one particular case.

Consider the next-to-leading-order (NLO) contribution to Compton scattering:

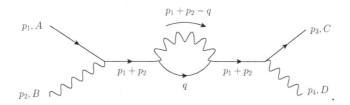

The amplitude for this diagram is given by

$$
\begin{aligned}
-i\mathcal{M} = \int \frac{\mathrm{d}^4 q}{(2\pi)^4}\, &\bar{u}_C(p_3)(ie\gamma^\nu)\frac{i(\not{p}_1 + \not{p}_2 + m)}{(p_1 + p_2)^2 - m^2}(ie\gamma^\rho) \\
&\times \frac{i(\not{q} + m)}{q^2 - m^2}(ie\gamma^\sigma) \\
&\frac{i(\not{p}_1 + \not{p}_2 + m)}{(p_1 + p_2)^2 - m^2}(ie\gamma^\mu)u_A(p_1)\left(\frac{-ig_{\rho\sigma}}{(p_1 + p_2 - q)^2}\right) \\
&\times \varepsilon^*_{D,\nu}(p_4)\varepsilon_{B,\mu}(p_2).
\end{aligned}
$$

$$(9.54)$$

Notice that the integral over the momentum q is unconstrained. It therefore has no finite value, instead growing without limit. How are we to make physical sense of a transition amplitude with an infinite value? To answer this, we must first realize that, as far as the initial and final states are concerned, there was a single electron of momentum $p_1 + p_2$ that acted as the virtual exchange particle in this situation. In effect, they do not "know" about the virtual photon emission and re-absorption in the center of the diagram. So whatever effect the loop has on the amplitude is through the effective electron propagator. We must write the full electron propagator, then, as a sum of approximations—one of which is the "bare" propagator that we have been using so far:

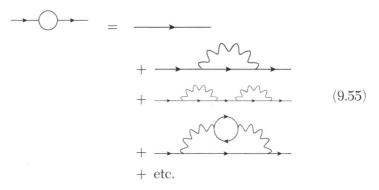

$$(9.55)$$

$+$ etc.

The form of the propagator given in the Feynman rules, then, which includes the physical electron mass, is the effective propagator when all of these internal corrections have been made. The bare propagator need not have the same form.

To see how this works, let's look at the simplest higher-order correction with a single loop—the same loop that appears in our correction to the previous Compton scattering. Taking the parts of the amplitude from that diagram that arise from the loop itself, and writing $p^\mu \equiv p_1^\mu + p_2^\mu$ for simplicity, we have

$$
\begin{aligned}
\text{Loop} &= \int \frac{\mathrm{d}^4 q}{(2\pi)^4} \left(ie\gamma^\rho \right) \left(\frac{i(\slashed{q} + m)}{q^2 - m^2} \right) \left(ie\gamma^\sigma \right) \left(\frac{-ig_{\rho\sigma}}{(p - q)^2} \right) \\
&= -e^2 \int \frac{\mathrm{d}^4 q}{(2pi)^4} \frac{\gamma^\rho \gamma^\tau \gamma_\rho q_\tau + m\gamma^\rho \gamma_\rho}{(q^2 - m^2)(p - q)^2}.
\end{aligned}
$$

$$(9.56)$$

We now use the first of several standard tricks to compute this integral: the Feynman parameter x makes use of the identity

$$\frac{1}{AB} = \int_0^1 \mathrm{d}x \, \frac{1}{(Ax + B(1-x))^2}, \tag{9.57}$$

to rewrite the denominator as

$$\mathrm{Loop} = -e^2 \int_0^1 \mathrm{d}x \int \frac{\mathrm{d}^4 q}{(2\pi)^4} \frac{\gamma^\rho \gamma^\tau \gamma_\rho q_\tau + m \gamma^\rho \gamma_\rho}{((q^2 - m^2)(1-x) + (p-q)^2 x)^2}. \tag{9.58}$$

Next we change variables according to $q^\mu = k^\mu + p^\mu x$:

$$\mathrm{Loop} = -e^2 \int_0^1 \mathrm{d}x \int \frac{\mathrm{d}^4 k}{(2\pi)^4} \frac{\gamma^\rho \gamma^\tau \gamma_\rho p_\tau x + m \gamma^\rho \gamma_\rho}{(k^2 + p^2 x(1-x) - m^2(1-x))^2}. \tag{9.59}$$

Notice that we have neglected the term in the numerator that is linear in k^μ, since an odd function integrated over all space will vanish anyway. In more general calculations, we can drop any odd powers in the numerator of the integrand. All k^μ-dependence is now contained in the denominator, so we will evaluate a simplified integral:

$$\int_0^1 \mathrm{d}x \int \mathrm{d}^4 k \, \frac{1}{k^2 - M^2}, \tag{9.60}$$

where $M^2 \equiv -p^2 x(1-x) + m^2(1-x))^2$, and then we will evaluate the numerator separately. This integral is simplified further by performing a "Wick rotation" to Euclidean space.[1] That is, we change variables again according to

$$k^\mu = (k^0, \mathbf{k}) \mapsto \ell^\mu = (\ell^0, \boldsymbol{\ell}) = (-ik^0, \mathbf{k}). \tag{9.61}$$

Notice that $k^2 = (k^0)^2 - k_i k_i = -(\ell^0)^2 - \ell_i \ell_i = -\ell^2$ as long as we understand ℓ^μ as a *Euclidean* four-vector. Also, note that $\mathrm{d}^4 k = i \mathrm{d}^4 \ell$. Once we find ourselves in a Euclidean space, we see that our

---

[1] The name comes from the fact that this change of variables is equivalent to rotating the integration contour for the k^0 component such that it lies along the imaginary axis rather than the real axis.

integral is actually spherically symmetric and we may make use of (4D) spherical polar coordinates. Our integral then becomes

$$i \int_0^1 dx \int d^4\ell \, \frac{1}{\ell^2 + M^2} = i \int_0^1 dx \int_0^\infty d|\ell| \, \frac{2\pi^2 |\ell|^3}{|\ell|^2 + M^2}$$

$$= i \int_0^1 dx \int_0^\infty d\xi \, \frac{\pi^2 \xi}{\xi + M^2}. \tag{9.62}$$

The factor of $2\pi^2 |\ell|^3$ appearing in the numerator here comes from converting to spherical polar coordinates and performing the angular part of the integral; it is equivalent to the factor $2\pi r$ coming from the angular integration in two dimensions or the $4\pi r^2$ in three dimensions. In the last line, we have used our final change of variables, $\xi = |\ell|^2$, $d\xi = 2|\ell| d|\ell|$, to reduce the power in the numerator.

Our final trick is to deal with the problematic infinite answer that will arise from this integral if we simply attempt to compute it at this point. The idea is to introduce a *regulator*: that is, a quantity that regularizes the integral to give it a finite value but which will reproduce the original result in some appropriate limit. There are many ways to regularize integrals like the previous one, some of which are better suited than others to particular situations or theories. The simplest regulator we can use, though, is a simple "momentum cut-off," Λ: a maximum allowed value of the momentum parameter ξ. The appropriate limit to obtain our original (infinite) integral is then of course $\Lambda \to \infty$. With the aid of partial fractions and standard integrals, the result is now easily shown to be

$$i\pi^2 \int_0^1 dx \left[\ln(\xi + M^2) + \frac{M^2}{\xi + M^2} \right]_{\xi=0}^{\Lambda}. \tag{9.63}$$

Having dealt with the denominator in our original loop integral, we now return to the question of the numerator. To evaluate this, we use the γ-matrix identities listed in Appendix C. In particular, we find

$$\text{numerator} = -e^2 \left(\gamma^\rho \gamma^\tau \gamma_\rho p_\tau x + m\gamma^\rho \gamma_\rho \right)$$
$$= 4e^2 \left(\not{p}x - m \right), \tag{9.64}$$

so that the full expression for the loop becomes

$$\frac{e^2}{4\pi^2} \int_0^1 \mathrm{d}x \; i \left(\not{p}x - m \right) \left[\ln(\xi + M^2) + \frac{M^2}{\xi + M^2} \right]_{\xi=0}^{\Lambda}. \qquad (9.65)$$

Notice that the result depends on the total momentum p^μ, through the combination M^2.

At the beginning of this section, it was asserted that this loop would merely alter our understanding of the electron propagator. Somewhat counterintuitively, this is most easily seen by summing a series of such loops. Consider a series of diagrams in which each is constructed from the last by adding on another one of the loops of the form we have just calculated (as in the first three contributions to Equation 9.55). If we denote the value of a single loop by $-i\Sigma(p)$ and the electron propagator by

$$iS(p) = \frac{i(\not{p}+m)}{p^2 - m^2} = i\left(\not{p} - m \right)^{-1}, \qquad (9.66)$$

where the factors of i are a standard convention, then the bare propagator diagram gives a contribution $iS(p)$, while a single loop (with propagators on either side) gives $iS(p)\Sigma(p)S(p)$. Similarly, higher-order diagrams give contributions $iS(p)\Sigma(p)S(p)\Sigma(p)\ldots\Sigma(p)S(p)$, with one factor of $\Sigma(p)S$ for each loop. Since each diagram contributes to the full effective propagator, we find we have a geometric series of contributions. Summing these, we have

$$iS(p) \sum_{k=0}^{\infty} \left(\Sigma(p)S(p) \right)^k = iS(p) \left(1 - \Sigma(p)S(p) \right)^{-1}$$

$$= i \left(S(p)^{-1} - \Sigma(p) \right)^{-1} \qquad (9.67)$$

$$= i \left(\not{p} - m - \Sigma(p) \right)^{-1}.$$

Now here comes the sneaky part: since we know that real particles must be on-shell, we demand that the full propagator for an on-shell electron have the physical electron mass and behave like the propagator of a free particle—that is, one with no interactions. If the electron did not interact, then infinite loop diagrams would not be an issue and the form of the propagator that we originally wrote down would have been sufficient. We should, then, expect the full propagator to have a singularity at the physical electron mass. Therefore, we may set

$$i \left(\slashed{p} - m_{\text{phys}} \right)^{-1} = i \left(\slashed{p} - m - \Sigma(p) \right)^{-1} \quad \text{for} \quad p^2 = m_{\text{phys}}^2 \quad (9.68)$$

or

$$\Sigma(p) = \slashed{p} - m \quad \text{when} \quad p^2 = m_{\text{phys}}^2. \quad (9.69)$$

Here m_{phys} is the physical mass of the electron. By constructing the full propagator out of a series of bare propagators and loops, we have hidden the underlying "bare mass" of the electron from experimental observation. The only value we can measure is the "dressed" mass m_{phys}. An alternative viewpoint is to consider $\Sigma(p)$ as affecting the energy of the electron rather than the mass, through the energy momentum relation $E^2 = \mathbf{p}^2 + m^2$, and for this reason $\Sigma(p)$ is known as the electron self-energy.

Since the bare mass is unobservable, it does not matter that the value of the self-energy, Σ, is infinite: we just assert that the value of the bare mass is also divergent and allow the two to cancel each other out, leaving us with a well-behaved physical electron mass. This is the renormalization process: each infinite quantity arising through loop diagrams is absorbed into one of the underlying but unobservable parameters of the theory in order to leave us with a finite and observable physical value. An interesting side effect of all this is found when we consider that the physical or renormalized mass is given by

$$m_{\text{phys}} = m + \Sigma(p), \quad (9.70)$$

but $\Sigma(p)$ is dependent on the total momentum transferred, and the above cancellation was only imposed at specific momenta. This means that at other values of momentum (when the electron is virtual and off-shell), the exact nature of the cancellation differs, and we find a physical mass that varies with the particular interaction being considered. Put another way, if we were to split $\Sigma(p)$ into a divergent part that cancels m and a finite remainder, then that remainder varies, giving us a variable value of m_{phys}.[2] This is a genuine property of interacting particles: the effective values of physical parameters vary with the energy scale of the particular interaction under consideration.

Just as corrections to the propagator produce an effective mass parameter, corrections to the simple QED vertex produce an effective vertex factor. That is, the bare coupling constant is also renormalized to produce a physical value. Crucially, the relationship between bare and effective coupling is also found to be dependent on the energy scale at which a process is considered. As with the mass, since the bare parameter cannot be known, we must find some way of fixing the value of the effective parameters so that the theory still has predictive power.

The renormalization procedure is really two procedures in one. First, regularization introduces some means to tame the infinite integrals appearing in Feynman diagrams so that we may make work with them in a meaningful way. This can be through a simple cutoff as we have seen, through introducing imaginary extra particles with large masses, or even through varying the number of space-time dimensions. As long as the regularized integral gives back the original integral in some limit, the exact nature of the regularization scheme does not really matter. Second, a set of renormalization conditions is introduced such that the effective parameters have their measured values at low energy (since this is typically where we measure them). In this way, the parameters' values at higher energies can be deduced. A particularly important case of this for later discussions is that the coupling constant for a theory of interacting particles is found to change as we measure it at different energy scales. In particular, the couplings, g_i, vary with the logarithm of the energy scale, μ, according to

$$\frac{\partial g_i}{\partial(\ln \mu)} = \beta_i(g_i), \tag{9.71}$$

---

[2] This separation of $\Sigma(p)$ into a finite and a divergent part can be seen directly from the square brackets of Equation 9.65, where it is easily verified that the first term is divergent and the second is finite.

where β_i is known simply as the coupling's "β-function." In the approximation that only one-loop corrections are considered, the β-functions are found to be fairly simple: the inverse couplings, $\alpha_i^{-1} = 4\pi/g_i^2$, varying linearly with $\ln \mu$. The gradients are different for each force and are given in Appendix A. All of this has the most notable consequences for the theory of strong interactions and is a point to which we will return when we consider quantum chromodynamics in Section 10.3.2.

The full details of these sorts of calculations and the renormalization process are somewhat beyond the scope of this text, but the interested reader is referred to any introductory text on quantum field theory. When first confronted with the renormalization procedure, many feel that they have somehow been duped, or that the infinities have been carelessly swept under the rug. This is not the case, however, and the procedure is built on a solid, mathematically rigorous foundation. Besides which, the procedure works! The renormalization process is directly responsible for the extraordinary agreement between experiment and theory discussed earlier in this chapter.

9.7 FORM FACTORS AND STRUCTURE FUNCTIONS

9.7.1 Electromagnetic Form Factors

The calculations considered so far have involved only fundamental particles. A complication arises when we wish to calculate the amplitude for an interaction involving one or more hadrons. The problem is that, while at low energy it appears that the hadron as a whole emits or absorbs photons, at a more fundamental level, such interactions are actually occurring in individual constituent quarks. For this reason, we have no reason to suspect that the effective Feynman rule for hadron-photon interaction should resemble that of, say, electron-photon interaction. What we can say, however, is that the interaction should be of a particular form, as constrained by Lorentz and gauge invariance. In particular, notice that the form of Equation 9.33

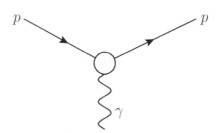

FIGURE 9.1 The interaction of a non-fundamental charged particle with a photon. The shaded blob shows that the details of the graph at this point are masked by higher-order contributions or by other interactions.

consists of three parts. The initial factor (apart from the charges which we have shifted here for convenience) is due to the exchanged photon. Each of the following factors is due to just one of the interacting fermions and takes the form of a symmetric rank-2 tensor. This forces us to conclude that, however complex the effective vertex factor is for a hadron, it must also lead us to the basic form of a symmetric rank-2 tensor for its contribution to the spin-averaged amplitude.

The details of the argument depend on the spin of the hadron in question. In the case of a proton, with spin $\frac{1}{2}$, we can proceed as follows. First, the lowest-order contribution is represented by the Feynman diagram shown in Figure 9.1, where the shaded blob denotes that we have hidden some of the details of this diagram, in the sense that zooming in would reveal a more fundamental picture of the interaction.

If the proton were point-like, we know that the current to which the photon couples would look like

$$j^\mu = (qe)\overline{u}_C(p_3)\gamma^\mu u_A(p_1), \qquad (9.72)$$

just as for the charged leptons. Since the subsequent steps in the calculation of the amplitude will be identical to the point-particle case, Lorentz invariance forces the *actual* hadron vertex to take the form of a four-vector. Furthermore, gauge invariance allows us to impose the Lorenz condition, thus requiring that the current have zero divergence, so $\partial_\mu j^\mu = 0$, or in momentum space, $k_\mu j^\mu = 0$, where k^μ is the momentum of the photon. So by symmetry arguments alone, we can be sure that the hadron-photon vertex must be given by a four-vector with the additional constraint that $k \cdot j = 0$. There are three momentum vectors that can contribute to j^μ, namely p_1^μ, p_3^μ and the photon momentum $k^\mu = (p_3 - p_1)^\mu$. Since only two of these are independent, we are free to choose any two linearly independent combinations. We choose k^μ and, for simplicity, $(p_1 + p_3)^\mu$, which is orthogonal to k^μ. Furthermore, there is only one independent scalar quantity that can be constructed from these momenta, since

$$
\begin{aligned}
k^2 + (p_1 + p_3)^2 &= (p_3 - p_1)^2 + (p_1 + p_3)^2 \\
&= 2m^2 - 2p_1 \cdot p_3 + 2m^2 + 2p_1 \cdot p_3 \qquad (9.73) \\
&= 4m^2 = \text{constant},
\end{aligned}
$$

so k^2 and $(p_1 + p_3)^2$ are not independent.

Since the only aspect of the Feynman rule we expect to change (compared with that for fundamental particles) is the vertex itself, the form of j^μ should be a group of terms sandwiched between two spinors. Recall from Section 8.8 that there are five independent Lorentz-covariant objects that may be constructed from spinors. Two of these have the wrong behavior under parity to contribute to j^μ, so we may construct a total of five independent vector quantities that can potentially contribute to j^μ, namely

$$
\begin{aligned}
&\bar{u}(p_3)\gamma^\mu u(p_1), \quad k^\mu \bar{u}(p_3)u(p_1), \quad (p_1 + p_3)^\mu \bar{u}(p_3)u(p_1), \\
&k_\nu \bar{u}(p_3)\sigma^{\mu\nu}u(p_1), \quad \text{and} \quad (p_1 + p_3)_\nu \bar{u}(p_3)\sigma^{\mu\nu}u(p_1).
\end{aligned} \qquad (9.74)
$$

Each of these may be scaled by an arbitrary coefficient function, but since each must maintain its Lorentz structure, this coefficient can depend at most on the one independent scalar quantity of interest, which we can take to be k^2. Also, since this scalar function can include constant terms such as $\sigma^{\mu\nu}$, we find that the fourth and fifth terms in the previous equation are accounted for by the second and third terms respectively. Thus we arrive at a current of the form

$$
j^\mu = qe\bar{u}(p_3)\left(a_1(k^2)\gamma^\mu + a_2(k^2)k^\mu + a_3(k^2)(p_1 + p_3)^\mu\right)u(p_1),
$$
$$(9.75)$$

where a_1, a_2, a_3 are the scaling functions.

We now impose the second constraint, $k_\mu j^\mu = 0$, known in this context as the Ward identity. Multiplying Equation 9.75 by p_μ, we find

$$
\begin{aligned}
k_\mu j^\mu &= qe\overline{u}(p_3)\left(a_1(k^2)\not{k} + a_2(k^2)k^2 + a_3(k^2)k \cdot (p_1 + p_3)\right)u(p_1) \\
&= qe\overline{u}(p_3)\left(a_2(k^2)k^2\right)u(p_1),
\end{aligned} \tag{9.76}
$$

where the final term vanishes due to the orthogonality of k^μ and $(p_1 + p_3)^\mu$, and the first term vanishes because

$$
\begin{aligned}
\overline{u}(p_3)\not{k}u(p_1) &= \overline{u}(p_3)\not{p}_3 u(p_1) - \overline{u}(p_3)\not{p}_1 u(p_1) \\
&= m\left(\overline{u}(p_3)u(p_1) - \overline{u}(p_3)u(p_1)\right) = 0.
\end{aligned} \tag{9.77}
$$

In order to satisfy the Ward identity, then, we see that the function $a_2(k^2)$ must be identically zero, reducing the independent terms in the current to just two:

$$
j^\mu = qe\overline{u}(p_3)\left(a_1(k^2)\gamma^\mu + a_3(k^2)(p_1 + p_3)^\mu\right)u(p_1). \tag{9.78}
$$

We now make use of the *Gordon identity*,

$$
\overline{u}(p_3)(p_1+p_3)^\mu u(p_1) = 2m\overline{u}(p_3)\gamma^\mu u(p_1) - i\overline{u}(p_3)\sigma^{\mu\nu}(p_3-p_1)_\nu u(p_1), \tag{9.79}
$$

to rewrite the second term in Equation 9.78. The end result, then, is that the proton-photon vertex is given by

$$
qe\overline{u}(p_3)\left(F_1(k^2)\gamma^\mu + \frac{i}{2m}F_2(k^2)\sigma^{\mu\nu}k_\nu\right)u(p_1), \tag{9.80}
$$

where F_1 and F_2 are known as the proton *form factors*. The factor of $i/2m$ is conventional. Using the Gordon identity again, this can also be written as

$$
qe\overline{u}(p_3)\left(F_1'(k^2)(p_1 + p_2)^\mu + \frac{i}{2m}F_2'(k^2)\sigma^{\mu\nu}k_\nu\right)u(p_1), \tag{9.81}
$$

which helps to make the interpretation of the terms a little clearer. To understand this expression, we should first note that we could perform a similar but rather more straightforward derivation for the interaction of a photon with a scalar particle, such as the charged pion. In this case, we find that there is just one form function, with the vertex given by

$$iqeF_\pi(k^2)(p_1 + p_3)^\mu. \qquad (9.82)$$

Notice that the form of the vertex factor is identical to that for a fundamental scalar particle, only scaled by the pion form factor. The first term of the expression for the proton vertex, then, behaves analogously to that of the pion vertex. The second term, on the other hand, has no equivalent in the scalar case, and so must be an artifact of the non-zero spin of the proton. This term is due to the proton's magnetic moment.

Notice that the vertex factor for a fundamental spin-$\frac{1}{2}$ particle contains only the first term of Equation 9.78 but both terms of Equation 9.81 if rewritten using the Gordon identity. It should come as no surprise that the magnetic moment in this case works out to give the g-factor of $g = 2$. However, we can take this a step further: suppose the fermion in Figure 9.1 is an electron and that the blob now denotes our ignorance, not of the hidden strong interactions, but of the higher-order electromagnetic interactions that manifest as loop corrections to the tree-level diagram. In this case, the second term of Equation 9.78 *is* present by exactly the same argument as above, and represents the *anomalous* magnetic moment of the electron. It is by computing the loop contributions to this term that one is able to arrive at the theoretical value for $(g - 2)$ that is in such impressive agreement with observation.

9.7.2 Structure Functions and the Quark Model

The interactions discussed in the preceding sections were for elastic scattering processes, but the same arguments can be applied to inelastic scattering, in which the final products are not the same as the initial particles. In particular, deep inelastic scattering processes involve the exchange of a photon between a high-energy electron and a proton, such that the proton produces hadronic jets. The vertex factor for such processes looks almost identical to Equation 9.80, but the scale-dependent coefficients are now referred to as *structure functions*, since they map the internal structure of the hadron. The detailed study of form factors and structure functions is worthy of a book of its own and many have been written, but it is worth mentioning here just a few points regarding their behavior. First, at intermediate energy scales, the structure functions are complicated, while at very high energy scales, they become essentially scale-invariant, notwithstanding some minor scaling violations. This is best explained by relating the structure functions to the distribution functions for a set of constituent particles within the hadron. In this context, these hypothetical constituents are known as partons, since they are the "parts" of the hadron. The scale-invariance is then a consequence of the finite time-scale of the interactions between these partons: at high energy, the interaction of the photon with a single constituent particle occurs over a shorter interval than the characteristic time-scale of the parton interactions. Second, the predicted form of the structure function from a parton distribution depends on the spin of the partons, and consistency with observation requires the parton distribution function for the charged components of the hadron to be spin-$\frac{1}{2}$ and account for only half of the total hadron momentum. In other words, the only electrically charged constituent particles of a hadron are required to be spinors, consistent with the quark model. The missing momentum is accounted for with the electrically neutral gluons. Furthermore, comparison of the structure functions for the proton and the neutron gives constraints on the charges of the constituents, placing them at $q = +2/3, -1/3, -1/3, \ldots$, again consistent with the quark model. It is this fact that provided the first clear evidence that quarks were physical particles rather than mere mathematical tools.

EXERCISES

1. **(a)** Take the adjoint of the Dirac equation and modify the derivative to a gauge-covariant derivative to arrive at the equation obeyed by the adjoint spinor in an electromagnetic field.

(b) Show that this equation along with the equations of motion for the photon and the electron follow from the Lagrangian given in Equation 9.3. (Hint: The Euler-Lagrange equation in which you differentiate with respect to ψ gives the equation for $\overline{\psi}$ and vice versa.)

2. Compare the list of Feynman rules against the QED Lagrangian. Identify which rule corresponds to which term and determine how you would go about deriving Feynman rules directly from a Lagrangian.

3. Following a derivation similar to Exercise 13, show that the full Pauli equation (Equation 3.56) is the low-energy limit of the Dirac equation in the presence of an electromagnetic field.

4. In ultra-high energy collisions, the mass of the colliding particles can be neglected to a first approximation.
   (a) For a high-energy collision between two distinguishable particles in the center-of-mass frame, write down the four-vectors for the initial and final states in the center-of-mass frame.
   (b) Hence show that the differential cross-section for electron-muon scattering is proportional to $(1 + (1 + \cos\theta)^2)/(1 - \cos\theta)^2$, where θ is the scattering angle.
   (c) Write this differential cross-section in terms of the Mandelstam variables.
   (d) Integrating over solid angle should now give the total cross-section for this process. However, the cross-section is found to be divergent. Identify the origin of this divergence and explain why it does not imply unphysical behavior. (Hint: consider which values of θ may cause the differential cross-section to diverge.)

5. The presence of delta functions in the differential cross-section formula allows it to be calculated analytically in the case of a two-particle final state (Equation 9.53). This applies equally to the decay rate in the case of a two-particle

final state. Derive the decay rate in this case, in terms of the invariant amplitude.

6. Consider the annihilation of an electron and a positron via $e^+ + e^- \to \gamma + \gamma$.

   (a) Why do on-shell electron-positron pairs never annihilate to produce a single photon?

   (b) Draw a Feynman diagram for the above process.

   (c) Calculate the spin-averaged invariant amplitude for this process.

7. (a) Draw two Feynman diagrams for one-loop corrections to elastic scattering of distinguishable particles.

   (b) One such correction takes the form of a correction to the photon propagator. Identify this diagram and calculate the photon's "self-energy" (more properly called the vacuum polarization when relating to photons).

8. (a) Find the invariant amplitude and differential cross-section for $\mu^- \text{-} \mu^+$ production in $e^- \text{-} e^+$ collisions in the ultra-relativistic limit.

   (b) Write this differential cross-section in terms of the Mandelstam variables. How is it related to your answer to Exercise 4?

9. Derive the γ-matrix identities listed in Appendix C.

10. Write down the invariant amplitude for Delbrück scattering to lowest order, and use Appendix C to evaluate its fermion trace structure.

11. Show that the Mandelstam variables obey the relation

$$s + t + u = p_1^2 + p_2^2 + p_3^2 + p_4^2.$$

CHAPTER 10

NON-ABELIAN GAUGE THEORY AND COLOR

Just as quantum electrodynamics may be derived from a localization of the $U(1)$ symmetry in the Dirac equation, we will find that localizing other global symmetries leads to the remaining forces of the standard model. In particular, we will show how the approximate isospin symmetry discussed in Section 6.2.1 leads to the weak interaction. However, it will also be shown that there are certain complications involved with this interaction, which are the underlying reason for the approximate nature of this symmetry. For this reason, this derivation is deferred until Chapter 11. In this chapter, we consider the larger but exact (and, therefore, simpler) color symmetry, and show that this leads to the strong interaction. We will see that the non-Abelian nature of these larger symmetry groups leads to behavior that is qualitatively different from that of the Abelian $(U(1))$ symmetry previously considered.

10.1 NON-ABELIAN SYMMETRY IN THE DIRAC EQUATION

10.1.1 $SU(3)$ and Color

We have seen in Section 6.3 that quarks come in three colors. This color degree of freedom is very different from the flavor degree of freedom, since all colors behave identically. Whereas different quark flavors have different masses and charges, there is no distinction between, say, a red up quark and a blue up quark. There is a perfect three-fold symmetry. Since the colors have the same mass, all three degrees of freedom for a given quark flavor may be described succinctly by a single Dirac equation

$$\left(i\partial\!\!\!/ - m \right) \Psi = 0, \tag{10.1}$$

where

$$\Psi = \begin{pmatrix} \psi_r \\ \psi_g \\ \psi_b \end{pmatrix} \tag{10.2}$$

is now a triplet of spinors. We can also write this in the form of an exterior product of one Lorentz spinor, ψ, and a *color* spinor, c_A: $\Psi = \psi c_A$. The color spinor is then a three-component $SU(3)$ spinor that accounts for the color degree of freedom of the quark. Specifically, red, green, and blue quarks respectively have color spinors

$$\begin{pmatrix} 1 \\ 0 \\ 0 \end{pmatrix}, \quad \begin{pmatrix} 0 \\ 1 \\ 0 \end{pmatrix}, \quad \begin{pmatrix} 0 \\ 0 \\ 1 \end{pmatrix}. \tag{10.3}$$

This symmetry goes deeper: since particles can be combined into linear superpositions, we could relabel, for example, r as $\frac{1}{\sqrt{2}} \left(|r\rangle + |b\rangle \right)$ and b as $\frac{1}{\sqrt{2}} \left(|b\rangle - |r\rangle \right)$, without affecting the physics. In fact, for any

$SU(3)$ matrix, M, a color transformation of the form

$$\begin{pmatrix} r \\ g \\ b \end{pmatrix} \mapsto M \begin{pmatrix} r \\ g \\ b \end{pmatrix} \tag{10.4}$$

is a symmetry of the system, as long as this relabeling is performed consistently on all quarks simultaneously. That is, there is a global $SU(3)$ symmetry in quark colors. Recall from Section 4.3.1 that there are eight generators, $T_1, \ldots, T_8$, of the $SU(3)$ group, each a traceless Hermitian 3×3 matrix. There is, of course, a great deal of freedom in choosing these generators, each choice leading to a different representation of the group. One possible representation is given by $T_i = \frac{1}{2}\lambda_i$, where λ_i are the Gell-Mann matrices:

$$\lambda_1 = \begin{pmatrix} 0 & 1 & 0 \\ 1 & 0 & 0 \\ 0 & 0 & 0 \end{pmatrix} \quad \lambda_2 = \begin{pmatrix} 0 & -i & 0 \\ i & 0 & 0 \\ 0 & 0 & 0 \end{pmatrix}$$

$$\lambda_3 = \begin{pmatrix} 1 & 0 & 0 \\ 0 & -1 & 0 \\ 0 & 0 & 0 \end{pmatrix} \quad \lambda_4 = \begin{pmatrix} 0 & 0 & 1 \\ 0 & 0 & 0 \\ 1 & 0 & 0 \end{pmatrix}$$

$$\lambda_5 = \begin{pmatrix} 0 & 0 & -i \\ 0 & 0 & 0 \\ i & 0 & 0 \end{pmatrix} \quad \lambda_6 = \begin{pmatrix} 0 & 0 & 0 \\ 0 & 0 & 1 \\ 0 & 1 & 0 \end{pmatrix}$$

$$\lambda_7 = \begin{pmatrix} 0 & 0 & 0 \\ 0 & 0 & -i \\ 0 & i & 0 \end{pmatrix} \quad \lambda_8 = \frac{1}{\sqrt{3}} \begin{pmatrix} 1 & 0 & 0 \\ 0 & 1 & 0 \\ 0 & 0 & -2 \end{pmatrix}, \tag{10.5}$$

where the $\sqrt{3}$ factor in the final matrix is to ensure the same normalization for all matrices. Specifically, these matrices are normalized such that $\text{tr}\,(T_i T_j) = \frac{1}{2}\delta_{ij}$ (or $\text{tr}\,(\lambda_i \lambda_j) = 2\delta{ij}$).

10.1.2 Localizing the SU(3) Symmetry

Localization of this $SU(3)$ symmetry works in essentially the same way as that of the $U(1)$ symmetry. We again need to construct a gauge-covariant derivative, that transforms according to

$$D_\mu M \Psi = M D_\mu \Psi, \tag{10.6}$$

for an arbitrary $SU(3)$ transformation matrix M. Since M is given by

$$M = \exp\left(\sum_{i=0}^{8} i\alpha_i(x)T_i\right), \tag{10.7}$$

where α_i are now a set of eight independent parameters, the covariant derivative must now have eight additional vector field terms $A_i, i = 1, \ldots, 8$. Each of these vector fields is included to absorb one of the unwanted terms arising from the partial derivative acting on one of the group parameters $\alpha_i(x)$. Therefore, each of these vector fields independently has gauge invariance. The proportionality constant for each vector field must be the same to retain the $SU(3)$ symmetry so we find a covariant derivative of the form

$$D_\mu = \partial_\mu + ig_3 \sum_{i=0}^{8} A_\mu^i T^i, \tag{10.8}$$

where g_3 is the strong coupling constant. This covariant derivative has the correct transformation property as long as we have the freedom to redefine A_μ according to the generalized gauge transformation

$$A_\mu^i \mapsto e^{i\alpha_j(x)T_j}\left(A_\mu^i + i\partial_\mu\alpha^i(x)\right)e^{-i\alpha_k(x)T_k}. \tag{10.9}$$

Notice, incidentally, that less care is taken with the positioning of $SU(3)$ indices than Lorentz indices. This is because there is no need to distinguish between contravariant and covariant vectors in the $SU(3)$ space.

Local $SU(3)$ symmetry, then, requires eight vector bosons, which we know as gluons. These mediate the strong force just as the photon mediates the electromagnetic force in the Abelian case. One key difference, however, is that the photon leaves the identity of the interacting fermion unchanged, while the gluons are capable of changing the color of the interacting quarks. To see why, consider the action of the generator T_1 on a quark in the red state. Writing again the quark wavefunction as a color triplet, a red state can be expressed as

$$|r\rangle = \begin{pmatrix} 1 \\ 0 \\ 0 \end{pmatrix}, \text{ so}$$

$$T_1 |r\rangle = \begin{pmatrix} 0 & 1 & 0 \\ 1 & 0 & 0 \\ 0 & 0 & 0 \end{pmatrix} \begin{pmatrix} 1 \\ 0 \\ 0 \end{pmatrix} = \begin{pmatrix} 0 \\ 1 \\ 0 \end{pmatrix} = |g\rangle. \qquad (10.10)$$

The generator T_1 has the effect of turning a red quark into a green quark. Similarly, the same generator converts green to red. If color is to be a conserved quantity (which, of course, it must be, since its conservation follows from the global $SU(3)$), then the gluons themselves must carry color. In particular, the gluon A_1^μ must be in the color state $\frac{1}{\sqrt{2}}(|r\bar{g}\rangle + |g\bar{r}\rangle)$. Likewise, in the chosen $SU(3)$ representation, the gluons can be written as

$$|1\rangle = \frac{1}{\sqrt{2}}\left(r\bar{g} + g\bar{r}\right)$$

$$|2\rangle = \frac{-i}{\sqrt{2}}\left(r\bar{g} - g\bar{r}\right)$$

$$|3\rangle = \frac{1}{\sqrt{2}}\left(r\bar{r} - g\bar{g}\right)$$

$$|4\rangle = \frac{1}{\sqrt{2}}\left(r\bar{b} + b\bar{r}\right)$$

$$|5\rangle = \frac{-i}{\sqrt{2}}\left(r\bar{b} - b\bar{r}\right) \qquad (10.11)$$

$$|6\rangle = \frac{1}{\sqrt{2}}\left(g\bar{b} + b\bar{g}\right)$$

$$|7\rangle = \frac{-i}{\sqrt{2}}\left(g\bar{b} - b\bar{g}\right)$$

$$|8\rangle = \frac{1}{\sqrt{6}}\left(r\bar{r} + g\bar{g} - 2b\bar{b}\right).$$

10.2   GLUON SELF-INTERACTIONS

Since the strong interaction is an interaction between colored particles and the gluons themselves carry a color charge, this leads to gluon-gluon interactions. These can also be derived by looking at the gluon kinetic term. Constructing the commutator of two covariant derivatives, as for the photon, we find

$$ig_3 G^i_{\mu\nu} \equiv [D_\mu, D_\nu]^i = \partial_\mu A^i_\nu - \partial_\nu A^i_\mu + if^{ijk}A^j_\mu A^k_\nu. \qquad (10.12)$$

The third term in this expression is specific to the non-Abelian case, and did not appear in the field strength tensor, $F_{\mu\nu}$, for the photon. It arises in this case because of the non-commutativity of the group generators appearing in D_μ. In fact, the presence of this term means that $G^i_{\mu\nu}$ is not gauge-invariant as $F_{\mu\nu}$ was. However, the term that will appear in the Lagrangian, $G^i_{\mu\nu}G^{i\mu\nu}$, is both Lorentz-invariant and gauge-invariant.

The equation of motion for a gluon in the presence of a colored fermion is then given by

$$D_\mu G^{i\mu\nu} = g_3 \overline{\Psi}\gamma^\nu T^i \Psi, \qquad (10.13)$$

which expands to

$$\partial^2 A^{i\nu} - \partial^\nu \partial \cdot A^i = 2g_3 f^{ijk}\partial_\mu \left(A^{j\mu}A^{k\nu} \right) - ig_3 A^\ell_\mu T^\ell \partial^\mu A^{i\nu}$$
$$+ ig_3 A^\ell_\mu T^\ell \partial^\nu A^{i\mu} + 2ig_3^2 A^\ell_\mu T^\ell f^{ijk} A^{j\mu}A^{k\nu}$$
$$+ g_3 \overline{\Psi}\gamma^\nu T^i \Psi. \qquad (10.14)$$

The first three terms on the right of this equation allow for three-gluon interactions, while the fourth term allows for four-gluon interactions. The final term of course accounts for gluon interactions with the fermion.

Another way to arrive at the same conclusion is to construct the Lagrangian for this system. The kinetic term for gluons is constructed in analogy with the kinetic term for photons:

$$
\begin{aligned}
\mathcal{L}_{\text{gluon}} &= -\frac{1}{4} G^i_{\mu\nu} G^{i\mu\nu} \\
&= -\frac{1}{4} \Big[\left(\partial_\mu A^i_\nu - \partial_\nu A^i_\mu \right) \left(\partial^\mu A^{i\nu} - \partial^\nu A^{i\mu} \right) \\
&\quad -2g_3 f^{ijk} A^j_\mu A^k_\nu \left(\partial^\mu A^{i\nu} - \partial^\nu A^{i\mu} \right) \\
&\quad +g_3^2 f^{ijk} f^{i\ell m} A^j_\mu A^k_\nu A^{\ell\mu} A^{m\nu} \Big].
\end{aligned}
\tag{10.15}
$$

The three-point and four-point interactions are clear in this formalism.

10.3   STRONG FORCE INTERACTIONS

10.3.1   Quantum Chromodynamics

The strong force, then, is explained in terms of a non-Abelian generalization of local phase symmetry. Theories of this nature are collectively known as gauge theories, since they all have a generalized notion of gauge invariance, as given in Equation 10.9. Since it is based around color, in analogy with quantum electrodynamics, the locally $SU(3)$-invariant theory that we have introduced is named quantum chromodynamics (QCD). The Feynman rules for QCD are similar to those for QED, with a few additional rules. In particular, the following alterations to the QED rules must be made.

- External (anti-)quarks carry an additional color spinor, c_A, which transforms as the (anti-)fundamental $(3/\bar{3})$ representation of $SU(3)$.

- External incoming/outgoing gluons carry an additional color polarization factor, $G_a/G_a^\dagger$, in the adjoint (8) representation of $SU(3)$, which serves to identify the particular combination of gluon basis states.

- Fermion-gluon vertices contribute an additional factor of the $SU(3)_C$ generator, T^a.

- Internal gluon propagators contribute an additional δ_{ab} where a, b are (adjoint representation) color indices.

In addition, there are now gauge boson self-interactions to consider, with their own set of Feynman rules. For a three-gluon vertex,

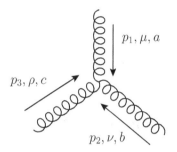

where $p_{1,2,3}$ are momenta, μ, ν, ρ are Lorentz indices, and a, b, c are color indices, we get a contribution of

$$-g_3 f_{abc} \left(g_{\mu\nu}(p_1 - p_2)_\rho + g_{\nu\rho}(p_2 - p_3)_\mu + g_{\rho\mu}(p_3 - p_1)\right), \quad (10.16)$$

where f_{abc} are the $SU(3)$ structure constants. With similar notation, we find

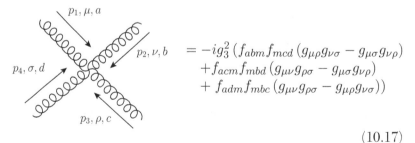

$$= -ig_3^2 \left(f_{abm}f_{mcd} \left(g_{\mu\rho}g_{\nu\sigma} - g_{\mu\sigma}g_{\nu\rho}\right)\right.$$
$$+f_{acm}f_{mbd} \left(g_{\mu\nu}g_{\rho\sigma} - g_{\mu\sigma}g_{\nu\rho}\right)$$
$$+ \left. f_{adm}f_{mbc} \left(g_{\mu\nu}g_{\rho\sigma} - g_{\mu\rho}g_{\nu\sigma}\right)\right)$$

$$(10.17)$$

A full list of all these rules may be found in Appendix B.

10.3.2 Scale-Dependence

Quantum chromodynamics is found to have some important scale-dependent features. In particular, it behaves very differently at the low-energy (large-distance) and high-energy (short-distance) scales. This difference is due in part to the scale-dependence of the coupling constant g_3: the inherent strength of QCD interactions depends on the energy of the particles taking part in the interaction, or equivalently on the length scales being probed. This may seem strange—after all, the strength of the interactions is measured by a coupling *constant*! However, such scale-dependence is a genuine feature of any theory of interacting particles, as we saw in Section 9.6. For an intuitive appreciation of its origin, it is easiest to consider first the analogous scale-dependence of the electromagnetic coupling. Consider a lone stationary electron and a test particle used to measure the electron's charge. If the test particle could be brought infinitesimally close to the electron, we would find some value, e_0. But now consider what happens as we move the test particle away from the electron. The intervening space, even when empty, is a sea of virtual photons and electron-positron pairs popping in and out of existence. Even though these particles are virtual, they have a real effect: the potential due to the real electron is attractive to the virtual electrons but repulsive to the virtual positrons, and so it polarizes the particle pairs. The net effect of this is that the space itself around the electron becomes polarized, screening the true value of the electron charge from the test particle. If the electron charge is measured at this larger distance scale, the measured value, e, is found to be less than the bare charge e_0. In fact, the bare charge is found to be formally infinite, and is one of the parameters of the theory that must be renormalized. The amount of this screening may be calculated through higher-order corrections to the tree-level graphs of QED, and is the origin of the β function, introduced in Section 9.6. This function gives the relative change in the coupling constant as the energy scale is increased, or equivalently as the theory is probed at shorter length scales. In the case of QED, the β function is positive: the coupling increases as the energy-scale is increased.

In QCD, the screening effect is reversed and the β function is negative. The force gets stronger at *longer* length scales. The reason for this is the gluon self-interactions. Essentially, a gluon exchanged between two quarks that are far removed from each other produces other gluons *en route*, increasing the coupling constant for the interaction. This means that the strong coupling at everyday energy scales (macroscopic length-scales) grows so high that its numerical value is greater than 1. At this point, the perturbative approach to calculating amplitudes breaks down, since increasing the number of vertices in a Feynman diagram *increases* that diagram's contribution to the amplitude. At shorter distances, the perturbative approach is still valid, leading to two very different regimes in strong interactions. The approximate scale-dependences of couplings are summarized in Appendix A.

10.4   HIGH-ENERGY QCD

10.4.1   Asymptotic Freedom

At short distance scales, the strong coupling constant is small. Indeed, its value decreases with decreasing length scales so that at arbitrarily small scales, the constant vanishes. This means that the quarks and gluons within a hadron behave more and more as free (non-interacting) particles as they are probed at smaller scales. This behavior is known as asymptotic freedom. In this regime, we may calculate scattering amplitudes through the perturbative approach. Such calculations allow us to verify that the baryon and meson color states are bound, while other quark combinations such as $qq\bar{q}$ are mutually repulsive. Perturbative calculations in the high-energy regime allow for the detailed analysis of the parton distribution functions. These in turn may be linked to the hadronic structure functions of Section 9.7.2 in order to match theory to experiment.

10.4.2   Perturbative QCD

Let us calculate the amplitude for quark-quark scattering via gluon exchange. The leading-order contribution to this amplitude comes from the diagram:

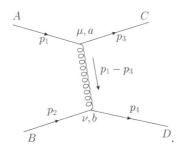

There is, of course, a crossed version of this diagram as well if the quarks are identical, but we will neglect this fact for now, since the argument we are going to use would still apply with two contributions. Using the QCD Feynman rules, we find that the amplitude is given by

$$-i\mathcal{M} = \overline{u}(p_3)c_C^\dagger \left(-ig_3\gamma^\mu(T^a)_{CA}\right) u(p_1)c_A \left(\frac{-ig_{\mu\nu}\delta_{ab}}{(p_1 - p_3)^2}\right)$$
$$\times\ \overline{u}(p_4)c_D^\dagger \left(-ig_3\gamma^\nu(T^b)_{DB}\right) u(p_2)c_B. \qquad (10.18)$$

Since the color spinors c_A and the Lorentz spinors $u(p)$ act in different spaces, they commute, and so we can factorize this expression as

$$-i\mathcal{M} = \overline{u}(p_3)\left(-ig_3\gamma^\mu\right) u(p_1) \left(\frac{-ig_{\mu\nu}}{(p_1 - p_3)^2}\right) \overline{u}(p_4)\left(-ig_3\gamma^\nu\right) u(p_2)$$
$$\times\ \left(c_C^\dagger T_{CA}^a c_A\right)\left(c_D^\dagger T_{DB}^a c_B\right), \qquad (10.19)$$

where the generators have been set equal by the Kronecker delta, δ_{ab}. The repeated "a" index implies summation over generators.

Apart from the obvious replacement $qe \mapsto g_3$, the first part of this amplitude is identical to the corresponding QED diagram for electron-electron scattering. The only additional complication for the QCD calculation, then, is the appearance of a *color factor*:

$$CF = \left(c_C^\dagger T_{CA}^a c_A\right)\left(c_D^\dagger T_{DB}^a c_B\right) = \frac{1}{4}\left(c_C^\dagger \lambda_{CA}^a c_A\right)\left(c_D^\dagger \lambda_{DB}^a c_B\right), \qquad (10.20)$$

where λ^a are the Gell-Mann matrices.

Let us now choose a specific color assignment: $q_r + q_r \to q_r + q_r$. That is, we consider two quarks in a color sextet state: the symmetric part of the product $3 \otimes 3 = 6 \oplus \overline{3}$. In this case, the color spinors are

$$c_A = c_B = c_C = c_D = \begin{pmatrix} 1 \\ 0 \\ 0 \end{pmatrix}, \qquad (10.21)$$

and so the only Gell-Mann matrices that will contribute to the color factor are λ^3 and λ^8, since these have diagonal entries. So we find:

$$CF = \frac{1}{4}\left(\begin{pmatrix} 1 \\ 0 \\ 0 \end{pmatrix}^T \lambda^3 \begin{pmatrix} 1 \\ 0 \\ 0 \end{pmatrix} \cdot \begin{pmatrix} 1 \\ 0 \\ 0 \end{pmatrix}^T \lambda^3 \begin{pmatrix} 1 \\ 0 \\ 0 \end{pmatrix} + \right.$$

$$\left. + \begin{pmatrix} 1 \\ 0 \\ 0 \end{pmatrix}^T \lambda^8 \begin{pmatrix} 1 \\ 0 \\ 0 \end{pmatrix} \begin{pmatrix} 1 \\ 0 \\ 0 \end{pmatrix}^T \lambda^8 \begin{pmatrix} 1 \\ 0 \\ 0 \end{pmatrix} \right)$$

$$= \frac{1}{3}. \qquad (10.22)$$

Since the amplitude is otherwise identical to the corresponding QED amplitude, we can deduce that the behavior will also be identical. In particular, the QED electron-electron scattering leads to a potential that falls off as $1/r$, and is positive (as we saw in Section 9.3). That is, the force between like charges is repulsive. So we can see that we have the same result here: two quarks in a symmetric state will repel. Although we have demonstrated this fact only in the case of red-red scattering, the result is necessarily the same for the remaining colors, since color symmetry is exact: we must be able to replace one color for another in a consistent manner, and arrive at the same result. It is worth checking, therefore, that the result really is the same for green-green and blue-blue scattering.

What about the color singlet state, the antisymmetric combination that we saw in Chapter 6? For quarks to form baryons, we require this state to be attractive. We can demonstrate that this is the case, again through the color factor. In the singlet state, the quarks' colors are fully antisymmetric. So a typical pair of quarks to consider might be in the state $|\psi\rangle = \frac{1}{\sqrt{2}}(|rg\rangle - |gr\rangle)$, for example. The scattering amplitude for quarks in this state can be expressed as

$$\left\langle \psi \left| \widehat{S} \right| \psi \right\rangle = \frac{1}{2}\left(\left\langle rg \left| \widehat{S} \right| rg \right\rangle - \left\langle rg \left| \widehat{S} \right| gr \right\rangle \right.$$
$$\left. - \left\langle gr \left| \widehat{S} \right| rg \right\rangle + \left\langle gr \left| \widehat{S} \right| gr \right\rangle \right), \quad (10.23)$$

where $\widehat{S}$ is the relevant scattering operator. So there are two distinct types of interaction to consider in this case: the elastic scattering processes,

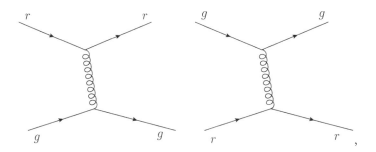

in which quarks retain their color identity, and the inelastic processes in which quark colors are swapped. Let's find the color factor for red-green elastic scattering first. In this case:

$$CF = \frac{1}{4}\left(\begin{pmatrix} 1 \\ 0 \\ 0 \end{pmatrix}^T \lambda^3 \begin{pmatrix} 1 \\ 0 \\ 0 \end{pmatrix} \cdot \begin{pmatrix} 0 \\ 1 \\ 0 \end{pmatrix}^T \lambda^3 \begin{pmatrix} 0 \\ 1 \\ 0 \end{pmatrix}\right.$$
$$\left. + \begin{pmatrix} 1 \\ 0 \\ 0 \end{pmatrix}^T \lambda^8 \begin{pmatrix} 1 \\ 0 \\ 0 \end{pmatrix} \cdot \begin{pmatrix} 0 \\ 1 \\ 0 \end{pmatrix}^T \lambda^8 \begin{pmatrix} 0 \\ 1 \\ 0 \end{pmatrix}\right)$$
$$= -\frac{1}{6}. \quad (10.24)$$

Thanks again to color symmetry, green-red scattering gives the same color factor. This leaves the color-changing processes

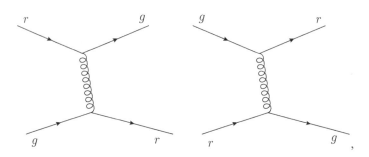

each giving a color factor of

$$CF = \frac{1}{4}\left(\left(\begin{pmatrix} 1 \\ 0 \\ 0 \end{pmatrix}^T \lambda^1 \begin{pmatrix} 0 \\ 1 \\ 0 \end{pmatrix} \right) \cdot \left(\begin{pmatrix} 0 \\ 1 \\ 0 \end{pmatrix}^T \lambda^1 \begin{pmatrix} 1 \\ 0 \\ 0 \end{pmatrix} \right) + \right.$$

$$\left. + \left(\begin{pmatrix} 1 \\ 0 \\ 0 \end{pmatrix}^T \lambda^2 \begin{pmatrix} 0 \\ 1 \\ 0 \end{pmatrix} \right) \cdot \left(\begin{pmatrix} 0 \\ 1 \\ 0 \end{pmatrix}^T \lambda^2 \begin{pmatrix} 1 \\ 0 \\ 0 \end{pmatrix} \right) \right) \qquad (10.25)$$

$$= \frac{1}{2},$$

where it is now the λ^1 and λ^2 matrices that contribute. Combining these results, we find from Equation 10.23 that the overall color factor for the scattering of a pair of quarks in the singlet state is

$$CF = \frac{1}{2}\left(2 \times -\frac{1}{6} - 2 \times \frac{1}{2} \right) = -\frac{2}{3}. \qquad (10.26)$$

Since this is negative, it implies that the force between differently colored quarks really is attractive.

Before performing similar calculations for the meson states, $q\bar{q}$, it is worth mentioning that there is an alternative method for finding the color factor for a process. This is to look at the gluons that can couple to a particular quark color. For example, to conserve color in the previous red-green elastic scattering, the only gluons that can couple to red-red and green-green currents are the $|3\rangle = \frac{1}{\sqrt{2}}(r\bar{r} - g\bar{g})$ and $|8\rangle = \frac{1}{6}(r\bar{r} + g\bar{g} - 2b\bar{b})$. The coupling at the red vertex is $\frac{1}{\sqrt{2}}$ for the

$|3\rangle$ and $\frac{1}{\sqrt{6}}$ for the $|8\rangle$. Similarly, the coupling at the green vertex is $-\frac{1}{\sqrt{2}}$ for the $|3\rangle$ and $\frac{1}{\sqrt{6}}$ for the $|8\rangle$. With this in mind, the color factor can also be expressed as

$$CF = \frac{1}{2} \sum_{\text{diagrams}} C_1 C_2, \qquad (10.27)$$

where $C_{1,2}$ are the "color-couplings" at each vertex. For the current example, we have

$$CF = \frac{1}{2} \left(\left(\frac{1}{\sqrt{2}} \right) \left(-\frac{1}{\sqrt{2}} \right) + \left(\frac{1}{\sqrt{6}} \right) \left(\frac{1}{\sqrt{6}} \right) \right)$$
$$= -\frac{1}{6}, \qquad (10.28)$$

which is in agreement with the more formal method presented above.

We will perform one more example before suggesting that the reader try some of these calculations themselves. The final example is quark-antiquark scattering in the singlet (meson) state, $\frac{1}{\sqrt{3}} \left(r\bar{r} + g\bar{g} + b\bar{b} \right)$. As before, this gives several contributions:

$$\left\langle \text{singlet} \left| \widehat{S} \right| \text{singlet} \right\rangle = \frac{1}{3} \left(\left\langle r\bar{r} \left| \widehat{S} \right| r\bar{r} \right\rangle + \left\langle r\bar{r} \left| \widehat{S} \right| g\bar{g} \right\rangle + \dots \right), \qquad (10.29)$$

with nine terms in total. Because of color symmetry, only two of these need to be calculated explicitly. We will choose red-antired elastic scattering and $r\bar{r} \to g\bar{g}$. The relevant diagram in this case is

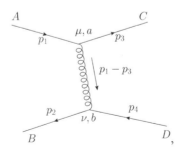

with amplitude

$$
\begin{aligned}
-i\mathcal{M} &= \overline{u}(p_3)c_C^\dagger \left(-ig_3\gamma^\mu(T^a)_{CA}\right)u(p_1)c_A \left(\frac{-ig_{\mu\nu}\delta_{ab}}{(p_1-p_3)^2}\right) \\
&\quad \times \overline{v}(p_2)c_B^\dagger \left(-ig_3\gamma^\nu(T^b)_{BD}\right)v(p_4)c_D \\
&= \overline{u}(p_3)\left(-ig_3\gamma^\mu\right)u(p_1)\left(\frac{-ig_{\mu\nu}}{(p_1-p_3)^2}\right)\overline{v}(p_2)\left(-ig_3\gamma^\nu\right)v(p_4) \\
&\quad \times \left(c_C^\dagger T_{CA}^a c_A\right)\left(c_B^\dagger T_{BD}^a c_D\right). \tag{10.30}
\end{aligned}
$$

Notice that, as far as the color factor is concerned, there is no difference between red-red and red-antired scattering. So for red-antired scattering, we find $CF = \frac{1}{3}$, and for $r\overline{r} \rightarrow g\overline{g}$, we have $CF = \frac{1}{2}$. Substituting these values into Equation 10.29, we get an overall color factor of

$$
CF = \frac{1}{3}\left(3\times\left(\frac{1}{3}\right) + 6\times\left(\frac{1}{2}\right)\right) = \frac{4}{3}. \tag{10.31}
$$

At first sight, this appears to have the wrong sign: a positive result seems to suggest a repulsive force, yet we know that the singlet is a stable configuration, since it is the state chosen by mesons. We must remember, however, that one of the particles in this case is an anti-quark, carrying an anticolor. In addition to the color spinor for each particle, we must also take account of its "color charge." While the red quark and antired antiquark are both described by the spinor c_1, one has a positive value and the other negative. Therefore, we expect the form of the potential deriving from the QED-like part of Equation 10.30 to resemble that of the electron-positron system rather than the electron-electron system. Therefore, in the quark-antiquark case, the attractive/repulsive nature of a force is determined by the *negative* of the color factor.

10.5 LOW-ENERGY QCD

10.5.1 Quark Confinement

At large scales (low energy scales), quarks and gluons do not behave as free particles: in fact, their behavior is probably as far from free as it could be. Instead, they are *confined* to the hadrons of which they are part, in the sense that they cannot be removed. Although the mechanism behind confinement is still not completely understood, its origins are becoming clearer thanks in part to the efforts of lattice QCD calculations (which we will discuss in Section 10.5.3). A qualitative argument for confinement is as follows, where it is instructive first to consider the analogous behavior of electromagnetic interactions.

The cloud of virtual photons emitted by an electrically charged particle radiates outward in all directions, with the density of photons at a given distance dropping off with the surface area of a sphere at that radius. In contrast, the gluons in a similar cloud are bound by their own self-interactions into string-like objects that end on the color-charged objects producing them. It's important to realize that these QCD strings are not the fundamental objects of string theory, though historically they were their precursors. We will revisit the origin of these QCD strings in Section 10.5.3 where we examine a slightly more rigorous argument for their formation.

QCD strings begin and end on colored objects as in Figure 10.1, in effect focusing the strong force between them. So unlike electromagnetic forces, which fall away with increasing distance according to an inverse-square law, the strong force is found to be constant with varying distance. This leads to a linearly increasing potential energy when two strongly interacting particles are separated. As such,

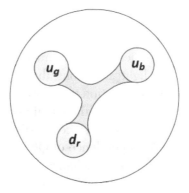

FIGURE 10.1 A schematic representation of QCD strings confining the constituent quarks in a baryon.

increasing the separation between two quarks, in an effort to remove one or more from a hadron, imparts an increasing amount of energy into the system. Before the quark is taken beyond the boundaries of the hadron, the amount of energy introduced is sufficient to produce a new quark-antiquark pair. In turn, this leads to the formation of new hadrons, rather than the isolation of a quark. Thus, the quarks are never observed outside of the confines of a hadron. The production of hadrons in this way is known as hadronization. The energy required to probe the internal structure of a hadron is immense, so hadrons produced through hadronization are not produced in isolation, but in collections of hadrons with roughly similar momenta, called a jet. Experimentally, then, the emission of a quark or gluon from a QCD process is not detected as such but must be inferred through the arrangement of jets in the detector.

10.5.2　The Residual Nuclear Force

The low-energy behavior of QCD is best described not in terms of quarks and gluons but in terms of the bound states of baryons and mesons. Since the quarks and gluons are confined within these composite particles, from the point of view of an observer at a distance from a hadron, the hadron appears to behave as a point particle. These composite particles then have their own dynamics. The residual nuclear force is the attractive force between baryons in the nucleus, and is mediated by mesons. As we saw in Section 9.3, it is a general result in quantum field theory that the force mediated by a boson with even-valued spin is attractive for like particles. In principle then, any meson is capable of mediating the nuclear force through a Yukawa-like interaction, but in practice it is carried by the pions. Since these are the lightest mesons, they are readily produced and so have the greatest contribution to the scattering amplitudes of baryons. Although they are not fundamental and there is a deeper theory underlying their interaction, since the deeper theory is effectively masked by the scale-dependence of QCD, it is useful to have a set of Feynman rules for pion exchange. The relevant rules are analogous to those given in Sections 7.3 and 8.8, but we have additional kinds of interaction. Specifically, since the hadrons are not the fundamental degrees of freedom for the theory, the Lagrangian for this model is only approximate. As such, we must take a perturbative approach even to the terms in the Lagrangian and include an entire series of interactions of increasing dimension. That is, we find that there are terms with increasing numbers of derivatives acting on the scalar particle, which manifest themselves as powers of momentum in the Feynman rules.

This approach to calculation is known as "effective field theory" and the general philosophy of the approach is to include all terms consistent with the symmetries of the model. Effective field theories are characterized by the presence of non-renormalizable interactions. However, this is not a problem for an effective theory, since it is only applied to a specific range of energy-scales, and so the high-energy limit where renormalizability becomes an issue is of no consequence. The theory is *expected* to break down at high energy, since it is a low-energy approximation, and so a hard momentum cutoff is imposed at a suitable scale. The particular effective theory used for describing residual hadronic strong interactions is known as "chiral perturbation theory," since the approximate symmetries to consider when writing down the Lagrangian are the chiral symmetries of the underlying quarks, which are considered in the massless limit. This is a valid approximation since the quark masses are much smaller than the typical energy scales of interest. The lowest-order contribution to the residual nuclear interaction is the exchange of a single pion, as in Figure 10.2, though the effective nature of the model means that the associated Feynman rules are somewhat complicated: in principle each vertex contributes an infinite series of terms with different powers of the transferred momentum. However, those terms with higher powers of momentum make smaller contributions, given that they are scaled down by the high-momentum cutoff, so in practice such series are summed to a specific order. Given the complicated nature of such calculations we will not explore them further here. Instead, we simply note that chiral perturbation theory has had much success in modeling the behavior of the residual nuclear force between pairs of nucleons in various bound states, as well as larger collections of nucleons. The methods may even be generalized to apply

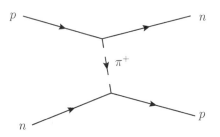

FIGURE 10.2 Pion exchange mediating the residual nuclear interaction between baryons.

to "hyperons"—those baryons with one or more strange quarks—by including the kaons as additional mediators.

10.5.3 Perturbative and Lattice QCD

In the short-distance regime, the strong coupling is smaller than 1. As such, increasing the number of vertices in a Feynman diagram reduces the magnitude of the diagram's contribution to the amplitude. This means that perturbation theory is valid and amplitudes may be calculated order-by-order in g_3.

At large distances, since g_3 is large, the perturbative approach is no longer valid. If one were to attempt a perturbative calculation, then diagrams with a large number of vertices would have a greater contribution to the amplitude than those with fewer vertices. The calculation would be divergent when calculated to higher and higher orders in g_3. Another approach is required when performing QCD calculations at low energy. An approach to such calculations that has had considerable success is the lattice approach. In particular, among other hadron properties, lattice calculations have correctly found the mass of the proton to within 2% accuracy. This approach to QCD involves discretizing space-time into a set of lattice points. A strongly interacting system is then modeled by placing the quarks on the lattice points, while gluons are placed on the edges between these vertices, as in Figure 10.3.

More precisely, lattice QCD is a method for calculating measurable quantities by direct approximate computation of the path integral (see Section 7.3.1). In principle, the path integral is a sum

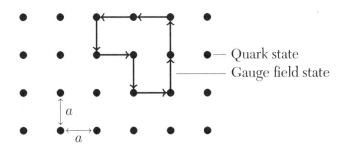

— Quark state

———— Gauge field state

FIGURE 10.3 Quarks placed on the points of a lattice of spacing a, with gauge fields placed on the edges connecting lattice points.

over all possible field configurations, but this sum may only be computed approximately in all but the simplest cases. Perturbation theory makes this approximation by expanding about an analytic solution in powers of one or more coupling constants. Lattice calculations take a different approach and approximate the path integral as a finite-dimensional integral in which the field is defined only at a finite set of discrete lattice space-time points, with lattice spacing a. A Fourier transform shows that the discrete lattice ensures a momentum cutoff. Similarly, the finite size of the lattice corresponds to a discrete set of momentum modes, thus rendering any calculation finite in both the infra-red (long-distance) and ultraviolet (short-distance) regions. Lattice QCD has, then, a built-in regularization scheme for dealing with the infinities that arise in quantum field theoretical calculations. To make contact with reality, calculations are considered in the continuum limit, in which $a \to 0$, and at which point any artifacts of the discretization should vanish. While the idea seems reasonably simple, the practicalities of such calculations are riddled with complications.

The first complication is the sheer size and cost of the calculations, even for relatively small lattices. For example, a field defined on a lattice with 32 lattice points in each direction (including time) requires a $32^4 = 1,048,576$-dimensional vector of field values, and manipulation of such vectors requires matrix multiplication using square matrices of this size: that is, with over 10^{12} elements. Lattice calculations thus require considerable computing power and run time. Second, even with a finite-dimensional integral, these calculations are not approached analytically, but must be approximated using Monte Carlo methods. That is, a selection, or ensemble, of field configurations is summed over as an approximation to the full (infinite) set of configurations. This leads to a further complication: as we said in Section 7.3.1, each path's contribution is weighted by its action, $S[\text{path}] = \int_{\text{path}} d^4x \, \mathcal{L}$, according to e^{iS}. For large values of S, this complex phase oscillates wildly for neighboring paths such that their contributions largely cancel out, whereas those paths close to the classical path (which has a stationary value of S) give similar contributions, leading to constructive interference in the weighting factor. This is a problem for lattice computations, since essentially *all* the configurations in a neighborhood must be included to achieve the correct interference. This problem is overcome by working in a four-dimensional Euclidean space, rather than Minkowski space-time. Notice that the only difference between a Minkowski-space and a Euclidean-space vector is the relative negative sign for the "time-time" component of the metric. So by moving to an imaginary time coordinate, $\tau = it$, via a Wick rotation, we find $x^\mu x_\mu = t^2 - x^2 - y^2 - z^2 = -\tau^2 - x^2 - y^2 - z^2$: with an imaginary time, the system becomes Euclidean with metric (-,-,-,-). The other effect of this imaginary coordinate is to convert the i in the exponent of the path integral to a negative sign. That is, in Euclidean space, the path integral is now weighted by e^{-S}. This has the effect that, while the paths with the largest contributions are still those with the smallest values of S, the contribution of an *individual* path far from the classical path is now heavily damped *without* the interference of other neighboring paths. This allows Monte Carlo methods to be used, since we may now choose a selection of paths with large contributions, and efficient algorithms exist for finding such an ensemble.

As stated previously that the gauge fields live on the links between lattice sites: why should this be? Well, consider that the gauge field's appearance in the theory is via the gauge-covariant derivative: $D_\mu = \partial_\mu + ig_3 A_{a\mu} T_a$. The derivative in a discrete space becomes a non-local operator: the derivative of a field at a site x is defined as some discrete difference, such as $\partial_\mu \psi(x) = \psi(x + a\widehat{\mu}) - \psi(x)$, where $\widehat{\mu}$ is the unit vector in the μ-th direction. That is, the action of the operator on a field at one point involves the value of the field at other points. Since the gauge field is incorporated into the definition of gauge-covariant derivative, it must also be non-local. Furthermore, it must be oriented, since it must distinguish between the location being acted upon by the derivative and the other (non-local) lattice point that is involved. For this reason, the gauge fields are conventionally shown as an arrow on the link between two lattice sites. It can then be shown that any closed path constructed of such links is gauge invariant. In this way, the gauge-invariance of the continuous theory is preserved on the lattice.

The final complication of lattice QCD that we will discuss is the fermion doubling problem. To see how this problem arises, it is worth first considering a bosonic field on the lattice. When the propagator for a bosonic field is computed on the lattice, it is found to be of the form

$$\frac{1}{\widetilde{p}^2 + m^2},$$
(10.32)

where

$$\widetilde{p}_\mu = \frac{2}{a} \sin\left(\frac{a p_\mu}{2}\right),$$
(10.33)

for lattice spacing a, whereas the propagator in a continuum Euclidean theory is

$$\frac{1}{p^2 + m^2}.$$
(10.34)

The momentum is thus deformed by the discretization process, with the discrete momentum and that in the continuum limit agreeing in the limit of small momentum. For larger values of momentum, near the maximum of $p_\mu = \pi/a$ at the boundary of the first Brillouin zone, the two diverge as shown in Figure 10.4. In the fermionic case, a similar procedure finds a propagator

$$i\widetilde{\slashed{p}}' + m^{-1},$$
(10.35)

where

$$\widetilde{p}'_\mu = \frac{1}{a} \sin\left(a p_\mu\right),$$
(10.36)

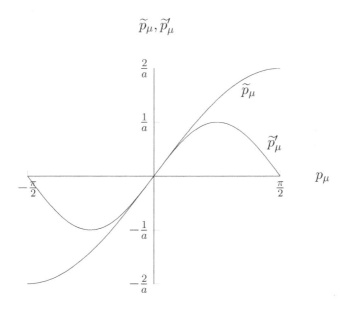

FIGURE 10.4 The relationship between continuum and lattice momenta in the first Brillouin zone.

compared with simply

$$i\not{p} + m^{-1} \qquad (10.37)$$

in the continuum. The problem is now clear from Figure 10.4: there are two distinct regions that appear to behave like the continuum, $p_\mu = 0$ and $p_\mu = \pm\pi/a$ (since we are restricted to the first Brillouin zone, the positive and negative extrema are identical).

This doubling of fermion states occurs once for each space-time dimension, giving 16 fermionic degrees of freedom for each physical fermion in four dimensions. A number of methods exist for removing these unphysical fermions, but each introduces its own side effects. For example, the Wilson fermion approach gives a mass to the unphysical degrees of freedom that scales as $1/a$. In this way, as the continuum limit is approached, the doubled fermions acquire an infinite mass and are removed from the system. The downside of this approach is that it also destroys the chiral symmetry of the underlying theory: that is, even in the continuum limit, with no mass terms, the left-chiral and right-chiral parts of the fermions are no longer independent. There are several other approaches to removing the "doublers," but it has been proven that no method is capable of doing so without introducing artifacts that remain present in the continuum limit.

The Origin of QCD Strings

The low-energy behavior of QCD is dominated by the complex nature of the QCD vacuum state. It is important to realize that "vacuum" does not necessarily translate as "empty" but simply denotes the lowest-energy state of the system. In the case of QCD, the ground state consists of a sea of quarks, antiquarks, and gluons. One of the successes of lattice QCD is the prediction of the QCD strings connecting the constituent quarks in hadrons. These manifest themselves as an emptying of the busy QCD vacuum state. The presence of a valence quark in a region suppresses the background fluctuations of the vacuum state. Somewhat counterintuitively, it is the resulting *lack* of virtual particles that raises the energy of the region above the vacuum state. It is the system's attempt to minimize its energy that leads to the formation of QCD strings. To see why, first consider the electromagnetic force. When two particles of opposite electric charge are placed in a small region, we can picture electric field flux lines connecting the two. In the space immediately between the charges, these flux lines are concentrated and have very little deviation from a series of straight lines. In the space outside this, however, the lines become more bent, and radiate from each charge in all directions as in Figure 10.5. If the color force behaved in this way, the vacuum suppression would extend into the space around the color charges, raising the system's energy. To minimize vacuum suppression, the flux lines are squeezed into a narrow tube directly connecting one charge to another, as in Figure 10.6: this is the origin of the QCD string. Lattice calculations have demonstrated this type of behavior for both mesons and baryons. Interestingly, in the case of baryons, they have shown that the strings do not connect each pair of quarks individually, but instead that the strings form a Y-shaped junction between all three quarks, as in Figure 10.1. This is the arrangement in which the least amount of space suffers vacuum

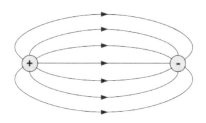

FIGURE 10.5 Lines of electric flux in the region between two charges.

FIGURE 10.6 Lines of color flux squeezed by the QCD vacuum into a flux tube connecting color charges.

suppression. This is true as long as the quarks are sufficiently separated. Since the "flux tube" is finite in cross-section, for very closely spaced quarks it is more energetically favorable for the suppressed region to form roughly a triangular prism around the quarks.

10.6   EXOTIC MATTER

The qqq and $q\bar{q}$ systems are not the only possible color singlets. It has long been suspected that more exotic hadrons may exist in the form of tetraquarks ($q\bar{q}q\bar{q}$) and pentaquarks ($qqqq\bar{q}$). In addition, due to gluon self-interactions, there is the possibility of a color-singlet collection of gluons, known as a glueball.

10.6.1   Pentaquarks and Tetraquarks

The tetraquark is a bound state of two quarks and two antiquarks, which would of course be a colorless combination. The possibility of such a state was proposed by Gell-Mann when he first put forward the quark model, and many specific examples have been predicted along with their calculated properties. Despite their long-theorized existence, though, tetraquark states have only been observed in recent years. Many possible tetraquark candidates have arisen over time as statistical bumps in the invariant mass distributions for various decays. However, further probing by other experiments has often smoothed out such bumps. In other cases, these fluctuations have remained but have not reached the necessary 5σ level of significance to be considered a genuine resonance. The first resonance to be confirmed, which may be a tetraquark state, is the $Z^-(4430)$, found in the decays of B mesons. This was first discovered by the Belle collaboration in 2007, but was only confirmed as a resonance by the LHCb collaboration in 2014, at the indisputably high significance level of 13.9σ. The state arises in the decay of B^0 mesons via the mode $B^0 \rightarrow \psi' + K^+ + \pi^-$, where ψ' is a charmonium $(c\bar{c})$ meson in the $n = 2$, $^3S_1$ state, where n is the principal quantum number. The $Z^-(4430)$ appears specifically in the invariant mass distribution for the ψ'-π^- pair, with a spin-parity of 1^+. Since this state is charged, the simplest quark configuration that can give rise to its decay is $c\bar{c}d\bar{u}$, discounting the possibility that it is an ordinary meson. What remains to be seen, however, is whether this is a tetraquark in the true sense—four quarks mutually bound by their color-charge—or a bound state of two mesons, maintained by the same residual nuclear attraction that binds baryons together in the nucleus. A third possibility is that the state is a different kind of "bound state of bound states," namely a diquark–anti-diquark pair. Further experiments will allow for the nature of the state to be pinned down, but in any case, it is clear that the $Z^-(4430)$ most likely really does represent an entirely new family of particles.

Another type of exotic hadron is the pentaquark. This is a kind of baryonic cousin of the tetraquark, consisting of four quarks and an antiquark. Again, there is now compelling evidence that such a state has been discovered, also by the LHCb collaboration in 2015. The evidence in this case comes from the decay of Λ_b^0 baryons to $J/\Psi + K^- + p$, in particular from the invariant mass spectrum for the J/Ψ and p. In fact, since the spectrum has a tall narrow peak *within* a broader flatter peak, two distinct pentaquark states are required to fit the data. These are the $P_c^+(4380)$ and $P_c^+(4450)$, both with quark content $uudc\bar{c}$, and together they fit the data to an incredible 15σ significance level, easily placing the discovery beyond

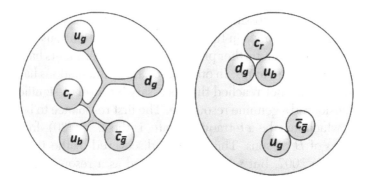

FIGURE 10.7 Possible arrangements of the constituents in a pentaquark. In (a), the quarks are mutually bound by their color charges through QCD strings. In (b), the pentaquark is a more loosely bound state of hadronic constituents, namely a baryon and a meson.

doubt. As with the observed tetraquark states, it is not yet known if these pentaquarks are combinations of five quarks strongly bound by their color-charge, or bound baryon-meson states, as in Figure 10.7. Further observation and analysis will no doubt make this clear.

Additional bumps in various invariant mass spectra have been identified as possible tetraquarks and pentaquarks, but only those in the previous examples have, to date, been definitively identified as such. Of course, as more data come in, this situation is likely to change and we may find that the exotic hadron family grows considerably in the near future.

10.6.2 Glueballs

Two- and three-quark states are commonplace, four- and five-quark states are now beginning to announce their presence, and in principle we could go even further, postulating six-quark states or more. On the other hand, a one-quark state remains forbidden, at least at low energy, by the confining nature of strong interactions. However, if we step over this forbidden state, we arrive at the possibility of a zero-quark state, in which gluons alone are bound by their self-interactions. Instead of quarks, these composite particles would consist of an overall color-neutral group of gluons. While there are no valence quarks in these glueball states, like other hadrons they contain a sea of virtual quark-antiquark pairs. Since the state must be color-neutral and gluons are not charged under other interactions, all glueball states must have all quantum numbers equal to zero other than angular momentum, parity, and C-parity. These unconstrained properties are determined by the particular arrangement of the valence gluons, and lattice QCD calculations have additionally predicted the properties of various states, including their masses, decay channels, and lifetimes. Several experimentally observed resonances are potential glueball candidates, though to date none has been definitively identified as such. In particular, the $f_0(1500)$ and $f_0(1710)$, along with several other f_0 states, have potential interpretations as the lightest predicted glueballs. Just as in the case of the flavor-neutral mesons, glueballs are capable of mixing with other neutral states. In this way, the glueballs may mix with ordinary mesons, complicating the identification of states.

10.6.3   Quark-Gluon Plasma

As stated previously, free quarks are forbidden at large distance scales. However, due to the asymptotic freedom of QCD, if the average energy density (that is, temperature) or particle density of a system is sufficiently high, then the system will undergo a phase transition to a state of matter in which both quarks and gluons become deconfined. That is, a new state of matter—the quark-gluon plasma or QGP—occurs, in which colored particles become free. Such a state would have dominated the universe for a fraction of a second after the Big Bang and was recreated artificially for the first time in 2010 at Brookhaven's Relativistic Heavy Ion Collider (RHIC). Quark-gluon plasmas have since also been created at the LHC and studied by the dedicated ALICE experiment as well as the general-purpose ATLAS and CMS experiments. In both cases, the plasma state has been achieved by colliding heavy nuclei in accelerators. Since each proton in an ionized nucleus experiences the accelerating force in a collider, the energy attainable in nucleus-nucleus collisions is considerably larger and spread over a greater volume than in collisions of individual hadrons, making such collisions ideal for producing the QGP state.

The QGP state is not straightforward to study experimentally, as it is not directly detected, since it exists only in a very localized region of both space and time. Instead, only the ordinary hadronic matter states radiating from the collapse of the state and into the surrounding detectors are observed. The experimental signatures of the state include a change in the expected momentum spectra of products due to the increase in degrees of freedom of the state over hadronic matter. At present, QGP physics remains relatively unknown, and the properties of the QGP state are neither precisely measured experimentally, nor universally agreed upon theoretically. However, with more powerful collisions, this is an area in which a great deal of progress is likely to be seen in the near future. One aspect of the nature of QGP that *is* clear is that, unlike the usual plasma state of ionized gas, the strong interactions between colored particles in QGP cause it to behave more like a liquid than a gas. In time, the properties of this intriguing exotic state of matter will be probed, allowing us to explore the QCD phase diagram and to deepen our overall understanding of QCD.

EXERCISES

1. Color $SU(3)$ symmetry demands that the color factor be the same for red-red scattering as it is for green-green or blue-blue (see Equation 10.22). Show this explicitly by considering the gluons that can contribute to each process.

2. Consider a higher-order process in which red and blue quarks scatter via exchange of *two* gluons. How many individual diagrams contribute and what is the overall color factor?

3. In principle, mesons other than the pions can contribute to the residual inter-nucleon interaction but will be very heavily suppressed. Use relevant Feynman diagrams to explain why this is.

4. The Lagrangian for QCD and a single quark flavor is given by

$$\mathcal{L} = \overline{\Psi}\left(i\gamma^\mu D_\mu - m\right) - \frac{1}{4}G^i_{\mu\nu}G^{i\mu\nu}. \qquad (10.38)$$

Show that this leads to the correct equations of motion for the quark and for the gluons.

5. (a) Due to gluon self-interactions, there are two distinct types of Feynman diagram that contribute to the QCD analogue of Compton scattering (quark + gluon → quark + gluon) at tree-level. What are they?

(b) Find the invariant amplitude for each diagram, making no assumption about the color state of the external particles.

(c) By factoring out all of the color information from these amplitudes, find a general expression for the color factor for each diagram.

(d) Hence find the color factor for each diagram in the case of elastic scattering between a red quark and a gluon in the $|1\rangle$ state.

SYMMETRY BREAKING AND THE HIGGS MECHANISM

11.1 THE WEAK FORCE AS A BOSON-MEDIATED INTERACTION

Fermi's original formulation of the weak interactions consisted of a four-point fermion interaction of the form $\frac{G_F}{\sqrt{2}}\overline{\psi}_1\psi_2\overline{\psi}_3\psi_4$, or

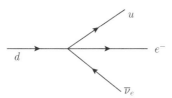

This vertex contributes a factor G_F to the amplitude, where G_F is the coupling for the interaction, known as the Fermi constant. The value of the Fermi constant is very much smaller than the couplings for the electromagnetic and strong nuclear interactions, at 1.166×10^{-5} Gev^{-2}, to account for the weakness of the interaction.

The theory was phenomenologically sound, and capable of reproducing the behavior of the weak interaction known at the time.

There was found to be, however, a serious theoretical issue with such an interaction: the four-fermion vertex led to a non-renormalizable theory, and was only viable to first order in G_F. That is to say that the inevitable infinities that arise in higher-order diagrams cannot be consistently removed in the case of the Fermi interaction. The first step toward a solution to this problem was the introduction of the weak intermediate bosons, the $W^\pm$ and the Z^0. The very small value of the Fermi constant was explained by giving these bosons a large mass; when they were later discovered experimentally at CERN's Large Electron-Positron Collider (LEP), they were indeed found to be massive particles at around 90 GeV.

11.1.1 $\mathbb{P}$ Violation

There are additional complications with the weak interactions, on top of the renormalizability issue. One of these is the experimentally determined fact that weak interactions violate parity. Before trying to build a theory that violates parity, let us first take a moment to better understand how a theory can *respect* the symmetry. If a theory is to remain invariant under a parity transformation, then the equations governing that theory must have all terms transform in the same way under parity. This means that a vector current like $\bar{\psi}\gamma^\mu\psi$ acting as a source for a vector boson is an example of a parity-invariant theory. Similarly, an axial current such as $\bar{\psi}\gamma^\mu\gamma^5\psi$ can act as a source for a pseudo-vector. What we cannot do, however, in a parity-invariant theory, is mix terms of these two opposing kinds. So another way of stating that weak interactions violate parity is to claim that the current responsible for the force contains both a vector component and an axial component. That is, these two types of current are mixed together in weak interactions. This mixing would manifest itself as an asymmetry in the helicities of weakly produced particles, and the amount of asymmetry tells us how much the currents mix. Again, it is an experimental fact that the particular amount of mixing found to occur is the maximal amount: there is an equal contribution from the vector and axial parts. This leads, for example, to an electron-neutrino current of the form

$$\frac{1}{2}\overline{e}\left(\gamma^{\mu} - \gamma^{\mu}\gamma^{5}\right)\nu_{e}, \tag{11.1}$$

with similar currents for other fermions. Notice that, for clarity, we have used e and ν_e to represent these particles' spinors, as opposed to, say, ψ_e and ψ_{ν}. This will help to reduce the number of necessary subscripts.

To see how to fit such a vertex into a theory, notice that we can factorize it as

$$\overline{e}\gamma^{\mu}\frac{1}{2}\left(\mathbb{1} - \gamma^{5}\right)\nu_{e}, \tag{11.2}$$

where we recognize the factor $\frac{1}{2}\left(\mathbb{1} - \gamma^{5}\right)$ as the left-chiral projection operator. Therefore, we can rewrite the vertex once more as

$$\overline{e_{L}}\gamma^{\mu}\nu_{e,L}, \tag{11.3}$$

and we see that it is only left-chiral fermions that take part in weak interactions. This explains the observation that all neutrinos are left-helical. Since the neutrino is (almost) massless, its helicity and chirality eigenstates (almost) coincide. The only interaction that the neutrino engages in is the weak interaction, so only left-chiral, and therefore left-helical neutrinos are produced. Even if a right-handed neutrino were to exist, it would not interact with the Standard Model particles through any of the three non-gravitational forces, and so we would have no way to infer its presence. Actually, this is somewhat of a simplification, and we will return later to the problem of neutrino masses and what they mean for right-helical neutrinos.

11.1.2 ℂ Violation

The charge conjugate of a left-chiral fermion is a left-chiral antifermion. The weak interaction only couples to left-chiral fermions and to right-chiral antifermions. So the ℙ violation found in the weak interactions also leads directly to a violation of charge conjugation symmetry, since a system with ℂ symmetry would treat left-chiral fermions and left-chiral antifermions on the same footing. This is the reason that many felt, after the discovery of $\mathbb{C}$ and $\mathbb{P}$ violation, that the combined symmetry of $\mathbb{CP}$ may still be respected, even by the weak interactions. In a sense, many felt in hindsight that $\mathbb{CP}$ was the "obvious" symmetry that they had meant *all along* when talking about charge conjugation.

11.2 RENORMALIZABILITY AND THE NEED FOR SYMMETRY

The approximate four-point nature of weak interactions, as well as the rather feeble coupling strength, indicate that the bosons that mediate the weak force are massive. However, the arguments we have used so far, based on symmetry principles, dictate that gauge bosons must be massless. Furthermore, the different particles that can interact with each other through the weak force are very different from each other (unlike the quark colors for the strong force which are essentially identical). For instance, the electron and the neutrino are clearly very different particles, with different charge and mass, yet they interact through the weak force. These problems might seem to suggest that the gauge symmetry approach is not suitable for the weak interaction, but there is a very compelling reason to persevere. Namely, gauge theories have been shown, by Gerard 't Hooft and Martinus Veltman, to be renormalizable, while other theories of spin-1 particles generally are not. The symmetry of a gauge theory in a sense protects the renormalizability. In fact, it is a general result that any theory involving particles of spin greater than $\frac{1}{2}$ inevitably leads to certain unphysical degrees of freedom unless those particles are coupled to a conserved current derived from an appropriate symmetry.

The solution to the previous problems is to notice that, at very high energies, the electron and the neutrino do begin to behave very similarly. The strong force becomes the dominant interaction, and neither of these particles participate in it. If we were only aware of the electron and neutrino through very high-energy experiments, we might well say that they look approximately identical! This is the key to building a theory of weak interactions. We assume that there really is a flavor symmetry—flavor $SU(2)$—linking the electron to the neutrino and the up quark to the down quark, but that this symmetry is "broken."

11.3   HIDDEN SYMMETRY

Symmetry breaking is any process by which a symmetric system appears to become less symmetrical, and is an important concept in several areas of physics. A classic example is that of phase transitions. A liquid is a highly symmetrical system: since there is no preferred orientation for any particular molecule, the system is isotropic. In fact, a liquid in a vacuum will form a sphere with $SO(3)$ rotation symmetry. The more familiar situation of a liquid in a gravitational field has, of course, a definite "top" and "bottom," but the symmetry in this situation is broken by external influences. In a situation like this, we say that the symmetry is explicitly broken, but this is not the type of symmetry breaking that interests us. Even in the absence of any external influence, as the liquid drop is cooled, it undergoes a phase transition to the solid state. At this point, the molecules become arranged in a crystalline form, and the symmetry is destroyed. Although there is no preferred direction, it is energetically favorable for the molecules to form a lattice, and so this occurs with the molecules in some specific but arbitrary orientation. The symmetry in this situation is said to be "spontaneously broken." As in this example, it is typically the case that symmetry breaking occurs as the energy of a system is lowered.

Although this symmetry is referred to as "broken," it is more accurate to call the symmetry "hidden," since the original symmetry is in fact still present. In the example of the liquid drop, there is still an $SO(3)$ symmetry in the sense that the crystal itself *could* still be in any orientation, even though the specific state that the system has chosen is anisotropic. More generally, spontaneous symmetry is characterized by a Hamiltonian that obeys some symmetry, but a ground state or vacuum that does not.

The Higgs mechanism was developed independently by several groups in around 1964, and is a means of spontaneously breaking the $SU(2)$ gauge symmetry required for renormalizability of the weak interactions. The mechanism is able to provide a mass to the weak bosons without destroying the renormalizability. We will build up to the full Higgs mechanism gradually by first looking at some similar but simplified models.

11.3.1 Toy Model 1: Z_2 Symmetry Breaking

Consider a real-valued scalar field, ϕ, which has a potential energy given by $V = -\frac{1}{2}\mu^2\phi^2 + \frac{\lambda}{4!}\phi^4$, depicted in Figure 11.1. Notice that the potential has a reflection (Z_2) symmetry in the value of ϕ. This leads, via the Euler-Lagrange equation or Hamilton's equations, to an equation of motion:

$$\left(\partial^2 - \mu^2\right)\phi = -\frac{\lambda}{3!}\phi^3, \tag{11.4}$$

which appears to describe spin-0 particles with a four-point self-interaction of coupling strength λ. There is a problem with the mass term, however. Notice that the μ^2 term in the Klein-Gordon equation has the wrong sign. This seems to suggest that the particles described by this equation have an imaginary mass of $i\mu$, which is clearly nonsensical. The problem arises because we can only interpret excitations of the field *around the vacuum* as particles, and in this case the vacuum is not at $\phi = 0$.

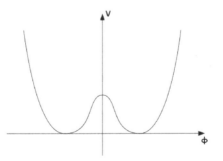

FIGURE 11.1 Potential energy of a real scalar with negative μ^2.

The stationary points of the potential are at $\phi = 0$ and $\phi = \pm v$ where $v = \sqrt{6\mu^2/\lambda}$. The first of these is a local maximum, while the others are local minima. This means that there are two ground states or vacua for this system: each of the local minima in the potential function is a valid vacuum state, and so the system will settle into one or the other. While the Hamiltonian has Z_2 symmetry, the ground state (whichever one is chosen) does not. Notice that it is precisely *because* the Hamiltonian is symmetric that the two minima are of equal energy. The field, then, will take a non-zero background value at all points in space: it is said to have acquired a non-zero vacuum expectation value, or vev. The field can then be split into the background portion and the residual value: $\phi = v + r$. The residual field r is still free to deviate from the background value, and excitations in r will give rise to particles. An important distinction, however, between the original ϕ field and the residual field r is that the r particles have a real mass. To see this, we can substitute $v + r$ into the potential:

$$
\begin{aligned}
V &= -\frac{1}{2}\mu^2 \left(v + r\right)^2 + \frac{\lambda}{4!}\left(v + r\right)^4 \\
&= -\frac{3}{2}\frac{\mu^2}{\lambda} + \mu^2 r^2 + \frac{\lambda v}{3!}r^3 + \frac{\lambda}{4!}r^4.
\end{aligned}
\tag{11.5}
$$

Between the first and second lines in Equation 11.5, use is made of the fact that the vev is given by $\sqrt{6\mu^2/\lambda}$. The first term is an unimportant constant, which will vanish from the equation of motion. The second term gives the mass of the particle, and the final two terms show that the particles will undergo both three-point and four-point self-interactions with couplings of λv and λ respectively. The equation of motion becomes

$$
\left(\partial^2 + \mu^2\right) r = -\left(\frac{\lambda v}{2!}r^2 + \frac{\lambda}{3!}r^3\right),
\tag{11.6}
$$

so we can see that the mass is now well-behaved. This equation of motion can also be arrived at by substituting $\phi = v + r$ directly into Equation 11.4.

This simple toy model demonstrates the underlying idea of spontaneous symmetry breaking in particle physics but is missing some of the most important features that we will require of the concept. To capture these, a slightly more complicated model is needed.

11.3.2 Toy Model 2: $U(1)$ Symmetry Breaking

We now consider a *complex* scalar field ϕ with a potential

$$V = -\mu^2 \phi^2 + \frac{1}{4}\lambda \left(\phi^* \phi\right)^2, \tag{11.7}$$

and corresponding wave equation:

$$\left(\partial^2 - \mu^2\right)\phi = -\frac{\lambda}{2}\left(\phi^* \phi\right)\phi. \tag{11.8}$$

It is important to recognize that there are two degrees of freedom in this system, though there are different ways to parametrize that freedom. The field can be split into real and imaginary parts, or into ϕ and ϕ^*, or even considered as polar coordinates $|\phi|$ and $\arg(\phi)$. This last parametrization is particularly useful for our purposes. It also reminds us that the degrees of freedom need not have physical significance. We have already seen, in Section 9.1, that complex-valued wave equations have a global $U(1)$ symmetry (allowing redefinition of the phase degree of freedom), and the above is no exception. In fact, the stationary points for this system are located at the origin, and in a continuous ring of fixed magnitude around the origin:

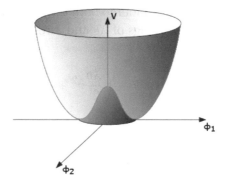

This potential is commonly known as the Mexican hat potential, though I prefer to think of it as the wine-bottle potential. The ring of minima occurs at a field value of magnitude

$$v = |\phi| = \sqrt{2\mu^2}\lambda, \qquad (11.9)$$

as can be seen by differentiating Equation 11.7. The field could "choose" any point on this ring to settle into as its background value. Also, because the Hamiltonian contains a shear term (the kinetic term), the background field will have the *same* value at all points, since a varying value would increase the overall energy of the system. We now write the field in terms of its vev, v, its phase, ξ, and a residual magnitude, h: $\phi = \frac{1}{\sqrt{2}}(v + h)e^{i\xi/v}$. The overall factor of $\sqrt{2}$ and the factor of v in the exponent allow both complex and real scalars to be simultaneously and consistently normalized. Substituting this into the wave equation gives:

$$\partial^2 h + \frac{2i}{v}\partial^\mu h \partial_\mu \xi + \frac{i}{v}(v + h)\partial^2 \xi$$
$$- \frac{(v + h)}{v}\partial_\mu \xi \partial^\mu \xi - \mu^2(v + h) = -\frac{\lambda}{4}(v + h)^3 \qquad (11.10)$$

Since h and ξ are both real-valued, we can separate this into real and imaginary components to arrive at two equations, one for each degree of freedom. Taking into account Equation 11.9, these can be written as

$$\left(\partial^2 + 2\mu^2\right) h = v\partial_\mu \xi \partial^\mu \xi - \frac{3\lambda}{4}vh^2 - \frac{\lambda}{4}h^3,$$
$$\partial^2 \xi = -2\partial^\mu \xi \partial_\mu h - h\partial^2 \xi. \qquad (11.11)$$

The residual magnitude h behaves very much like the r particle in the case of discrete symmetry. Having fallen into the potential well, radial perturbations now increase the energy in the system and the h field has gained a mass $m_h = \sqrt{2}\mu$. The ξ particle, however, is very different. The ξ degree of freedom traverses the bottom of the well, with all values of ξ giving the same potential energy. Put another way, perturbations in the ξ direction around the vev do not alter the energy in the system since the potential is flat in this direction. So ξ

particles behave as massless particles: the equation of motion contains no linear term and so no mass. This is a general result: when a continuous global symmetry is spontaneously broken, excitations of the field that change the value away from the vev in a direction tangential to the set of minima correspond to massless particles known as Goldstone bosons. All of this can also be shown directly via the Lagrangian, which after substitution of the real fields h and ξ contains no quadratic term in ξ, which would correspond to a mass (see Exercise 1). The following types of interactions have also appeared:

Before continuing with the main thrust of the argument, it is worth taking a short detour to look at an interesting application of these ideas. We have seen that a Goldstone boson arises if an exact global symmetry is broken. But if the symmetry that is broken is only approximate, then the boson related to "angular" variations is no longer massless. However, since there is an approximate symmetry, the system must behave almost as though it has a Goldstone boson. In this case, then, a "pseudo-Goldstone" boson is found, which has a small but finite mass. A good example of this process is in the strong interactions: the up and down quarks have an approximate chiral $SU(2)_L \otimes SU(2)_R$ symmetry due to their small masses, with left- and right-chiral parts transforming independently. This symmetry is then broken by the non-zero vev of the QCD vacuum, down to isospin $SU(2)$, where the two chiralities now mix. However, since the quarks are not perfectly massless and the chiral symmetry is therefore not exact, its spontaneous breaking leads to a set of three (massive) pseudo-Goldstone bosons: the pions. It is this spontaneous breaking of (approximate) chiral symmetry that guarantees that the pions have a small mass when compared with the other mesons. In the case that we consider three quark flavors to be almost massless, then there is an approximate $SU(3)_L \otimes SU(3)_R$ symmetry that is broken to flavor $SU(3)$, resulting in the light pseudo-scalar meson octet.

Local $U(1)$ Symmetry Breaking

Suppose now that the $U(1)$ symmetry of the previous example is promoted to a local symmetry. We have seen how this results in electromagnetic interactions (or at least interactions that behave similarly to electromagnetic interactions but possibly with a different set of charges—for simplicity, we will refer to them as electromagnetic). The symmetry breaking occurs as before with two differences. First, the derivative in the scalar wave equation must be replaced with a covariant derivative, leading to modified equations for the real fields

$$\left(\partial^2 + 2\mu^2\right) h = g^2 \left(v + h\right) A_\mu A^\mu + \frac{1}{v^2} \left(v + h\right) \partial_\mu \xi \partial^\mu \xi$$

$$-\frac{\lambda}{4} \left(3vh^2 + h^3\right) + \frac{2g}{v} \left(v + h\right) A^\mu \partial_\mu \xi, \quad (11.12)$$

$$\text{and } \partial^2 \xi = -\frac{h}{v} \partial^2 \xi - \frac{2}{v} \partial_\mu h \partial^\mu \xi - 2g A^\mu \partial_\mu h - g \left(v + h\right) \partial_\mu A^\mu.$$

The most important difference, though, is in the Maxwell equation for the photon. Recall that the current due to a scalar particle is given by

$$j^\mu = iqe \left(\phi^* \partial^\mu \phi - \phi \partial^\mu \phi^*\right) - 2(qe)^2 A^\mu \phi^* \phi, \quad (11.13)$$

so the Maxwell equation becomes:

$$\partial^2 A^\mu - \partial^\mu \partial \cdot A = iqe \left(\phi^* \partial^\mu \phi - \phi \partial^\mu \phi^*\right) - 2(qe)^2 A^\mu \phi^* \phi$$

$$= \frac{iqe}{2} \left[(v + h) \left(\partial^\mu h + \frac{i}{v} (v + h) \partial^\mu \xi \right) \right.$$

$$\left. - (v + h) \left(\partial^\mu h - \frac{i}{v} (v + h) \partial^\mu \xi \right) \right]$$

$$-(qe)^2 A^\mu (v + h)^2$$

$$= -(qe)^2 (v + h)^2 \left(A^\mu + \frac{1}{qev} \partial^\mu \xi \right). \quad (11.14)$$

Notice that ξ only appears with a derivative: it is of the right form to act as a gauge transformation on the photon. We are thus able to "gauge away" this object, effectively absorbing it into the photon.

This leaves us with

$$\partial^2 A^\mu - \partial^\mu \partial \cdot A = -(qe)^2 (v+h)^2 A^\mu$$
$$= -(qe)^2 v^2 A^\mu - 2(qe)^2 vh A^\mu - (qe)^2 h^2 A^\mu. \tag{11.15}$$

The term involving A^μ and v^2, however, demonstrates that in absorbing the phase term ("eating the Goldstone boson"), the photon has now acquired a mass. That is, the term would sit more comfortably on the left of the equation where we can identify it with the mass term of the Proca equation. A massive spin-1 particle, of course, has an additional longitudinal polarization state over its massless counterpart, and so requires an extra degree of freedom from somewhere. This is no problem, since the phase degree of freedom of the complex scalar is now unphysical and effectively lost to the photon. This is the essence of the Higgs mechanism: the symmetry is still intact (albeit hidden), protecting the renormalizability of the theory, but the gauge boson has "stolen" a longitudinal polarization from the scalar field to become a massive particle.

So far, we have seen the argument for the Goldstone boson providing a mass to the gauge boson by looking directly at the equations of motion. This was in order to ease any readers unfamiliar with Lagrangian mechanics into the concepts and to make clear the physical interpretation of these ideas. The same argument can be made, however, by looking just at the Lagrangian for the system, and in fact becomes a little simpler. This will prove valuable when breaking larger symmetry groups, since working with the equations of motion for non-Abelian symmetries quickly becomes unwieldy. The Lagrangian for the Abelian symmetry we have just considered is given by

$$(D_\mu \phi)^* D^\mu \phi + \mu^2 \phi^* \phi - \frac{\lambda}{4} (\phi^* \phi)^2 - \frac{1}{4} F_{\mu\nu} F^{\mu\nu}, \tag{11.16}$$

where $D_\mu = \partial_\mu + iqe A_\mu$. This expands to

$$\mathcal{L} = (\partial_\mu \phi)^* \partial^\mu \phi + iqe A^\mu (\phi \partial^\mu \phi^* - \phi^* \partial^\mu \phi) + (qe)^2 \phi^* \phi A_\mu A^\mu$$
$$+ \mu^2 \phi^* \phi - \frac{\lambda}{4} (\phi^* \phi)^2 - \frac{1}{4} F_{\mu\nu} F^{\mu\nu}. \tag{11.17}$$

When substituting in the real fields, we find

$$
\begin{aligned}
\mathcal{L} &= \frac{1}{2}\partial_\mu h \partial^\mu h + \frac{1}{2}(qe)^2 (v+h)^2 \left(A_\mu + \frac{1}{qev}\partial_\mu \xi \right) \\
&\quad \times \left(A^\mu + \frac{1}{qev}\partial^\mu \xi \right) + \frac{qe}{v} A^\mu (v+h)^2 \partial_\mu \xi \\
&\quad + \frac{1}{2}(qe)^2 (v+h)^2 A_\mu A^\mu + \frac{1}{2}\mu^2 (v+h)^2 \\
&\quad - \frac{\lambda}{16}(v+h)^4 - \frac{1}{4}F_{\mu\nu}F^{\mu\nu}.
\end{aligned}
\tag{11.18}
$$

A gauge transformation $A_\mu \mapsto A_\mu - \frac{1}{qev}\partial_\mu \xi$ simplifies the second term to

$$
\frac{1}{2}(qe)^2 (v+h)^2 A_\mu A^\mu = \frac{1}{2}(qev)^2 A_\mu A^\mu + \dots,
\tag{11.19}
$$

and we see that a mass of qev has again appeared for the gauge boson, since a quadratic term in the Lagrangian will lead to a linear term in the equation of motion, and so provides a mass.

Combining the μ^2- and λ-dependent terms, we arrive at

$$
\mathcal{L}_{\text{scalar}} = \frac{1}{2}\mu^2 v^2 - \mu^2 h^2 - \frac{\lambda}{16}v^4 - \frac{3\lambda}{8}v^2 h^2 - \frac{\lambda}{4}vh^3 - \frac{\lambda}{16}h^4.
\tag{11.20}
$$

The first term, being constant, has no effect on the equation of motion, as it will vanish when the Lagrangian is differentiated. The second term shows, as before, that the real particle h has acquired a real-valued mass of $\sqrt{2}\mu$, while the remaining terms again give the interactions of the h particle.

It should be stressed that this is *not* the Higgs mechanism as realized in the Standard Model, but a simple model to illustrate the concepts: the $U(1)$ symmetry in the Standard Model is *not broken*. The symmetry that must be broken to account for the weak interactions is the $SU(2)$ flavor symmetry, which we are now ready to tackle.

11.3.3 The Higgs Mechanism: *SU*(2) ⊗ *U*(1) Breaking

The symmetry that must be broken in the Standard Model is $SU(2)$ flavor symmetry. For this we require a pair of complex scalars, $\Phi = \begin{pmatrix} \phi_1 \\ \phi_2 \end{pmatrix}$, transforming as a doublet under $SU(2)$. The gauge bosons of this symmetry are the W_1, W_2 and W_3. Breaking this $SU(2)$ is not, by itself, enough to reproduce the properties of the weak interactions. Recall that the fermions we wish to pair together into weak doublets, such as the electron and neutrino, must also have different electric charges. This problem can be overcome by having the scalar doublet charged under an additional $U(1)$ symmetry. However, this is *not* the electromagnetic force: it is another $U(1)$ mediated by a new gauge boson, the B. The relevant charge for this group is not, therefore, the electric charge but some new quantity, known as the weak hypercharge.[1] This gives a covariant derivative of the form

$$D_\mu = \partial_\mu + ig_2 W_\mu^a T^a + \frac{i}{2} g_1 Y B_\mu, \qquad (11.21)$$

where g_1 and g_2 are the couplings under the $U(1)_Y$ and $SU(2)$ groups respectively. The factor of $\frac{1}{2}$ in the $U(1)_Y$ term is merely convention, and is included so that the matrix manipulations are more straightforward: since $T^a = \frac{1}{2}\sigma^a$, an overall factor of $\frac{1}{2}$ can be pulled out from the gauge terms of the derivative. Also, now that we have fixed our conventions, we can choose a hypercharge for the scalar doublet. We assign $Y_\Phi = 1$.

With $SU(2) \otimes U(1)_Y$ symmetry, there is now a lot of freedom in how to choose the vacuum state. However, because of that same symmetry, it does not matter how we choose it: different choices of vacuum really just correspond to re-parametrizations of the system. The standard convention is to choose a vacuum in which the scalar

---

[1] We will subsequently refer to this simply as hypercharge. However, the reader should be aware that it should strictly be "weak hypercharge" to distinguish it from an unrelated quantity relating to quark flavors.

doublet has an expectation value

$$\Phi = \frac{1}{\sqrt{2}} \begin{pmatrix} 0 \\ v \end{pmatrix}, \tag{11.22}$$

with $v = \mu/\sqrt{\lambda}$.

The scalar doublet can then be parametrized as

$$\Phi = \frac{1}{\sqrt{2}} e^{i\xi_a T_a} \begin{pmatrix} 0 \\ v + h \end{pmatrix}, \tag{11.23}$$

and the Lagrangian is given by

$$\mathcal{L} = (D_\mu \Phi)^\dagger (D^\mu \Phi) + \mu^2 \Phi^\dagger \Phi - \lambda \left(\Phi^\dagger \Phi \right)^2 - \frac{1}{4} W_{\mu\nu}^a W^{a\mu\nu}. \tag{11.24}$$

Assuming for a moment that all four fields h and ξ_a are zero, the first term of this Lagrangian becomes

$$\frac{1}{8} \begin{pmatrix} 0 & v \end{pmatrix} \begin{pmatrix} g_1 B + g_2 W_3 & g_2 (W_1 - iW_2) \\ g_2 (W_1 + iW_2) & g_1 B - g_2 W_3 \end{pmatrix}^2 \begin{pmatrix} 0 \\ v \end{pmatrix} \tag{11.25}$$

which reduces to

$$\frac{1}{8} g_2^2 v^2 \left(W_1^2 + W_2^2 \right) + \frac{1}{8} v^2 \left(g_2 W_3 - g_1 B \right)^2. \tag{11.26}$$

This provides mass terms as in the Abelian case. The situation is complicated here, however, by the presence of the linear combination $g_1 B^\mu - g_2 W_3^\mu$. This combination is a sign that the physical states do not coincide with the particular representation of $SU(2)$ that we have chosen. It is the linear combination as a whole that acquires a vev through this symmetry breaking, and so we identify the combination as a particle species: Z_μ^0. To ensure standard normalization of both the group generators and this new massive particle, we require

$$Z^0 = \frac{g_2 W_3 - g_1 B}{\sqrt{g_1^2 + g_2^2}}. \tag{11.27}$$

Introducing a parameter, θ_w, defined by $\tan\theta_w = g_1/g_2$, we can write this as

$$Z^0 = \cos\theta_w B - \sin\theta_w W_3. \qquad (11.28)$$

The angle θ_w parametrizes the amount of mixing between the $SU(2)$ and $U(1)_Y$ groups in producing the Z^0 particle, and is known as the weak mixing angle. The remaining physical states are $W^\pm = \frac{1}{\sqrt{2}}(W_1 \mp iW_2)$ and the remaining orthogonal combination

$$A = \sin\theta_w B + \cos\theta_w W_3. \qquad (11.29)$$

Notice that this last particle is massless, since the Lagrangian developed no quadratic term proportional to it. This particle corresponds to the unbroken part of the original symmetry, a residual $U(1)$ symmetry that gives rise to electromagnetism. In other words, the remaining linear combination A is the photon.

The masses of the $W^\pm$ bosons can be read off directly from the quadratic terms in the Lagrangian as

$$m_W = \frac{g_2 v}{2}, \qquad (11.30)$$

while the mass of the Z^0 is given by

$$m_Z = \frac{g_2 v}{2\cos\theta_w}. \qquad (11.31)$$

Three of the four gauge bosons are massive, which means that three of the four scalar degrees of freedom have been gauged away. This leaves one remaining real scalar field that can interact with these massive vector bosons. Since this scalar must be invariant under the remaining unbroken symmetry (it would have broken the symmetry otherwise), it must be electrically neutral. This is the Higgs boson, predicted in 1964 and discovered experimentally almost 50 year later at the Large Hadron Collider in 2012.

11.4 ELECTROWEAK INTERACTIONS

The electromagnetic and weak interactions have been described in terms of a gauge group $SU(2) \otimes U(1)_Y$, with the massive weak bosons arising from the symmetry breaking, and the electromagnetic interactions being the residual unbroken part of the group. The two interactions have been unified into a larger gauge theory known as electroweak theory. We must now consider how the fermions fit into this theory. Since we wish only left-chiral fermions to participate in weak interactions, we must temporarily strip the fermions of their masses in order to treat their chiral components differently. Notice that, as a bonus, this has also solved the problem of different members of a doublet having different masses: the masses are equal if all set to zero! The fermions, then, are grouped into $SU(2)$ doublets if left-chiral but are left as $SU(2)$-invariant singlets if right-chiral. This gives the Lagrangian separate fermionic kinetic terms for left- and right-chiral fermions of the form:

$$\mathcal{L} \supset \left(\begin{array}{cc} \overline{\nu_{e,L}} & \overline{e_L} \end{array} \right) i\slashed{D} \left(\begin{array}{c} \nu_e \\ e \end{array} \right) + \overline{e_R} i \slashed{\partial} e_R + \ldots, \qquad (11.32)$$

and allows for interaction with the $SU(2)$ generators only if left-chiral. For this reason, the flavor $SU(2)$ group is commonly denoted $SU(2)_L$ with the L standing for "left-chiral."

Since the $W^\pm$ are linear combinations of W^1 and W^2, they couple only to left-chiral fermions. In particular, they couple the two components of a left-chiral flavor doublet, such as the electron and its associated neutrino. Also, since $W^\pm$ emission or absorption involves a change in charge of the fermion, such interactions are referred to as charged-current processes. The Z^0, on the other hand, is composed partly of W^3, which couples to left-chiral fermions, and partly of the B, which couples to anything with a non-zero hypercharge. As such, the Z^0 interaction is not purely a left-chiral one as the charged current interactions are.[2] In particular, it couples to a current of the

---

[2] The exception to this statement is that, since right-chiral neutrinos have $Y = 0$, The neutrino-Z^0 interaction *is* purely left-chiral.

form:

$$g_z \overline{f} \gamma^\mu \left(I_3 - Y_R \sin^2 \theta_w \gamma^5 \right) f, \qquad (11.33)$$

where f is any fermion, I_3 is the weak isospin of the left-chiral component, and Y_R is the hypercharge of the right-chiral component. The coupling in this case is given by $g_z = g_1 / \sin \theta_w$. Notice that the interaction arising from this current is the emission or absorption of a Z^0 by a fermion that retains its identity afterward. This is in contrast with the charged-current interactions, in which the fermion changes to its weak doublet partner. Since Z^0 interactions involve no change in the charge of the fermion, they are known as neutral currents. As for the remaining gauge boson, we can also show that the electromagnetic coupling is related to the weak and hypercharge couplings by

$$e = g_1 \cos \theta_w = g_2 \sin \theta_w. \qquad (11.34)$$

11.4.1 Hypercharge and Weak Isospin

The left-chiral fermions are arranged into doublets under the $SU(2)_L$ symmetry group, while the right-chiral fermions are singlets under $SU(2)_L$, which is to say that they do not transform. The two components of a left-chiral weak doublet must be distinguished in some way: that is, we require a flavor analogue of the color states of QCD. This is termed weak isospin, since it is similar to the isospin concept introduced back in Section 6.2.1. It is not quite the same, however, since it is only the left-chiral species that have non-zero values of weak isospin. Also, our original version of isospin applied only to hadrons, whereas weak isospin applies also to leptons. The weak doublets have a weak isospin of $\frac{1}{2}$ with the upper and lower components of the doublet taking different values of the third component, I_3. On the other hand, the right-chiral weak singlets have a weak isospin of 0.

We must also consider the hypercharges of the fermions under the $U(1)_Y$ part of the group. We know that the hypercharges must be equal for the two components of a doublet, and the Gell-Mann–Nishijima formula provides a relationship between electric charge, hypercharge, and weak isospin that fits the bill:

$$Y = 2\left(Q - I_3\right). \qquad (11.35)$$

The factor of 2 is a result of the convention used for the B term in the covariant derivative (Equation 11.21). The following table summarizes the properties of the left- and right-chiral fermions:

	Left-chiral				Right-chiral		
Fermion	I	I_3	Y	Fermion	I	I_3	Y
ν_L	$1/2$	$+1/2$	-1	ν_R	(absent)		
e_L^-	$1/2$	$-1/2$	-1	e^-	0	0	-2
u_L	$1/2$	$+1/2$	$+1/3$	u_R	0	0	$+4/3$
d_L	$1/2$	$-1/2$	$+1/3$	d_R	0	0	$-2/3$

Notice that the left- and right-chiral parts of, say, the electron do not have the same properties. This is perfectly acceptable, since the electroweak theory treats these as independent massless particles. The left-chiral electron and the right-chiral electron are not (yet) related, and so we should not expect them to have the same properties. In fact, the only reason we associate the left- and right-chiral electrons with each other at all is because of the mass term that links them as we will see in Section 12.2. But as far as the gauge group itself is concerned, there is no connection between these independent particles. On the other hand, notice that the hypercharges are the same for members of the same weak doublet. This all leads us to a set of Feynman rules that are summarized in Appendix B.

EXERCISES

1. Starting from the Lagrangian

$$\mathcal{L} = \partial_\mu \phi^* \partial^\mu \phi + \mu^2 \phi^2 - \frac{1}{4} \lambda \left(\phi^* \phi\right)^2 ,$$

and expanding ϕ as $\phi = (v + h)e^{i\xi}$, re-derive the properties of the spontaneously broken $U(1)$ model in the Lagrangian formalism.

2. Find expressions for the couplings in the three-point and four-point Higgs boson self-interactions in terms of the Higgs boson and $W^\pm$ masses.

3. Consider a spontaneously broken model in which a triplet of real scalars transforms as the 3 of $SU(2)$. What properties would this model have at low energy and why is it not suitable as a theory of weak interactions?

4. By considering a low-energy decay process, find an expression for the Fermi constant G_F in terms of g_2 and M_W.

5. Show that $\sin \theta_w = g_1 / \sqrt{g_1^2 + g_2^2}$ and $\cos \theta_w = g_2 / \sqrt{g_1^2 + g_2^2}$.

12

THE STANDARD MODEL OF PARTICLE PHYSICS

12.1 PUTTING IT ALL TOGETHER

The Standard Model of Particle Physics was developed gradually during the 1960s and 1970s, and is at its core simply a combination of the $SU(3)_C$ symmetry of QCD and the electroweak theory of the previous chapter. As such, the Standard Model is described by the gauge group $SU(3)_C \otimes SU(2)_L \otimes U(1)_Y$. However, there is more to the model than just its gauge group: we must also specify the fermion content and a few other parameters that we will meet in this chapter. In particular, the Standard Model consists of three generations of fermionic matter, each of which contains similar representations of the gauge group. Specifically, the fermion content for the first generation is given by

$$
\begin{aligned}
Q_L &= (3, 2)_{+^1/_3} \\
u_R &= (3, 1)_{+^4/_3} \\
d_R &= (3, 1)_{-^2/_3} \\
L_L &= (1, 2)_{-1} \\
e_R &= (1, 1)_{-2},
\end{aligned}
\tag{12.1}
$$

where the numbers in parentheses are the representation under $SU(3)_C$ and $SU(2)_L$, while the subscript is the hypercharge. Q_L and L_L here refer to the left-chiral quarks and leptons respectively. Notice that there is no right-chiral neutrino in the Standard Model. With the discovery of neutrino oscillations, we now know that neutrinos have mass, and so we must include a right-chiral component,

$$\nu_R = (\mathbf{1}, \mathbf{1})_0, \tag{12.2}$$

though we defer discussion of this point until Sections 13.1–13.2. We now turn to the problem of giving the fermions in the theory mass, which so far they are lacking.

12.2   FERMION MASSES

The electroweak theory solves several issues associated with weak interactions. It provides a means of giving mass to gauge bosons while retaining renormalizability. It also allows for chiral interactions, and for particles within a weak doublet to have different properties. However, to achieve all of this, we were forced to assume that all of the fermions of the Standard Model are massless. This leaves us with a final problem: how can we give masses back to the fermions without undoing all the work that we have put into building a consistent theory? Recall that the different electromagnetic charges of the electroweak theory, along with the weak boson masses, only appeared *after* symmetry breaking. We are going to perform a similar trick here: the fermion masses will be a by-product of symmetry breaking. In order to achieve this, we must include a type of interaction in our theory that we have not yet considered; namely, direct interactions between the scalar and fermion sectors of the model. What sort of interactions could we have between these two? As in Section 8.8, the interaction is a Yukawa-type interaction.[1] However, if the theory is to

---

[1] Recall that Yukawa interaction now means something much broader than its original sense. Initially, it referred to the specific interaction of nucleons and pions, but we now use it to refer to any interaction in which a fermion emits or absorbs a scalar.

retain the full $SU(3)_C \otimes SU(2)_L \otimes U(1)_Y$ and Lorentz symmetries that we have carefully constructed, we find that only certain combinations of fermions and scalars are allowed. Let us find, then, the allowed interactions by examining a number of possibilities. First, we know that a left-chiral spinor obeys an equation of the form:

$$i\partial\!\!\!/\psi_L = \text{(source)}. \tag{12.3}$$

Recall that the left side of this equation transforms as a *right-chiral* spinor, since the gamma matrices in $\partial\!\!\!/$ inter-convert left and right chirality. Therefore, the right side must also transform as a right-chiral spinor. So the left-chiral electron cannot have as a source any left-chiral fermion, as this would violate Lorentz invariance. Something that we might attempt, then, is to include interactions of the form:

$$i\partial\!\!\!/\psi_L \propto \phi(\psi_L)^c, \tag{12.4}$$

since we know that the charge conjugate spinor transforms with the opposite chirality. Here we run into a problem, though, since the two sides of the equation must also transform identically under *gauge* transformations. Specifically, under a $U(1)_Y$ transformation through θ, each field transforms as $\phi \mapsto e^{iY\theta/2}\phi$, so for the equation to be gauge-invariant, the sum of hypercharges on each side must be equal. In general, this is not true of the conjugate spinor, although there is an important caveat to this statement that we will consider in Section 13.2. This constraint, then, is what determines the allowed Yukawa interactions.

Put another way, the interaction terms in the Lagrangian must be scalars under both Lorentz and gauge transformations. So terms such as $\overline{e}_L \Phi e_R$ are allowed, along with its Hermitian conjugate, where e_L and e_R are the wavefunctions of the left- and right-chiral electrons, and Φ is the scalar. Since the theory must treat both members of a weak doublet equally, we find that the allowed interaction terms in the Lagrangian for one fermion generation take the form

$$\mathcal{L}_{\text{int}} = -y_e \overline{L}_L \Phi e_R - y_d \overline{Q}_L \Phi d_R - y_u \overline{Q}_L \tilde{\Phi} u_R, \tag{12.5}$$

where the y's are a set of independent Yukawa couplings. The third term here contains $\tilde{\Phi} = i\sigma^2\Phi^*$ in analogy with the charge conjugate spinor $\psi^c = i\gamma^2\psi^\dagger$, since this construct has the correct transformation under $SU(2)_L$. Note that there is no $\overline{L}_L\tilde{\Phi}\nu_R$ term, since right-chiral neutrinos are not a part of the Standard Model. When kinetic terms of the form $\mathcal{L} = \overline{L}_L\left(i\not{D}\right)L_L + \ldots$, are included, this Lagrangian allows interaction through source terms, for example, of the form

$$i\not{\partial}e_L = y_e\Phi e_R \to \frac{y_e}{\sqrt{2}}(v + h)e_R. \tag{12.6}$$

In this way, we can see that after symmetry breaking, when Φ is given a vev of magnitude $v/\sqrt{2}$, the electron gains a mass of $m_e = y_e v/\sqrt{2}$. The residual field that we see as the Higgs boson is also coupled to the fermion in this way, with the *same coupling* as that which provides the mass. This means that the Higgs must interact with all the Standard Model fermions with a coupling proportional to each fermion's mass. For this reason, one of the common decay channels of the Higgs boson is via a top loop, such as

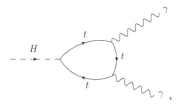

since the top mass is so high. A similar diagram with gluons in place of photons also acts as the dominant production channel in hadronic collisions.

In the previous chapter, we saw that the Higgs mechanism is capable of giving mass to gauge bosons in a manner that allows us to retain renormalization. Now we see that with the introduction of Yukawa couplings, the same mechanism can provide fermion masses despite the chiral nature of the weak interactions. In this way, then, the Standard Model can accommodate all of observed masses for fundamental particles. One will often hear the statement that the Higgs boson is responsible for giving other particles their mass. However, I hope that the preceding sections have demonstrated that this is rather a misreading of what is really going on. Particle masses come from the vacuum expectation value of the full Higgs field, not from the Higgs boson itself. The Higgs does not *give* particles their mass: it is merely what is left over *after* particles have gained their mass.

12.3   QUARK MIXING AND THE CKM MATRIX

The overall hypercharge of each term in Equation 12.5 vanishes, so each is invariant under $U(1)_Y$. Notice, however, that there is actually no reason to constrain ourselves to fermions from the same generation. While the second term in the Lagrangian, for example, allows for interactions of the form $i\bar{\partial}d_L = y_d\Phi d_R$, a similar term, $-y_{ds}\overline{Q}_L\Phi s_R$, leads to $i\bar{\partial}d_L = y_{ds}\Phi s_R$. That is, the left-chiral down quark can couple to the right-chiral down but equally to the right-chiral strange or bottom quarks. The possible interactions can be summarized in matrix form, with the left-chiral down-type quarks coupling to the right-chiral according to

$$
i\bar{\partial}
\begin{pmatrix} d'_L \\ s'_L \\ b'_L \end{pmatrix}
=
\begin{pmatrix} y_{dd} & y_{ds} & y_{db} \\ y_{sd} & y_{ss} & y_{sb} \\ y_{bd} & y_{bs} & y_{bb} \end{pmatrix}
\phi
\begin{pmatrix} d'_R \\ s'_R \\ b'_R \end{pmatrix},
\qquad (12.7)
$$

where the reason for the primed quark names will become clear. We also find, of course, a similar equation for the right-chiral components. So far, this just tell us that cross-generational interactions can occur with the scalar fields in the theory. The twist comes when we break the symmetry and impose the scalar vev. Now, each of the entries in the previous matrix behaves as a mass term, leading to a mass matrix:

$$
\begin{pmatrix} m_{dd} & m_{ds} & m_{db} \\ m_{sd} & m_{ss} & m_{sb} \\ m_{bd} & m_{bs} & m_{bb} \end{pmatrix},
\qquad (12.8)
$$

with a similar matrix for the up-type quarks.

We have masses that apparently link all three flavors of down-type quark. Rather than giving a single simple mass to each quark, we have managed to couple each left-chiral quark flavor to a complicated combination of right-chiral quarks. It makes little sense to talk about different masses for the same particle, so what does this matrix represent? Well, if we measure the masses of the down-type quarks, we will find three well-defined values: one for each of the three degrees of freedom for the system. These will be the eigenvalues of the mass matrix. The objects that we generally refer to as quarks, then, are really linear combinations of the underlying degrees of freedom, but they are the combinations for which we find well-defined masses. That is, the physical quarks are the mass eigenstates, as opposed to the flavor eigenstates. In order to find the mass states, we must diagonalize the mass matrix M_d by means of a similarity transformation: $M_d = U_d^\dagger M_d^{(D)} U_d$ where $M_d^{(D)}$ is the diagonalized matrix and U_d is some unitary matrix. The matrix U_d transforms between the flavor states d', s', b' seen by the weak interaction, and the states of definite mass d, s, b. A similar matrix, M_u, and transformation, U_u, exist in principle for the up-type quarks. Since the weak interactions are rare, we generally consider the mass eigenstates, which is why we place the primes on the flavor states. The only physical significance of the flavor states is in weak interactions.

The weak currents to which the $W^\pm$ bosons couple are given by

$$
\begin{aligned}
J^\mu &= \begin{pmatrix} \bar{u}'_L & \bar{c}'_L & \bar{t}'_L \end{pmatrix} \gamma^\mu \mathbb{1}_3 \begin{pmatrix} d'_L \\ s'_L \\ b'_L \end{pmatrix} \\
&= \begin{pmatrix} \bar{u}_L \bar{c}_L & \bar{t}_L \end{pmatrix} U_u^\dagger \gamma^\mu U_d \begin{pmatrix} d_L \\ s_L \\ b_L \end{pmatrix},
\end{aligned}
\tag{12.9}
$$

and its Hermitian conjugate. The γ matrices and the unitary matrices U_u, U_d act in different spaces. As such, they commute and we can write the weak interactions as

$$
J^\mu = \begin{pmatrix} \bar{u}_L & \bar{c}_L & \bar{t}_L \end{pmatrix} \gamma^\mu V \begin{pmatrix} d_L \\ s_L \\ b_L \end{pmatrix},
\tag{12.10}
$$

where $V = U_u^\dagger U_d$. The individual matrices, U_u and U_d, have no physical significance by themselves then. They only have physical meaning in the combination V. For this reason, there is freedom in how we choose to interpret V. Conventionally, we take V to act on the down-type quarks, effectively parametrizing the quark-mixing in such a way that it all takes place in the down-type quarks. By convention, the up-type quarks are taken to be both mass and flavor eigenstates. The matrix V is known as the Cabibbo-Kobayashi-Maskawa (or CKM) matrix. This is the reason for the non-conservation of quark flavor in weak interactions: it is not so much that the weak interactions mix quark flavors as it is that the mass states mix the weak flavors. A third quark generation in the mixing matrix brings with it additional complications, so we will first look at the case of just two generations.

12.3.1 The Cabibbo Hypothesis

In the case of only two generations of matter, we have a 2×2 unitary CKM matrix that mixes d and s states. Unitary $N \times N$ matrices in general have N^2 parameters ($2N^2$ real parameters that are constrained by the unitarity condition), so for two generations, we might expect the mixing matrix to have four parameters. However, three of these degrees of freedom can be transformed away by global relative phase transformations of the spinors for the flavor and mass eigenstates, and so are not physical. This only leaves one degree of freedom, so the mixing matrix for two generations is defined by a single real parameter. For simplicity, we choose the matrix to be real valued, reducing the unitarity condition to an orthogonality condition: $V^T V = \mathbb{1}_3$. Therefore, the mixing matrix is reduced to a simple rotation matrix:

$$V = \begin{pmatrix} \cos\theta_C & \sin\theta_C \\ -\sin\theta_C & \cos\theta_C \end{pmatrix}, \tag{12.11}$$

where θ_C is the Cabibbo quark-mixing angle, $\theta_C \sim 13°$. The model is named after Nicola Cabibbo, who was the first to propose it in 1963, as a means of explaining the cross-generational nature of weak interactions. We can also see why it is conventional to shift all mixing into the down-type quarks: when the model was proposed, only three quark flavors were known. Cabibbo's model only mixed the down-type quarks, since there was no second up-type quark to mix.

We find, then, that vertices such as

are possible, but so are:

though these interactions are weaker than the others, since the coupling for the first carries a factor of $\cos\theta_C$, while the second carries a factor of $\sin\theta_C$. We say that the first type are "Cabibbo allowed" processes, while the second are "Cabibbo suppressed." By comparison, a weak interaction-lepton vertex simply gives a factor of g_2, since lepton states do not mix in the same way.

As an example, the decay of a D^0 meson $(c\bar{u})$ can proceed via

$$D^0 \rightarrow K^- + \pi^+ + \pi^0 \quad \text{or} \quad D^0 \rightarrow \pi^- + \pi^+ + \pi^0, \quad (12.12)$$

among other decay modes. The first mode here is Cabibbo allowed, since c and s are in the same generation. However, the second mode is Cabibbo suppressed. Therefore, we expect the first process to be the dominant decay mode despite the smaller phase space (K^- is more massive than π^-).

The charged weak currents necessarily change flavor, since they must couple quarks of different charges. A flavor-changing neutral current, on the other hand, would require the transformation of say, a strange quark to a down quark. Such processes are found to be heavily suppressed in nature. In fact, one of the successes of the electroweak theory is its lack of flavor-changing neutral processes at tree level. However, it is easy to see that such processes can occur through

higher-order interactions, such as

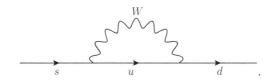

Of course, if quarks were free, a process like the one above would be forbidden anyway on kinematic grounds (it cannot conserve four-momentum), but it serves to demonstrate the *principle* that flavor-changing neutral currents are present at higher order. This was initially a problem for any model of weak interactions, since it predicted that decays such as

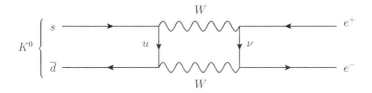

should be allowed, with a coupling of $(g_2 \cos \theta_C) \cdot (g_2 \sin \theta_C)$. This was problematic because the observed rate of such decays was found to be greatly suppressed.

The solution to the problem came from Glashow, Iliopoulos, and Maiani, in the form of the (then still hypothetical) charm quark. They saw that the inclusion of a second up-type quark allowed mixing to be described by the above matrix. In turn, this means that there is a second diagram contributing to kaon decay:

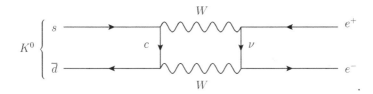

In this case, though, the charm *flavor* state couples to the strange *flavor* state, whereas the up couples to the down. So, in terms of *mass* states, the coupling in this case is $(-g_2 \sin \theta_C) \cdot (g_2 \cos \theta_C)$. If the charm quark were equal in mass to the up quark, these two amplitudes would thus cancel, forbidding this mode of kaon decay. As it is, the up and charm masses are different, and the cancellation is incomplete. Hence the process is merely suppressed, rather than forbidden.

12.3.2 Neutral Mesons

Neutral Meson Mass States

Quark mixing leads in turn to mixing of the neutral mesons. The classic example of this is the mixing of kaons via the interactions depicted in Figure 12.1.

Here, a $\overline{K}^0$ has transformed into a K^0, so we can see that a system prepared in the $\overline{K}^0$ state later has a non-zero probability of being found in the K^0 state. Put another way, since the quark flavors need not be conserved, we know that the flavor symmetries are not exact and that the Hamiltonian is not invariant under flavor transformations. So the mass eigenstates (physical states) are not the K^0 and $\overline{K}^0$ but some linear combination of these flavor states. How, then, are we to find the physical kaon states? The answer (almost!) comes from the combined symmetries of charge conjugation and parity. While the weak interactions violate these individually, it is a reasonable as-

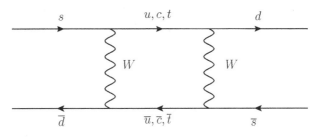

FIGURE 12.1 Neutral kaon mixing.

sumption that the combined symmetry is respected. If $\mathbb{CP}$ is an exact symmetry, then $\mathbb{CP}$ must commute with the Hamiltonian, and the mass eigenstates and $\mathbb{CP}$ eigenstates would be identical. That is, we would expect physical states to be eigenstates of $\mathbb{CP}$.

Consider, then, the combined action of $\mathbb{C}$ and $\mathbb{P}$ on the flavor eigenstates of the kaon system:

$$
\begin{aligned}
\mathbb{CP} \left| \overline{K}^0 \right\rangle &= \mathbb{CP} \left| s\overline{d} \right\rangle \\
&= \left| \overline{s}d \right\rangle \\
&= \left| K^0 \right\rangle \\
\text{similarly,} \ \ \mathbb{CP} \left| K^0 \right\rangle &= \left| \overline{K}^0 \right\rangle .
\end{aligned}
\tag{12.13}
$$

It is easy then to check that

$$
\left| K_1 \right\rangle = \frac{1}{\sqrt{2}} \left(\left| K^0 \right\rangle + \left| \overline{K}^0 \right\rangle \right) \text{ and } \left| K_2 \right\rangle = \frac{1}{\sqrt{2}} \left(\left| K^0 \right\rangle - \left| \overline{K}^0 \right\rangle \right)
\tag{12.14}
$$

are $\mathbb{CP}$ eigenstates with $\mathbb{CP} \left| K_1 \right\rangle = \left| K_1 \right\rangle$ and $\mathbb{CP} \left| K_2 \right\rangle = - \left| K_2 \right\rangle$. If $\mathbb{CP}$ is conserved, we expect these to be the physical states.

The decay rates of these two particles are very different. Since the K_1 has positive "CP-parity," it can only decay to $CP = 1$ states: the dominant decay mode is $K_1 \rightarrow \pi^+ + \pi^-$. On the other hand, the K_2 can only decay to $CP = -1$ states, the dominant mode being $K_2 \rightarrow \pi^+ + \pi^- + \pi^0$. Since the mass of the kaon is only slightly larger than the mass of three pions, the phase space of available states is much smaller for K_2 decay than it is for K_1 decay, pushing up the decay rate for the K_2. Experiments confirm that there is a short-lived and a long-lived neutral kaon, K_S and K_L. So we identify

$$
K_S = K_1 \text{ and } K_L = K_2 \quad (\mathbb{CP}),
\tag{12.15}
$$

if $\mathbb{CP}$ is an exact symmetry.

Flavor Oscillations

In the mass basis, the behavior of the kaons is straightforward. We have two orthogonal physical states, so if it does not interact with anything, a particle prepared in one state will remain in that state at all future times, described by a rest-frame wavefunction of the form $|\psi(t)\rangle = |\psi(0)\rangle \, e^{-(\frac{\Gamma}{2}+im)t}$, where Γ is the decay rate and m the mass (see Equation 5.42). However, interactions of this particle take place in the flavor basis, so we must also consider how the flavor states evolve in time. Inverting the relationship in Equation 12.14 to express the flavor states in terms of mass states, we find

$$\left|K^0\right\rangle = \frac{1}{\sqrt{2}}\left(|K_1\rangle + |K_2\rangle\right) \text{ and } \left|\overline{K}^0\right\rangle = \frac{1}{\sqrt{2}}\left(|K_1\rangle - |K_2\rangle\right).$$
(12.16)

We can now substitute in the wavefunction of the mass states to find the time-evolution of the flavor states:

$$
\begin{aligned}
\left|K^0(t)\right\rangle &= \frac{1}{\sqrt{2}}\left(|K_1(t)\rangle + |K_2(t)\rangle\right)\\
&= \frac{1}{\sqrt{2}}\left(|K_1\rangle \, e^{-\left(\frac{\Gamma_1}{2}+im_1\right)t} + |K_2\rangle \, e^{-\left(\frac{\Gamma_2}{2}+im_2\right)t}\right)\\
&= \frac{1}{\sqrt{2}}\left(\frac{1}{\sqrt{2}}\left(|K^0\rangle + |\overline{K}^0\rangle\right) e^{-\left(\frac{\Gamma_1}{2}+im_1\right)t}\right.\\
&\qquad\qquad \left. +\frac{1}{\sqrt{2}}\left(|K^0\rangle - |\overline{K}^0\rangle\right) e^{-\left(\frac{\Gamma_2}{2}+im_2\right)t}\right)\\
&= \left|K^0\right\rangle \times \frac{1}{2}\left(e^{-\left(\frac{\Gamma_1}{2}+im_1\right)t} + e^{-\left(\frac{\Gamma_2}{2}+im_2\right)t}\right)\\
&\quad +\left|\overline{K}^0\right\rangle \times \frac{1}{2}\left(e^{-\left(\frac{\Gamma_1}{2}+im_1\right)t} - e^{-\left(\frac{\Gamma_2}{2}+im_2\right)t}\right).
\end{aligned}
$$
(12.17)

From this, we can see that a particle initially in the $|K^0\rangle$ state has a non-zero $|\overline{K}^0\rangle$ contribution at later times. However, we can be more specific than this. The transition amplitude for finding a particle in the $|\overline{K}^0\rangle$ state at a time t, if initially in the $|K^0\rangle$ state, is given by

$$\mathcal{M} = \left\langle \overline{K}^0 \,\middle|\, K^0(t)\right\rangle = \frac{1}{2}\left(e^{-\left(\frac{\Gamma_1}{2}+im_1\right)t} - e^{-\left(\frac{\Gamma_2}{2}+im_2\right)t}\right), \quad (12.18)$$

and so the probability of finding the particle in the $\left|\overline{K}^0\right\rangle$ state at time t is given (up to an overall normalization) by

$$
\begin{aligned}
|\mathcal{M}|^2 &= \frac{1}{4}\left(e^{-\left(\frac{\Gamma_1}{2}+im_1\right)t} - e^{-\left(\frac{\Gamma_2}{2}+im_2\right)t}\right) \\
&\quad\times\left(e^{-\left(\frac{\Gamma_1}{2}-im_1\right)t} - e^{-\left(\frac{\Gamma_2}{2}-im_2\right)t}\right) \\
&= \frac{1}{4}\left(e^{-\Gamma_1 t} + e^{-\Gamma_2 t} - e^{-\frac{\Gamma_1}{2}-\frac{\Gamma_2}{2}}\left(e^{i(m_2-m_1)t} + e^{-i(m_2-m_1)t}\right)\right) \\
&= \frac{1}{4}\left(e^{-\Gamma_1 t} + e^{-\Gamma_2 t} - 2e^{-\frac{\Gamma_1}{2}-\frac{\Gamma_2}{2}}\cos\left((m_2-m_1)t\right)\right),
\end{aligned}
$$

$$(12.19)$$

and we see that, not only do the flavor states mix, but in fact the probability of a transition from the initial state to the opposite state oscillates periodically over time. A similar calculation shows that the probability of measuring the particle in its original state also oscillates over time.

12.3.3   More General Quark Mixing

In the case of three generations, we have a 3×3 unitary mixing matrix: the CKM matrix. By the same argument that we employed for the two-generation case, this suggests nine degrees of freedom, but five of these can be transformed away, leaving four real parameters. An orthogonal matrix only has $\frac{1}{2}N(N-1) = 3$ degrees of freedom, so we cannot accommodate all of the CKM parameters with a real matrix. We will worry about precisely what this means in Section 12.4. For now, let's consider the problem of how best to parametrize the matrix. First, notice that three parameters could have been accommodated in a 3×3 rotation matrix, so we have three mixing angles that parametrize the mixing between generations 1 and 2, generations 2 and 3, and generations 1 and 3. These are referred to as Euler angles in analogy with the angles used to describe three-dimensional orientation. The fourth degree of freedom must be a complex phase. There are infinitely many ways to represent these

Euler angles and phase in the CKM matrix, but a common way is

$$
\begin{pmatrix}
c_{12}c_{13} & s_{12}c_{13} & s_{13}e^{-i\delta} \\
-s_{12}c_{23} - c_{12}s_{23}s_{13}e^{i\delta} & c_{12}c_{23} - s_{12}s_{23}s_{13}e^{i\delta} & s_{23}c_{13} \\
s_{12}s_{23} - c_{12}c_{23}s_{13}e^{i\delta} & -c_{12}s_{23} - s_{12}c_{23}s_{13}e^{i\delta} & c_{23}c_{13}
\end{pmatrix},
$$
$$(12.20)$$

where c_{ij} and s_{ij} are the cosine and sine, respectively, of the mixing angle between generations i and j, and δ is the complex phase.

The parametrization that we choose has no effect, of course, on the measured values of the CKM matrix entries. Ignoring the matter of the complex phase for now, the magnitudes of the entries are found experimentally to be

$$
\begin{pmatrix}
|V_{ud}| & |V_{us}| & |V_{ub}| \\
|V_{cd}| & |V_{cs}| & |V_{cb}| \\
|V_{td}| & |V_{ts}| & |V_{tb}|
\end{pmatrix}
=
\begin{pmatrix}
0.974 & 0.225 & 0.003 \\
0.225 & 0.973 & 0.041 \\
0.009 & 0.040 & 0.999
\end{pmatrix}.
\quad (12.21)
$$

It is important to emphasize that these results are obtained experimentally: there is nothing in the Standard Model that can fix these values for us.

Notice that the entries that account for mixing of third-generation quarks with other generations have small values. This accounts for the relative stability of hadrons composed of b-quarks despite their high masses. In fact, interpreting these values in terms of the Euler angles above, we find $\theta_{12} = \theta_C = 13.04°$, $\theta_{23} = 2.38°$, and $\theta_{13} = 0.201°$. So the mixing of generations 1 and 2 is only lightly suppressed, while mixing of generations 2 and 3 is more heavily suppressed, and the first and third generations barely mix at all.

12.4 $\mathbb{CP}$ VIOLATION IN THE WEAK SECTOR

We turn now to the question of the complex phase in the CKM matrix. This phase is responsible for the violation of $\mathbb{CP}$ symmetry in weak interactions. In fact, it is a general result that a complex coupling constant leads to $\mathbb{CP}$-violation if it cannot be phased away. We can sketch an argument for this in the specific case of the CKM matrix as follows. The terms responsible for quark masses in the Lagrangian, or equivalently in the Hamiltonian, are of the general form

$$\frac{v}{\sqrt{2}}\left(y\bar{q}_L q_R + y^*\bar{q}_R q_L\right),\qquad(12.22)$$

where q_L, q_R are left- and right-chiral quarks, and y is some Yukawa coupling. Note that the couplings must be related by complex conjugation, as the Lagrangian (Hamiltonian) must be Hermitian, and the constructs $\bar{q}_L q_R$ and $\bar{q}_R q_L$ are Hermitian conjugates. Under a $\mathbb{CP}$ transformation, the above mass terms become

$$\frac{v}{\sqrt{2}}\left(y\bar{q}_R q_L + y^*\bar{q}_L q_R\right),\qquad(12.23)$$

and we can see that the Lagrangian (or Hamiltonian) is invariant under $\mathbb{CP}$ if and only if $y = y^*$. That is, a real coupling respects $\mathbb{CP}$, while a complex coupling violates it, if the complex component cannot be phased away.

Since it requires three quark generations, $\mathbb{CP}$ violation manifests itself only in situations where the three generations are all present. In particular, $\mathbb{CP}$-violation is observable in the decays of neutral mesons. We have already seen that the neutral meson flavor states mix to produce the physical states. Previously, the physical states we constructed for the kaon system were based on the assumption of exact $\mathbb{CP}$ symmetry. In particular, we identified the $\mathbb{CP}$ eigenstates $K_{1,2}$ with the mass eigenstates $K_{S,L}$. However, experiments by Cronin and Fitch in 1964 showed that, if a beam of neutral kaons is left to travel through empty space such that the short-lived K_S component decays away, there are still $CP = +1$ decays. This demonstrates that $\mathbb{CP}$ cannot be conserved. Notice that it does not tell us exactly where the $\mathbb{CP}$ violation takes place, though. It could be in the kaon mixing, or it could be in the decay processes themselves. If the violation takes place in the kaon mixing, then the identification we made above is no longer accurate. We find instead that K_S is only *mostly* K_1, but has a little K_2 mixed in. This additional $\mathbb{CP}$-violating mixing is parametrized as

$$|K_S\rangle = \frac{1}{\sqrt{1 + \varepsilon^* \varepsilon}} (|K_1\rangle - \varepsilon |K_2\rangle)$$
$$|K_L\rangle = \frac{1}{\sqrt{1 + \varepsilon^* \varepsilon}} (|K_2\rangle + \varepsilon |K_1\rangle),$$

(12.24)

where ε is a small quantity that parametrizes the mixing. As stated above, though, this "indirect violation" is not the only possible source of $\mathbb{CP}$ violation in kaon systems. It is possible that further violation occurs during the decay of the kaon (direct violation), through the so-called "penguin" graphs:[2]

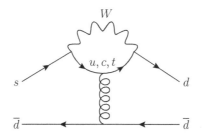

It is now known that both types of violation occur and experiments have been able to identify that direct violation is on the order of 10^3 times weaker an effect than indirect violation. In fact, there is also a third type of $\mathbb{CP}$ violation that arises from interference effects of direct and indirect violation.

The neutral kaons are the system in which $\mathbb{CP}$ violation was originally observed, but they are not the only system that exhibits violation, since the same mixing can apply to any neutral mesons. When investigating $\mathbb{CP}$ violation, however, the D mesons are found not to be terribly fruitful, since the D^0 and $\overline{D}^0$ states mix much less than their strange counterparts. The B mesons, on the other hand, are found to be the ideal sector for studying $\mathbb{CP}$ violation, and there are several detectors at various experimental facilities dedicated to the

---

[2] Apparently named after their resemblance to a penguin.

study of B physics. There are several reasons that B mesons are so well suited to this application. First, they have a surprisingly long lifetime for such massive particles, due to the weakness of quark mixing in the third generation. Second, the $\mathbb{CP}$-violating differences in decay rates are sufficiently large that accurate measurements may be taken. The $\mathbb{CP}$ violation that is observable in the B sector is qualitatively different from that in the kaon system. Whereas the kaons show very distinct mass eigenstates due to the lucky coincidence that the K_L mass is so close to the mass of three pions, the decay rates for the equivalent B mass states are so similar that they cannot (yet) be resolved. We cannot hope to observe the mass states decaying to the "wrong" final states if we cannot be sure which we are looking at! For this reason, studies of B mesons commonly work instead with the flavor eigenstates B^0 and $\overline{B}^0$. Of course, as in the kaon case, these mix and so we must find a way to distinguish them. One trick is to observe B mesons produced in particle-antiparticle pairs. Although we do not know which meson was which at the time of their production, we can use the fact that the subsequent evolution of each is correlated with the other. If one meson undergoes decay to a leptonic state, the lepton pair produced is sufficient to identify the B meson that produced it *at the time of decay*. This allows us to identify the meson that has not yet decayed, and since the rate of oscillation is known, we can extrapolate to find the quantum state of the second particle at the moment of *its* decay. Also, we should note that the B^0 and the $\overline{B}^0$ are capable of decaying to the same final states, again due to mixing, but this time in the resulting kaon, rather than in the initial B meson. For example, both B^0 and $\overline{B}^0$ can decay to $J/\Psi + K_L$, via

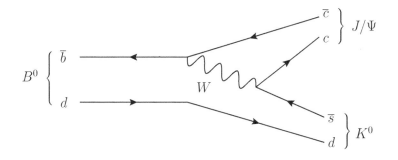

and

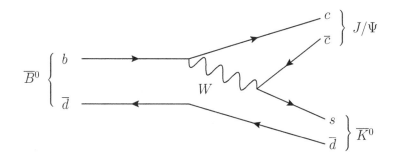

but the K^0 and $\overline{K}^0$ are *measured* as the same state, either K_L or K_S. In this way, we can measure the asymmetry in the decays of the B^0 and the $\overline{B}^0$, when decaying to the same final states, and hence determine the degree of $\mathbb{CP}$ violation.

We have already seen that the individual $\mathbb{P}$ and $\mathbb{C}$ symmetries are violated maximally in the weak sector. How might we quantify the degree of weak $\mathbb{CP}$ violation so that we may ask if the same is true of the combined symmetry? The answer comes from a quantity, J, known as the Jarlskog invariant. This is defined as

$$J = \left| \mathrm{Im} \left(V_{ij}^* V_{ik} V_{\ell k}^* V_{\ell j} \right) \right|, \tag{12.25}$$

where $(i, \ell) = (u, c, t)$, $(j, k) = (d, s, b)$ and $i \neq \ell$, $j \neq k$. Unitarity of the CKM matrix demands that taking the imaginary part of the product of four matrix elements in this way gives the same value regardless of which particular set of elements we choose. The quantity is invariant under phase shifts of the quarks and is guaranteed to take a value in the interval $[0, 1]$, vanishing when there is no $\mathbb{CP}$ violation present. As such, it acts as a useful measure of the degree of $\mathbb{CP}$ violation, which is independent of how we might choose to parametrize the CKM matrix. Far from being maximal, the amount of $\mathbb{CP}$ violation found to occur in the weak interactions is small at around $J = 3 \times 10^{-5}$.

12.5 SUCCESSES OF THE STANDARD MODEL

The Standard Model is a hugely successful theory of the fundamental particles and their interactions. We have already seen that the predictions of QED and their corresponding experimental results agree to a precision unmatched by any other theory in the history of science. Furthermore, the multitude of observed hadrons can be explained in terms of bound states of quarks, with the hadron properties at least partially deducible from a simple static model of the constituent quarks. The high-energy perturbative regime of QCD allows for calculation of the form factors for hadrons, while the lattice approach has had remarkable successes in modeling the behavior of the strong force at low energy scales. While these are technically the most impressive feats of the Standard Model, it may be argued that possibly the most *striking* examples of its success come from the weak sector. The electroweak theory predicted the existence of not just one, nor indeed two nor three, but *four* new particles whose existence have now been confirmed, along with their estimated properties. These are, of course, the W^+, W^-, Z^0 and the Higgs. The model is also capable of explaining the chiral nature of the weak interactions, and the origin of the $\mathbb{C}$-, $\mathbb{P}$-, and $\mathbb{CP}$-violation also observed in nature. Despite these successes, it is universally accepted that the Standard Model is merely a work in progress on the way to a deeper understanding of the particle universe. This is due, in part, to a number of problems and drawbacks with the model. Before discussing some of the less successful aspects of the Standard Model, we will examine one final point of the model that is particularly pleasing.

12.5.1 Anomaly Cancellation

Any quantum-mechanical theory consists of two parts: the Hamiltonian (or Lagrangian), and some means of quantizing the theory. Early in the history of quantum physics, it was implicitly assumed that any symmetry of the Hamiltonian would be a symmetry of the full theory; after all, this was certainly true of classical physics. Unfortunately, this assumption is not valid, since the quantization procedure must also respect any symmetry of the Hamiltonian in order for the symmetry to remain in the quantum theory. Since this came as something of a shock at the time, symmetries violated by the quantization procedure in this way are referred to as anomalies. It should be noted, though, that this name is almost as unfortunate as the term "imaginary" for those complex numbers that are not real: it gives the impression that there is something mysterious about their appearance. In fact, "anomalous" symmetries are common and can be innocuous in quantum theories.

A simple example of an anomalous symmetry is the invariance of a theory under a change in the scale at which it is viewed. In a classical theory, there is no reason to suspect that the fundamental strength of a force would depend on the distance scale under consideration. That is, from a classical point of view, the coupling constant for a theory should be just that: constant. This invariance of the coupling under scale transformations is then a symmetry of the classical theory. In contrast, we have already seen how the screening effect of quantum fluctuations (or equivalently, the renormalization process) leads to a scale-dependence of the coupling constant. Scale invariance is thus an anomalous symmetry. The preceding example is of an anomaly in a global symmetry.

Recall, though, that the structure of the Standard Model relies on gauged, or local, symmetries. Whereas a global anomaly is an interesting but harmless artifact of the quantization procedure, the same cannot be said of an anomalous *local* symmetry. Since the consistency of a theory of spin-1 bosons *requires* gauge symmetry, the loss of that symmetry through an anomaly would be catastrophic for the theory. Therefore, any consistent gauge theory must contain no gauge anomalies. It turns out that, for QED and QCD, there is nothing to worry about, since no anomalies appear. This is because the contribution to the anomaly carries opposite signs in left-chiral and right-chiral fermions. Since QED and QCD are left-right symmetric, the left- and right-chiral contributions must be equal and so cancel. However, things are not quite so straightforward once we introduce the electroweak theory. The chiral nature of the theory does potentially allow for the emergence of gauge anomalies, but the theory is saved by a very neat result.

The anomaly in electroweak theory (known as the chiral anomaly), is found to be generated by triangular Feynman diagrams of the form:

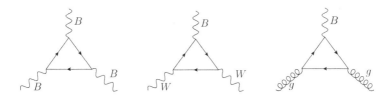

where the closed loop must sum over all fermions that may contribute. The first diagram is proportional to the sum of the cubes of the hypercharges of all the fermions with the contributions of left-chiral and right-chiral fermions having opposite sign. Since these charges repeat for different fermion generations, we need consider only one generation. We find:

$$
\sum_{\substack{\text{left-}\\\text{chiral}}} Y^3 - \sum_{\substack{\text{right-}\\\text{chiral}}} Y^3 = \left[(-1)^3 + (-1)^3 + 3\left(+\frac{1}{3}\right)^3 + 3\left(+\frac{1}{3}\right)^3 \right]
$$

$$
- \left[(-2)^3 + 3\left(+\frac{4}{3}\right)^3 + 3\left(-\frac{2}{3}\right)^3 \right]
$$

$$
= 0, \tag{12.26}
$$

where the factors of 3 account for the quark colors. The particular set of hypercharges for a single generation of fermions ensures that the anomaly vanishes, and the theory is safe! A similar cancellation is found in the second and third diagrams, though verification of this is left to the reader. Notice that we did not consider diagrams with, for example, one gluon. This is due to the fact that such diagrams necessarily vanish: each vertex in one of these triangle diagrams contributes a generator from the associated gauge group (an identity for $U(1)_Y$). The closed fermion loop then requires us to trace over these generators, and since the generators for a special unitary group are traceless, a single contribution from a given group causes the diagram to vanish.

12.6 DRAWBACKS OF THE STANDARD MODEL

Despite its successes, the Standard Model suffers from a number of problems. First, there is the rather unsatisfactory point that the parameters of the theory, such as masses, couplings, mixing angles, and the $\mathbb{CP}$-violating phase, must be introduced by hand. With no way of determining their values *a priori*, this is a total of 19 parameters with apparently no underlying explanation. In addition, the Standard Model fails to predict some observed phenomena. For instance, it has no candidate for dark matter, and no way to explain dark energy. Even its treatment of neutrinos is found to be inadequate, as we will see in Section 13.1.

In the following sections, we will look specifically at three problems with the Standard Model: matter-antimatter asymmetry, the hierarchy problem, and the strong $\mathbb{CP}$ problem. The second of these arises from the electroweak sector of the model, while the latter arises from a combination of the electroweak and strong sectors. However, even the simplest part of the Standard Model has its issues. In particular, as the energy scale is increased, and the $U(1)$ coupling rises, so too does the value of the β function. That is, not only does the coupling increase but so too does the *rate* of increase. At a finite energy scale known as the Landau pole, the coupling is found to take on an infinite value. This may suggest that there should be some new physics below this scale to prevent the pole being reached. One way out of this issue is to embed the $U(1)$ group into a larger group with a negative β-function. This is an example of Grand Unification, which will be explored in Section 13.3. However, it should be noted that the β-function is calculated using perturbation theory and so may only apply when a coupling is small. If a coupling tends to infinity, then it can certainly no longer be called small! As such, it may be that the behavior of the β-function changes at scales where $g > 1$, and that the Landau pole is not a genuine effect but a result of our attempting to apply perturbative predictions to non-perturbative systems.

12.6.1 Baryogenesis

Grand unification is also a possible solution to the problem of matter-antimatter asymmetry, though some would argue that this is not a shortcoming of the Standard Model. If baryon number is truly conserved in fundamental interactions, it begs the question of where the baryons came from originally. In particular, we expect matter and antimatter to have been produced in equal measure at the big bang, and yet we see essentially only matter today. Andrei Sakharov showed that a matter-antimatter asymmetry requires several conditions in order to develop from an initially symmetric universe. These are the violation of $\mathbb{C}$, $\mathbb{CP}$, and baryon number, B, together with a departure from thermal equilibrium. The requirement for violation of baryon number is self-explanatory, and the remaining conditions allow for any matter-preferential processes not to be automatically balanced by analogous antimatter-preferential processes. We have already seen that the Standard Model does indeed violate both $\mathbb{C}$ and $\mathbb{CP}$, and departure from thermal equilibrium is achieved during any phase transition, for example, the electroweak symmetry breaking.

In fact, the Standard Model is also capable of violating baryon number at sufficiently high energy through the sphaleron process. The details of this can get a little tricky, but essentially, the space of possible vacuums in the $SU(2)$ flavor sector is sufficiently complex that there are different field configurations that look identical *locally* but are distinguished *globally* by their topology. In any one of these energy wells, the local concept of baryons and leptons is different from neighboring wells. If this idea seems alien, recall that what we mean by "Higgs boson" is similarly dependent on the vacuum state we find ourselves in. A particle is an excitation of a quantum field about its minimum value, so the concept of a "particle" depends on that minimum. With enough energy, the universe could climb over the energy barrier (known as a sphaleron) into a different well, violating baryon and lepton number along the way. Estimates of the energy required for direct scaling of the sphaleron barrier are around 9 TeV, explaining why no baryon-violating processes have yet been directly observed in nature. Since scaling this barrier requires a large deviation from the vacuum state, the sphaleron process is a *non-perturbative* process and hence cannot be depicted through Feynman diagrams.[3]

In this way, then, the Standard Model appears to contain all of the ingredients necessary for baryogenesis—that is, the production of a matter-dominated universe from an initial symmetric state, in which matter and antimatter were present in equal measure. However, the sphaleron process actually acts to *undo* baryogenesis, rather than promote it. In the unbroken electroweak phase, sphaleron processes would be plentiful and destroy any existing baryon asymmetry. It is only during the symmetry breaking phase transition that any net baryogenesis could occur, after which the temperature is too low for further sphaleron processes, "freezing in" the asymmetry at that point. For this reason, some feel that electroweak baryogenesis is incapable of producing the *extent* of asymmetry we see today. However, it should be stressed that opinion is divided on this point.

When we discuss grand unification in Section 13.3, we will see how it can give rise to B-violating processes that are perturbative, in which exchange particles couple directly to B-violating currents.

12.6.2 The Hierarchy Problem

The hierarchy problem refers to a discrepancy between the Higgs mass and the expected energy scale for "new physics." In particular, while a massless particle will remain massless under the renormalization process, a non-zero particle mass receives quantum corrections through the particle's self-interactions. These corrections are found to be divergent for a scalar particle. Clearly, an infinite particle mass is nonsensical, so how are we to ensure consistency? The answer is that the divergence is again a sign of our ignorance of the underlying high-energy theory. In Section 9.6, we saw that, during renormalization, it is necessary to introduce a finite energy scale in order to evaluate loop integrals. This "cutoff" is then allowed to diverge after evaluation, and physical quantities are extracted along the way.

---

[3] In fact, since it does not involve the exchange of a virtual particle, the sphaleron process is arguably beyond the realm of particle physics.

However, this process assumes that the theory is valid up to arbitrarily large energy scales. If the standard model is really just an effective theory up to some physical energy, then the cutoff need only run up to this point. This implies that the mass of the scalar particle should be of the order of the physical cutoff, which is problematic if we only expect the Standard Model to break down at very high energy. As we will see in Section 13.3, the expected scale of a grand unified theory is typically on the order of 10^{16} GeV, but even this ludicrously large scale is better than the alternative: the Planck scale, at which gravitational effects can no longer be neglected in a quantum theory, is of the order of 10^{19} GeV! The scalar particle discovered at the LHC in 2012, which is widely believed to be the Standard Model Higgs, is found to have a mass of 125 GeV, respectively 14 and 17 orders of magnitude too small for either of these scales. In fact, the hierarchy problem was recognized long before this experimental discovery, since there were already theoretical bounds on the Higgs mass.

In order for a scalar particle to retain a small mass after renormalization, the "bare" mass appearing as an unobservable parameter in the underlying theory must cancel off the large corrections, leaving only the observable quantity. With the orders of magnitude involved, this is equivalent to measuring the lengths of two objects roughly the size of the Milky Way, and finding that they differ by only a meter. Needless to say that, without some underlying reason for this cancellation, such a phenomenal coincidence is a little hard to swallow. This puts the hierarchy problem in the class of fine-tuning problems, of which there are other examples. Ultimately, then, the problem is one of naturalness: there must be an underlying reason that such a precisely balanced system is in some way natural or expected. The obvious answer is that the scale of new physics is much lower than 10^{19} GeV, but what is the nature of this new physics? One possibility will be considered in Section 13.4.

A very good question at this point is why this is considered a problem only for the Higgs mass: why not for the fermions or gauge bosons of the Standard Model? The answer that is usually given to this question is that the masses of these particles are "protected" by a symmetry: chiral symmetry in the case of the fermions and gauge symmetry in the case of the vector bosons. This answer is somewhat esoteric, so let us attempt to clear up its meaning. As shown in Section 9.6, the corrections to a particle's mass are generally of the form

$$\delta m = k_1 \Lambda^n + k_2 m \ln \left(\frac{m}{\Lambda} \right), \qquad (12.27)$$

where k_1 and k_2 are constants, Λ is the cutoff, and n is 1 for fermions and 2 for vector bosons. In the case of a massless particle, a vector boson or fermion has, respectively, gauge symmetry or chiral symmetry (that is, left-chiral and right-chiral components do not mix). Since the appropriate symmetry is non-anomalous, it is guaranteed to be respected in the renormalized theory as well, so a massless particle cannot be given mass by quantum corrections. Clearly the second term above vanishes in this case, which in turn implies that $k_1 = 0$. For a massive particle, the second term is non-zero and does indeed provide corrections to the mass, but notice that a small mass means small corrections since the term is proportional to m. The first term must again be zero, since the results must agree in the massless limit; it is in this sense that the symmetry "protects" the mass. In the case of scalar particles, there is no such guardian symmetry, and so the corrections to the mass are typically of the order of the cutoff.

12.6.3 The Strong $\mathbb{CP}$ Problem

Finally, we consider the strong $\mathbb{CP}$ problem. We have seen that $\mathbb{CP}$ violation is a natural consequence of the electroweak theory. In fact, it is also found to be a natural part of the theory of strong interactions. Essentially, this is because $\mathbb{CP}$ is anomalous in QCD: there is a term in the QCD Hamiltonian that, even if initially set to zero, is reintroduced by quantum corrections, and this term violates $\mathbb{CP}$ symmetry. The term in question takes the form

$$i\theta \varepsilon_{\mu\nu\rho\sigma} G^{\mu\nu} G^{\rho\sigma}, \qquad (12.28)$$

where $G^{\mu\nu}$ is the gluon field strength tensor, and $\varepsilon_{\mu\nu\rho\sigma}$ is the totally anti-symmetric tensor, while $\theta \in [0, 2\pi]$ is a parameter. Since this term is purely imaginary, it is clear that it violates $\mathbb{CP}$ for any $\theta \neq 0$, by the arguments of Section 12.4. The problem is that no $\mathbb{CP}$ violation has been detected in the strong sector, which seems to suggest that θ is at least very close to, if not equal to, 0. We have another fine-tuning problem: θ may take any value in its range with equal likelihood and yet appears to have chosen a value so small that its departure from 0 is (as yet) undetectable. How are we to find a natural explanation for such a small value? One possible solution is a hypothetical particle called the axion, whose properties will be explored in Section 13.6. The astute reader may be wondering why there is no such problem for the $SU(2)_L$ and $U(1)_Y$ parts of the theory: the answer is that the behavior of a fermion under these parts of the theory depends on the fermion's chirality. In particular, the right-chiral fermions do not transform under $SU(2)_L$ and so an appropriate phase rotation of the right-chiral fermions may remove the analogous term in the weak sector.

From the preceding sections it should be clear that, while the Standard Model is a remarkable achievement, it is still a work in progress, and is certainly not the final theory of particle physics. The next chapter will explore some of the current ideas in particle physics that take us beyond the Standard Model.

EXERCISES

1. Show that each term in the Yukawa part of the Lagrangian (Equation 12.5) has no net hypercharge.

2. **(a)** Draw Feynman diagrams for the processes
   $K^- \to \mu^- + \overline{\nu}_\mu$ and $\pi^- \to \mu^- + \overline{\nu}_\mu$ at the quark level. Without performing a full calculation, estimate the ratio of the transition amplitudes of these two processes.

   (b) The above decays are the dominant modes for these mesons. Why isn't a decay to $e^- + \overline{\nu}_e$ favored? (Hint: This is a tough one! Some things to consider are the chirality and helicity of the decay products as well as how mass mixes these concepts.)

3. Consider a transition amplitude $\mathcal{M}$ for a process that can occur through two distinct channels: $\mathcal{M} = \mathcal{M}_1 + \mathcal{M}_2$. If a complex phase is introduced in the couplings for these channels, it shows itself in the amplitudes as

$$\mathcal{M} = \mathcal{M}_1 e^{i\delta_1} + \mathcal{M}_2 e^{i\delta_2},$$

where $\mathcal{M}_i$ now refer to the amplitude for each channel *if the couplings were real*.

(a) Write down the amplitude, $\mathcal{M}^{(\mathbb{CP})}$, for the $\mathbb{CP}$-conjugate process.

(b) Hence show that

$$|\mathcal{M}|^2 - \left|\mathcal{M}^{(\mathbb{CP})}\right|^2 = 2i\left(\mathcal{M}_1^*\mathcal{M}_2 - \mathcal{M}_1\mathcal{M}_2^*\right)\sin(\delta_2 - \delta_1).$$

(c) By writing $\mathcal{M}_{1,2}$ in terms of their magnitudes, $|\mathcal{M}_{1,2}|$, and complex arguments, $\phi_{1,2}$, show that the above expression is equal to

$$-4\left|\mathcal{M}_1\right| \cdot \left|\mathcal{M}_2\right|\sin(\phi_2 - \phi_1)\sin(\delta_2 - \delta_1).$$

(d) Hence show that $\mathbb{CP}$ symmetry is violated in the case of complex couplings.

4. (a) Write down the individual matrices that would account for a simple mixing of the first and second, second and third, and first and third quark generations. Show that these may be multiplied together to give the magnitudes of the elements of the CKM matrix as parametrized in Equation 12.20. How should the complex phase be included in one of the three rotation matrices to give the full parametrization?

(b) For the parametrization of the CKM matrix in Equation 12.20, show that the Jarlskog invariant is given by $J = c_{12}c_{23}c_{13}^2 s_{12}s_{23}s_{13}\sin(\delta)$.

5. Find the invariant amplitude for the process $W^- \to e^- + \bar{\nu}_e$. (For simplicity, neglect the mass of the electron.) Hence use the two-body decay rate formula found in Exercise 5 to estimate the partial decay rate of the W^- boson via this channel.

6. As well as the diphoton decay via a top loop, the Higgs boson can decay in several other ways. By considering the couplings between various particle species and the necessary final states these would lead to, find some other likely decay modes.

7. Show that the potential gauge anomalies of the Standard Model vanish.

13

BEYOND THE STANDARD MODEL

Some of the issues with the Standard Model have been addressed by theorists introducing extensions to the model. However, many of these extensions remain speculative, with little or no experimental verification. Their acceptance among (portions of) the physics community is based instead on the elegance of the proposed models, and their ability to solve many of the problems highlighted previously. This chapter will explore some of these theories that go "beyond the standard model." The first extension that we consider is the least contested, since it is similar in form to the quark mixing of the previous chapter, and is introduced to accommodate an experimental observation: namely, that neutrinos must have mass.

13.1 NEUTRINO OSCILLATIONS AND THE PMNS MATRIX

One of the products of the nuclear processes powering our sun is a large number of neutrinos. In the 1960s, observations of the solar neutrinos seemed to suggest that the number emitted was too low. Neutrinos are, of course, notoriously difficult to observe at all, but even taking this into account, the number observed was found to be only around one third of the value predicted by models of solar evolution. This became known as the solar neutrino problem and its solution came with the realization that neutrino detectors only detect *electron*-neutrinos: mu-neutrinos and tau-neutrinos are not detected. However, only electron-neutrinos should be emitted by weak nuclear processes in the Sun's core, so this then raises another problem: how are the electron-neutrinos emitted by the Sun changing to other neutrino flavors?

We can find the answer to this puzzle by considering again the kaon oscillations that we saw in the previous chapter. As long as no conservation laws are violated and there is an appropriate mixing channel for two particle species, the flavor state will oscillate in time. That neutrino flavors oscillate in this way was definitively demonstrated in the 1990s. In a beam of neutrinos, the probability of measuring neutrinos of a particular flavor fluctuates over time. We saw in the case of kaons, though, that the rate of this oscillation is determined by the difference in mass between the two particles. So if neutrinos are all massless, then no oscillations take place. Put another way, if neutrinos were massless, then their flavor and mass states would be identical: there would be no leptonic equivalent of the CKM matrix, and therefore no means of mixing leptons of different generations.

To accommodate neutrino oscillations, then, we require a leptonic analogue to the CKM matrix, known as the Pontecorvo-Maki-Nakagawa-Sakata (PMNS) matrix. So we now know neutrinos to be massive particles, even though no direct measurement of neutrino masses has ever been made. In fact, current experimental constraints tell us that the sum of the masses of the three neutrinos must be less than 1.2 eV. For most purposes, then, we can still approximate the neutrinos as being massless. As stated above, the introduction of the PMNS matrix is considered standard, and some would even argue that this is now a part of the Standard Model itself. However, it raises another problem, as we will see in the next section, and the possible solution presented there does not, as yet, have any observational backing.

13.2   THE SEE-SAW MECHANISM

The problem raised by the introduction of neutrino masses is this: why should the neutrino mass be non-zero but so small that it has not yet been directly measured? This is another example of a fine-tuning problem, similar to the strong CP and hierarchy problems discussed previously. One possible answer is the see-saw mechanism. In fact, there are a few variations on this idea, and the version that we will consider is the Type I see-saw.

It was stated in Section 12.2 that there are restrictions on the kind of interaction and mass terms we can include because of gauge-invariance and Lorentz-invariance. The argument put forward there suggested that an equation such as

$$i\partial\!\!\!/ f_L = m f_L^c \qquad (13.1)$$

is disallowed because of the different transformations of f_L and f_L^c under gauge transformations. There is, however, one exception to this rule. If the particle we are describing is neutral with respect to all charges, then both sides of the equation are unaffected by gauge transformations. There appears, therefore, to be nothing stopping us from including such a mass term for the right-chiral neutrino. The only charge that would be violated is lepton number. But, in fact, there is no reason to suspect that the individual lepton numbers are conserved, at least since the discovery of neutrino oscillations. This allows for two very different types of mass for neutrinos: a Dirac mass, m_D, that couples left-chiral neutrinos to right-chiral neutrinos (which is the type we have considered so far), and a Majorana mass, m_M, that couples right-chiral neutrinos to left-chiral antineutrinos. Notice that we cannot include the same kind of term for the left-chiral neutrino, though, since this has a non-zero hypercharge, and also transforms under $SU(2)_L$.

Including all possible contributions to the mass, we find we have a 2×2 matrix of mass terms (per generation):

$$\begin{pmatrix} 0 & m_D \\ m_D & m_M \end{pmatrix} \begin{pmatrix} \nu_L \\ \nu_R \end{pmatrix}. \qquad (13.2)$$

The natural scale for m_D is the energy scale of electroweak symmetry breaking, since this is where the Dirac mass terms come from. The natural scale for m_M, however, is at the scale of whatever new physics underlies the Standard Model, say somewhere around the 10^{16} GeV mark. The exact value is not important for this argument, as long as $m_M \gg m_D$.

The physical neutrinos would be the eigenstates of the previous matrix with masses equal to the eigenvalues, which are found to be approximately m_M and $-\frac{m_D^2}{m_M}$. One of the mass eigenstates has a mass close to the high mass scale mentioned above, while the other has a vanishingly small mass. It doesn't matter, incidentally, that the smaller eigenvalue is negative. Only the magnitude is physically meaningful anyway, since a redefinition of one of the neutrino's wavefunctions via a global phase shift of π will make this eigenvalue positive. The eigenvectors corresponding to these masses are approximately

$$\begin{pmatrix} 0 \\ 1 \end{pmatrix} \text{ and } \begin{pmatrix} 1 \\ 0 \end{pmatrix} \tag{13.3}$$

respectively. So the low mass neutrino is almost entirely left-chiral while the high mass neutrino is almost entirely right-chiral. As well as explaining the low mass of neutrinos, this also provides a possible dark matter candidate: the sterile neutrino. We also can see, incidentally, why this is known as the see-saw mechanism: as the right-chiral neutrino mass goes up, the left-chiral mass is driven down. At low energy, the apparent effect of all this will be very different. Recall from Section 11.1 that the low-energy view of massive boson-mediated interactions is a direct four-fermion interaction. Essentially, we have not "zoomed in" enough to the interaction region to see that the two vertices do not actually coincide. A similar process occurs with neutrino masses, and the low-energy view is of a purely left-chiral neutrino as in the Standard Model, with a mass derived from *two* interactions with the Higgs field: the individual mixed-chirality interactions at high-energy blend into one fixed-chirality interaction as we zoom out, as in Figure 13.1.

FIGURE 13.1 At high energy (left), a left-chiral neutrino interacts with the Higgs field and changes chirality, before a second interaction changes it back. As we zoom out (right), the right-chiral component propagates a negligible distance, transforming the process into an effective four-point interaction at low energy.

13.3   GRAND UNIFICATION

The history of physics has been one of gradual unification: the building of connections between disparate ideas and phenomena, and their integration into the same few frameworks. The once separate ideas of electric current and static charge were initially combined into a unified understanding of electricity. Later, this understanding would be further unified with magnetism and optics by Faraday, Ampere, Maxwell, and others into a single theory of electromagnetism. This, of course, was not the end of the story, however, as we have also seen that electromagnetism is itself combined with the weak interaction to give the electroweak theory. This theory, in conjunction with QCD, gives us the Standard Model. It seems strange, then, to stop here: why not continue to unify and see if we can explain strong and electroweak forces in terms of a single underlying theory?

In Section 13.2, it was stated that the natural scale for a Majorana neutrino mass might be somewhere around 10^{16} GeV. Where did this value come from? The β-functions (see Section 10.3.2) for the individual groups $SU(3)_C$, $SU(2)_L$ and $U(1)_Y$ show that the couplings for these groups' corresponding forces converge at high energy (around 10^{16} GeV). In fact, the convergence is not exact and the couplings "miss" each other, but the trend is strong enough to suggest that there may be new physics in the intervening energy region that ensures convergence (see Figure 13.2). There is no good reason for these couplings to converge unless they are somehow related at high energy. In the same way that we broke the electroweak sector of this theory to give the low-energy electromagnetic theory, we can ask the question: is there a larger simple group that contains the Standard Model? Such models are known as Grand Unified Theories (GUTs) and there are several candidates. The first example of a GUT was $SU(5)$, proposed by Georgi and Glashow. This group is capable of breaking via a Higgs-like mechanism to the Standard Model group. Importantly, though, not only does the $SU(5)$ *group* contain the Standard Model group, but a simple combination of $SU(5)$ *representations* is also capable of giving all the particles of the Standard Model. Not all unified models are GUTs, since some are still semi-simple: that is, they still contain a product of several simple Lie groups. However, even such a partial unification brings with it an immediate payoff. Recall from the previous chapter that the $U(1)$ part of the Standard Model has an apparent sickness in the form of the Landau pole: the divergent value of its coupling constant at a finite energy. If we embed $U(1)_Y$ along with the other subgroups of the Standard Model into a simple non-Abelian group at some energy below the scale of the Landau pole, then this problem is solved, since the coupling of a non-Abelian group *decreases* with increasing energy scale.

For a clearer understanding of grand unification, let's take a closer look at the Georgi-Glashow model. There are 24 generators of $SU(5)$, each of which is a 5×5 traceless Hermitian matrix, normalized such that $\mathrm{tr}(T_i T_j) = 2\delta_{ij}$. Eight of these have the general form

$$
T = \begin{pmatrix} & & & 0 & 0 \\ & \lambda & & 0 & 0 \\ & & & 0 & 0 \\ 0 & 0 & 0 & 0 & 0 \\ 0 & 0 & 0 & 0 & 0 \end{pmatrix}
\tag{13.4}
$$

and three have the form

$$
T = \begin{pmatrix} 0 & 0 & 0 & 0 & 0 \\ 0 & 0 & 0 & 0 & 0 \\ 0 & 0 & 0 & 0 & 0 \\ 0 & 0 & 0 & & \\ 0 & 0 & 0 & & \sigma \end{pmatrix},
\tag{13.5}
$$

where λ is a Gell-Mann matrix and σ a Pauli matrix. Another matrix is diagonal:

$$T = \frac{1}{\sqrt{15}} \begin{pmatrix} 2 & 0 & 0 & 0 & 0 \\ 0 & 2 & 0 & 0 & 0 \\ 0 & 0 & 2 & 0 & 0 \\ 0 & 0 & 0 & -3 & 0 \\ 0 & 0 & 0 & 0 & -3 \end{pmatrix}. \tag{13.6}$$

These twelve block-diagonal matrices give the generators of the Standard Model, while the remaining twelve, with non-zero elements in the off-diagonals, are broken at high energy. The $SU(5)$ group is the smallest simple group that can contain the full Standard Model as a subgroup. This is not enough, however, to suggest its validity as a possible unified group: when a group breaks to a collection of subgroups, the representations of the larger group must likewise break to representations of the subgroups. The rules governing such breaking are known as the group's branching rules. While it is possible to calculate the branching rules for a given group, most often the branching rules that theorists are interested in have already been calculated and may be found in tables. An important point about such branching rules, though, is that a group typically has many possible subgroups to which it can break, depending on the particular vacuum occupied by the scalar field responsible. We must also check, then, that the observed representations of the Standard Model gauge group embed into representations of $SU(5)$. The fact that they do so, and in fact do so without the need to postulate new unobserved fermions, is a remarkable property of the model. In particular, when $SU(5)$ breaks to $SU(3) \otimes SU(2) \otimes U(1)$, two of its representations, the $\bar{5}$ and 10, break as follows:

$$\begin{aligned} \bar{5} &\mapsto (\bar{3}, 1)_{+2/\sqrt{15}} \oplus (1, 2)_{-3/\sqrt{15}}, \\ 10 &\mapsto (3, 2)_{+1/\sqrt{15}} \oplus (\bar{3}, 1)_{-4/\sqrt{15}} \oplus (1, 1)_{+2/\sqrt{15}}, \end{aligned} \tag{13.7}$$

where the numbers in brackets are the $SU(3)$ and $SU(2)$ representations, and the subscript gives the charge under $U(1)$. Comparing this with the fermionic Standard Model representations given in Section 12.1, we see that the $U(1)$ charges are consistently out by a factor of $\sqrt{5/3}$. This is not a problem, though, as we are free to redefine the hypercharge as long as we compensate with a similar rescaling of the $U(1)_Y$ coupling. It is the *relative* $U(1)_Y$ charges that matter, and these are exactly as we would hope. So we can identify the $\overline{5}$ as

$$
\overline{5} = \begin{pmatrix} \overline{d}_r \\ \overline{d}_b \\ \overline{d}_g \\ e^- \\ -\nu_e \end{pmatrix}, \tag{13.8}
$$

with a similar identification for the **10**. Another bonus of grand unification is an explanation of the anomaly cancellation from the previous chapter. There, we saw that the particular charge assignments for a single fermion generation were necessary for the cancellation of chiral anomalies, but no underlying reason was given. With a GUT, we can see that the charges *must* work out the way that they do. In turn, this explains the quantization of charge, which again has no explanation within the standard model. While we are free to redefine the coupling and charges for the $U(1)$ subgroup, the same cannot be said of the other subgroups, since the charges are determined by a normalization condition in this case. In fact, it is only when this relative $U(1)$ normalization factor is taken into account that the couplings appear to converge, as in Figure 13.2.

The off-diagonal matrices mix parts of the $SU(5)$ representations that do not mix in the Standard Model subgroup. Each of these matrices corresponds to a gauge boson that does not survive the symmetry breaking and, just as in the electroweak theory, these bosons gain a mass at the breaking scale. Despite this high mass, these new bosons should have a non-zero contribution to scattering amplitudes, though the contribution will be hugely suppressed at low energy. An example of such an interaction would be

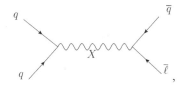

,

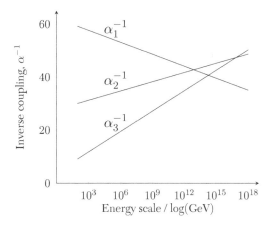

FIGURE 13.2 A plot of the inverse gauge couplings against energy scale. The initial values on the left are the observed values at an energy scale equal to the mass of the Z^0 boson. The gradients are determined from perturbation theory. The $U(1)_Y$ coupling's initial value and gradient are both corrected by the appropriate factor of $\sqrt{3/5}$.

where q is a quark, ℓ is a lepton, and X is one of the GUT bosons. In this way, we see that unified theories are capable of violating baryon number conservation and thus satisfying the third Sakharov condition.

Interactions like the one above also allow for the proton to decay to mesons and leptons, making the proton unstable. We can even quantify the proton lifetime approximately with a rather neat argument. The amplitude for the previous diagram is of the order

$$\mathcal{M} \approx \frac{g^2}{M_X^2}, \tag{13.9}$$

where g is the unified coupling and M_X is the X boson mass, since the mediator's momentum in such a process would be negligible compared with M_X. Using Equation 5.45, the decay rate of the proton must look something like

$$\Gamma = |\mathcal{M}|^2 \rho(E), \tag{13.10}$$

where $\rho(E)$ is the available state space. Since g is dimensionless, $\mathcal{M}$ has units of $[\text{Mass}]^{-4}$. The decay rate in natural units has units of $[\text{Mass}]$, so the state space must have units of $[\text{Mass}]^5$. Since the only mass-scale available to constrain the available states is the proton mass, we can deduce that $\rho(E)$ must be on the order of m_p^5. Thus we find

$$\Gamma = \frac{g^4}{M_X^4} m_p^5. \tag{13.11}$$

Since g is the value of the coupling at the unification point, its value is the value at which the three Standard Model couplings meet. Hence its value, or at least its order of magnitude, is known. Substituting in the relevant values, then, leads to a proton lifetime on the order of 10^{31} years. While this means we certainly do not need to worry about protons disappearing any time soon, it also rules out $SU(5)$ as a viable theory, since the current experimental constraints on the proton lifetime are around 10^{34} years. Although $SU(5)$ is ruled out in its simplest form, its supersymmetric extension is still viable. Additionally, there are many larger GUT groups that are still viable, such as E_6 and the Pati-Salam model, among others. One group in particular that has received much attention is $SO(10)$, which again has an elegant structure, not least because a whole Standard Model fermion generation with the addition of a right-chiral neutrino fits into a *single* 16 spinor representation.

13.3.1 Magnetic Monopoles

An interesting prediction of Grand Unified Theories is the existence of magnetic monopoles. Classically, of course, magnetic fields are produced by dipoles and there is no object that has a net magnetic charge. Cutting a bar magnet in two gives two new dipoles rather than isolated north and south magnetic poles. However, the possible existence of monopole is not ruled out by Maxwell's equations—it would merely require a slight modification of them. Monopoles arise naturally in GUTs as topological entities frozen into the vacuum state during symmetry breaking. Specifically, the electromagnetic field is related to one of the remaining group generators after the symmetry breaking phase transition. If this symmetry breaking "chooses" different vacuums in different regions of space, such that the differences wrap around a point, then there is no way to untwist the situation without restoring symmetry. The simplest example of this is the breaking of a Z_2 symmetry in a universe with only one spatial dimension. If the vacuum state $\langle\phi\rangle = v$ is chosen to the right of the origin but $\langle\phi\rangle = -v$ is chosen to the left, then continuity of the value of the field requires a high-energy state at the origin itself, in order to connect these two half-spaces. Similar situations exist with larger groups in larger spaces. In the case of the electromagnetic field, the energetic state is the monopole, and its magnetic charge is an artifact of us attempting to treat the field equally at all points, despite its actually having very different values as we move around the pole.

13.4   SUPERSYMMETRY

One of the most widespread additions to the Standard Model is supersymmetry (SUSY). There are several ways to introduce the concept of supersymmetry, all leading to the same set of ideas. One way is to postulate a symmetry between particles of different spin: in the same way that the quark colors are interchangeable through $SU(3)_C$ interactions, supersymmetry transformations would interchange bosons with fermions. My preferred way to introduce the concept, however, is as the answer to a question: can we combine the internal gauge symmetries of the Standard Model with the space-time symmetries of the Poincaré group in a non-trivial manner? Originally, the response to this question was thought to be "no." In fact, a key result due to Coleman and Mandula in 1967 rigorously proved that such a combination is impossible based on a few simple assumptions. That would seem to suggest that the question had been definitively answered and that no further work in this area were necessary. However, over the next few years, supersymmetric theories were formulated, initially in two dimensions and then in four, followed by the Haag-Łopuszański-Sohnius theorem in 1975, which modified the Coleman-Mandula result to allow for more general models. In particular, it relaxed one of the assumptions of its predecessor, allowing the generators of symmetry groups to obey anticommutation relations, $\{T_i, T_j\} = i f_{ijk} T_k$, generalizing the notion of a Lie algebra. What the generalized theorem showed was that the only way to include a non-trivial relationship between Poincaré generators and gauge-group generators was to have each independently obey anti-commutator relations with a set of new generators Q_α. These new generators are referred to as fermionic generators, since they are found to carry non-zero spin. In fact, these objects are Weyl spinors, and the index α runs over the spinor components. Notice that there is still no *direct* non-trivial interaction between the bosonic generators: this type of algebraic structure is known as a graded Lie algebra.

Recall that the Poincaré algebra consists of momentum operators P_μ and angular momentum operators $J_{\mu\nu}$. The supersymmetry algebra is essentially unique, up to the number of independent Q_α generators, and is given *schematically* by:

$$\begin{aligned}
\{Q, Q\} &\propto R, \\
\{Q, \overline{Q}\} &\propto P, \\
[Q, P] &= 0, \\
[Q, J] &\propto Q,
\end{aligned}$$
(13.12)

where P and J still obey any pre-existing commutation relations, and R is either 0 or some gauge-group generator that commutes with all others, depending on the particular theory in question. This demonstrates the essential properties of the graded Lie structure. Notice that the R-symmetry is the only part of the algebra that connects space-time and internal symmetries in a non-trivial way.

How, then, does this relate to the idea mentioned earlier of each boson having a fermionic partner? The answer to that comes from applying the new generator, Q, to a particle. Since Q itself carries a non-zero spin, its effect is to alter the spin of the particle by $\frac{1}{2}$, changing a boson to a fermion, and vice versa. If Q is a symmetry of the theory (that is, if Q can be applied to the Hamiltonian or Lagrangian without altering it), then apart from spin the properties of these partners must be identical. In the same way that we placed quarks of different colors into multiplets earlier, supersymmetry allows us to place particles of different spin into "supermultiplets."

Since the particles within a given multiplet should have the same properties (apart from spin in this case), and no such pairs of similar particles are known with different spin, supersymmetry predicts the existence of an as-yet undiscovered "superpartner" for each Standard Model particle. Each Standard Model fermion is twinned with a scalar, while each vector boson is twinned with a spin-$\frac{3}{2}$ "gaugino." This implies that the superpartners have masses sufficiently high that we have not yet seen evidence of them in collider experiments, and this is where we run into a potential problem with supersymmetry. One of the properties that the particles of a supermultiplet should share is mass. It's clear, then, that if supersymmetry does exist, it must be broken, much like the electroweak symmetry.

Given the lack of direct evidence for the idea, why is supersymmetry considered so enticing? The answer is that it is capable of solving some of the problems with the Standard Model that we discussed in the previous chapter. Supersymmetric theories are mathematically very elegant, and compelling for a number of reasons. Not least of these is their ability to solve the hierarchy problem. The reason for this is that the problematic divergent Feynman diagrams that result in the inflated Higgs mass are joined by a set of similar diagrams involving the superpartners. Since the relevant couplings are necessarily the same for particles of the same supermultiplet, and since the contribution to the amplitude from particles of differing spin is of opposite sign, these diagrams cancel each other off. For example, a diagram such as

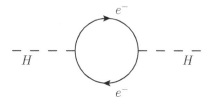

with an electron circulating in the loop, is canceled off by

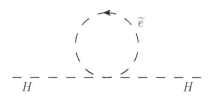

where the particle in the loop is now the electron's superpartner, the "selectron."[1] Some of the motivation for SUSY requires it to be broken at a fairly low energy. Specifically, introduction of SUSY at a sufficiently low energy alters the Standard Model β functions in such a way as to solve the "near miss" problem of the previous section. Since SUSY has not yet been observed at the LHC, some feel the future is beginning to look bleak for the theory, at least in its simplest form. However, these issues can be resolved with more sophisticated (or, depending on one's view, convoluted) models, and this is still very much an active area of research.

13.5   GRAVITONS

We have seen the particles of the Standard Model and described them with a set of equations: the Klein-Gordon equation for spin-0 particles, the Dirac equation for spin-$\frac{1}{2}$ particles, and the Maxwell equation for spin-1 particles. Supersymmetry predicts the existence of spin-$\frac{3}{2}$ partners of the gauge bosons, described by their own equation. So can we find equations for particles of higher spin? The answer is "yes" but only in a limited sense.

Just as the equation for spin-1 particles is simply the classical Maxwell equation for the electromagnetic field reinterpreted as a quantum wave-equation, we can attempt to reinterpret the equations of gravity. Since we are interested in relativistic wave equations, however, the gravitational equations we must consider are Einstein's equations of General Relativity. Without delving into too much detail, the general theory of relativity says that the geometry of space-time itself is affected by the presence of matter and energy, quantified in this case by a rank-2 symmetric tensor called the stress-energy-momentum tensor. That is, the space-time metric is deformed as

---

[1] The supersymmetric partners of the Standard Model particles are named in a systematic way: bosonic superpartners of Standard Model fermions are given an initial "s" as in selectron, squark, and sneutrino, while fermionic superpartners are given an "-ino" suffix, as in gaugino, wino, and photino. This leads to some of the more interesting names in all of physics!

$g_{\mu\nu} = \eta_{\mu\nu} + h_{\mu\nu}$, where $\eta_{\mu\nu}$ is the Minkowski metric of flat space-time and $h_{\mu\nu}$ is some deviation from flat space. This leads to a curvature in space-time that we perceive as gravity, with the amount of curvature quantified by $h_{\mu\nu}$. However, the equation describing the curvature is non-linear, and as such cannot be treated as simply as the Maxwell equation. In particular, the presence of non-linear terms means that self-interactions are automatically incorporated, which in turn prevents us from using simple plane waves as a basis. For small perturbations in free space, though, we can *approximate* the equation with a linear equation:

$$\partial^2 h_{\mu\nu} + \partial_\mu \partial_\nu h - \partial^\rho \left(\partial_\mu h_{\nu\rho} + \partial_\nu h_{\mu\rho} \right) = 0, \qquad (13.13)$$

where $h = \eta^{\mu\nu} h_{\mu\nu}$.

Reinterpreting this as a quantum wave-equation leads us to the prediction of gravitons. These are the particles that mediate the gravitational force, in the same way that the gauge bosons mediate the non-gravitational forces. Gravitons are found to be spin-2 particles. A particularly pleasing result, as we saw in Section 9.3, is that the Feynman rules for even-spin-mediated interactions between particles of like charge are always attractive, the relevant "charge" in this case being mass or energy.

A spin-2 particle is described by a symmetric rank-2 tensor (just as a spin-1 particle is described by a vector) and so has 10 components. Recall that the Maxwell equation had a gauge freedom that reduced the physical degrees of freedom to two, but that the introduction of a mass in the Maxwell equation required the Lorenz condition and removed gauge freedom, only reducing the degrees of freedom to 3. If we were to introduce a mass to the above equation, we would find additional constraints that reduced the degrees of freedom to 5, as we would expect for a spin-2 particle. However, the graviton is massless, and has its own version of "gauge" symmetry known as general covariance. In particular, we can make a transformation of the form:

$$h'_{\mu\nu} = h_{\mu\nu} + \partial_\mu \chi_\nu(x) + \partial_\nu \chi_\mu(x), \qquad (13.14)$$

for an arbitrary vector-valued function $\chi_\mu(x)$, without altering the equation of motion. When working with the Maxwell equation, we chose to impose a condition such that all but the d'Alembertian term ($\partial^2 A^\mu$) vanished. Can we do something similar here using the general covariance symmetry? That is, can we choose a gauge in which

$$\partial_\mu \partial_\nu h = \partial^\rho \left(\partial_\mu h_{\nu\rho} + \partial_\nu h_{\mu\rho} \right) \tag{13.15}$$

holds? Since both sides are symmetric in μ and ν, we can rewrite the left-hand side as

$$\frac{1}{2} \left(\partial_\mu \partial_\nu h + \partial_\nu \partial_\mu h \right) = \partial^\rho \left(\partial_\mu h_{\nu\rho} + \partial_\nu h_{\mu\rho} \right), \tag{13.16}$$

and it is then clear that the answer is "yes" if $2\partial^\rho \partial_\nu h_{\mu\rho} = \partial_\nu \partial_\mu h$, or

$$2\partial^\rho h_{\mu\rho} = \partial_\mu h. \tag{13.17}$$

Substituting in Equation 13.14, we find that this is satisfied by the transformed tensor, $h'_{\mu\nu}$, as long as we choose χ_μ as a solution to

$$\partial^2 \chi_\mu = \frac{1}{2} \partial_\mu h - \partial^\rho h_{\mu\rho}. \tag{13.18}$$

Since such a solution always exists, but requires a condition for each component of χ_μ, we are free to simplify the wave equation to

$$\partial^2 h_{\mu\nu} = 0, \tag{13.19}$$

reducing the degrees of freedom to six in the process.

This simplified equation has the obvious plane-wave solutions

$$h_{\mu\nu} = \varepsilon_{\mu\nu} e^{-ik_\rho x^\rho}, \tag{13.20}$$

where k_ρ is a massless four-momentum and $\varepsilon_{\mu\nu}$ is a polarization tensor. However, we find that there is a residual gauge invariance of the form:

$$h'_{\mu\nu} = h_{\mu\nu} + \left(k_\mu \xi_\nu + k_\nu \xi_\mu \right) e^{-ik_\rho x^\rho}, \tag{13.21}$$

since $h'_{\mu\nu}$ in this case still obeys the condition in Equation 13.17. So the degrees of freedom are further reduced to two. These two degrees of freedom correspond to the two possible helicity states

(± 2) of a graviton. Classically, of course, these solutions correspond to gravitational waves: deviations from the flat background metric, which propagate through space. These manifest themselves as a warping of space as the wave passes through, with space being alternately compressed and stretched along perpendicular axes, both also perpendicular to the direction of propagation. The two helicity states of the graviton correspond to the two independent polarizations of these waves, which as linear polarizations can be thought of as having their axes arranged in a "+" configuration and a "×." For many years since first predicted in the early twentieth century, these waves were to General Relativity what the Higgs boson was to the Standard Model: namely, a key prediction of the theory that had yet to be observed experimentally. This changed, of course, with their detection by the Laser Interferometer Gravitational-Wave Observatory (LIGO) experiment in 2016.

13.5.1 Can We Go Further than Spin-2?

The Feynman rule for the propagator of a spin-1 particle contains a part that remains constant with increasing energy and momentum.[2] This would suggest that high momentum transfers between particles are just as likely as low momentum transfers, leading to a divergent amplitude and cross-section. The situation is even worse for higher spins, where the corresponding term actually increases with increasing momentum transfer, making infinitely energetic interactions infinitely likely. This is clearly nonsensical. The only thing that saves a theory of a spin-1 particle is a conserved current for the particle to couple to: for example, the electromagnetic current for the photon. This conserved current allows us to gauge away the unphysical behavior. Likewise, a spin-$\frac{3}{2}$ particle, such as a gaugino, only gives a sensible theory if there is an appropriate current for the particle to couple to: in this case the appropriate current is that associated with

---

[2] For simplicity, this has been left this part out of the Feynman rules listed in this book, since it is gauge-dependent: the form of the rules given in this book are computed in a particular gauge. This does not affect their validity, of course, since any physical result must hold regardless of the chosen gauge.

supersymmetric transformations, known as the supercurrent. A spin-2 particle requires a symmetric traceless rank-2 conserved current, which is provided in the form of the stress-energy-momentum tensor. For this reason, if a (fundamental) spin-2 particle is discovered, it *must* couple to the stress-energy-momentum tensor: in other words, it *must be* the graviton. Since no conserved currents are known that have more than two Lorentz indices, it seems that we cannot have a fundamental particle of spin greater than 2.

13.5.2 Problems with Gravity

All of the above is not to say that the theory of the graviton is perfect, however. There are additional problems relating to the high-energy behavior of the theory (it is not renormalizable) that mean that the previous description of the graviton cannot be the correct or full description of gravity. This is why a full quantum description of gravity is still not yet understood. There are candidates, such as string theory and loop quantum gravity, but none of these is yet considered complete or mainstream.

13.6 AXIONS

As discussed in Section 12.6.3, there is no way in the Standard Model to explain the smallness of the $\mathbb{C}\mathbb{P}$-violating parameter in the QCD Lagrangian. If a chiral fermion undergoes a phase rotation, the value of the θ parameter is altered. This does not make θ unphysical, however, since such a rotation shifts the value into the Yukawa terms. We cannot simultaneously rotate away θ and retain the appropriate degrees of freedom in the Yukawa matrices. This is not a problem if one or more of the quarks is massless, since θ can effectively be rotated into the zero component of the Yukawa matrix, but this is ruled out experimentally as all quark flavors have non-zero mass. However, this gives a hint as to a possible solution to the strong $\mathbb{C}\mathbb{P}$ problem: a massless fermion has a global axial $U(1)$ symmetry, $e^{i\gamma^5 \xi}$, and it is this that allows for free phase rotation. Peccei and Quinn proposed additional fields to allow such an axial symmetry without modifying the fermion sector of the theory. Specifically, a new complex scalar field is introduced with the appropriate symmetry, which is then spontaneously broken. While we do not intend to delve into the math here, it can be shown that this is sufficient to restore CP-invariance in QCD. Since the symmetry is global, this leads to a Goldstone boson: a pseudo-scalar particle called the axion. In fact, because chiral symmetry is anomalous, the new $U(1)$ is not actually a symmetry of the full quantum theory and so the axion is a pseudo-Goldstone boson with a non-zero mass. The properties of the axion are found to be related to each other. As such, phenomenological constraints give estimates for all the axion properties: the particle is hypothesized to be very light ($m \sim 10^{-3}$ eV) with a weak coupling to fermions and a long lifetime.

13.7   DARK MATTER

Galactic rotation requires that each galaxy be composed of around six times more matter than is currently observed, and the missing mass is known as dark matter, since it emits no light. Dark matter, then, is that matter in the universe that we do not detect directly through electromagnetic interactions but whose existence is inferred through its gravitational effects. An obvious candidate for dark matter is black holes in the galactic halo or other "Massive Compact Halo Objects" (MACHOs). However, this possibility has been essentially ruled out in more ways than one. First, a direct search for such objects in our own galaxy has ruled out the majority of possible MACHOs of a wide range of sizes. Second, cosmological models of the formation of the first atoms, which agree with the observed ratios of the lightest elements, require the amount of matter of a type we would consider "ordinary" to be equal to the amount observed. Thus the nature of dark matter is widely considered to be some sort of new particle. Furthermore, models of galactic formation require that these particles be "cold." Cold in this context has less to do with actual temperature than it does with speed: dark matter particles must be moving at non-relativistic speeds. What we need, then, is a particle or particles with a non-zero mass that interacts only weakly with ordinary matter. Such particles are thus known as "Weakly Interacting

Massive Particles" or WIMPs. Crucially, Standard Model neutrinos are ruled out as their mass is too small, making them relativistic and only viable as a "hot dark matter" candidate.

There are many candidates for dark matter, and we have already seen some of them. Constraints on the axion mass also require it to be very weakly interacting. Its mass is predicted to be extremely small but non-zero. We have already stated that neutrinos are ruled out by their small mass and the axion mass is predicted to be considerably smaller. However, a crucial difference is that the QCD phase transition in the early history of the universe would have produced large numbers of axions that were not in thermal equilibrium with the rest of the universe and were already non-relativistic, despite their mass. This makes the axion a very appealing dark matter candidate, as many pieces appear to fit. The possibility that axions make up dark matter is currently being actively tested in the Axion Dark Matter Experiment (ADMX). This experiment uses a resonant cavity tuned to the possible axion mass together with a strong magnetic field to prompt the decay of axions to pairs of photons. In this way, a range of axion masses has already been ruled out and it is hoped that the experiment will ultimately either confirm the axionic nature of dark matter or rule out all otherwise possible masses.

We have already mentioned the right-chiral cousins of the Standard Model neutrinos, and have seen that they would be neutral under all Standard Model interactions. Furthermore, the mass eigenstate consisting almost entirely of a right-chiral component would have a large mass. At first sight, this makes these "sterile" neutrinos the obvious candidate for a dark matter particle. However, this idea is not without its problems. Specifically, the heavy neutrino states have a small left-chiral component, allowing for weak interactions. This means that the sterile neutrinos can decay by photon emission via channels like

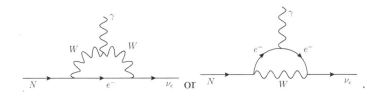

A sterile neutrino with too large a mass will decay at an appreciable rate due to the same 5th power argument that we used for proton decay in Section 13.3. Since dark matter is required in abundance, such decays would be detectable as an excess of X-ray photons of energy equal to the mass of the sterile neutrino (since the light neutrino mass is negligible). That these decays have not been detected provides an upper bound for the mass of the sterile neutrino if this really is the nature of dark matter. A mid-scale neutrino with a mass on the order of a few keV still provides a viable candidate for so-called "warm" dark matter: that is, dark matter of a range of speeds intermediate between those of hot and cold matter. Opinion on the validity of such models is divided, however, since they typically require more tuning than alternative models, with too small or large a mass leading to further problems.

In order to explore a final dark matter candidate, we return to the concept of R-symmetry briefly introduced in Section 13.4: the part of the internal symmetry group that has a non-trivial interaction with the space-time generators. In some SUSY models, the $U(1)$ R-symmetry is broken to a Z_2 symmetry, with different particles of a supermultiplet taking different values under this "R-parity." Specifically, the Standard Model particles take the value $+1$ and their superpartners take the value -1. If R-parity is conserved, this means that any allowed Feynman diagram vertices necessarily include an even number of superpartners. In turn, this means that the lightest of the superpartners must be a stable particle. If this lightest supersymmetric particle (LSP) is electrically neutral and color neutral, then it is a candidate for dark matter.

Each of the particles mentioned previously is theorized for another reason and plays a secondary role as a dark matter candidate. There is another possibility: that the true nature of dark matter is simply some other particle that has the necessary properties of mass and weak interaction with "ordinary" matter. We already know of three particles—the neutrinos—that have no interaction with the $U(1)_Y$ or $SU(3)_C$ parts of the Standard Model, so there is of course nothing to prevent the existence of whole "hidden sectors" of additional particles, not interacting with *any* of the Standard Model gauge groups. It is tempting to imagine that such hidden sectors could have their own gauge groups and interactions and even their own version of atoms, molecules, and so forth. However, the evidence for dark matter suggests that it is arranged only in diffuse uniform halos throughout galaxies, ruling out this rather fanciful possibility.

We can see, then, that there are multiple potential explanations for dark matter. However, there is no reason to believe that only one of these possibilities is realized in nature. It is plausible (indeed, likely) that dark matter is actually composed of a combination of different particle types. As such, many models have been proposed that incorporate a variety of dark matter constituents, and which seek to place constraints on the contribution of each component.

We should also mention that there is an alternative to dark matter. Some have proposed a modification of Newtonian gravity on large distance scales to account for the observed phenomena most commonly ascribed to dark matter, such as the galactic rotation curves. Specifically, a gravitational field that drops off faster than $1/r^2$ at long distances can also explain the observed rotation curves. Others have argued against the naturalness of such modified gravitational theories, pointing out that many still require additional particle content to make them work, which may seem to undermine the reason for their introduction. There is no reason to suppose that all particles in the universe should interact through the same combination of gauge groups as our own atoms. It almost seems obvious that there should be additional particle species that we have not yet detected, making dark matter the more natural addition.

13.8 DARK ENERGY AND INFLATION

As is clear from the preceding sections, the study of the universe at the smallest scale through particle physics has a surprising overlap with its study at the largest scales through cosmology. This relationship between apparently disparate areas of physics goes further when we consider dark energy and inflation. One could very reasonably argue that these are areas beyond the scope of particle physics, but it is worth pushing boundaries a little to highlight some of the common ground between disciplines.

13.8.1 Inflation

There are a number of problems with the Big Bang theory that may be solved by the idea of inflation. First, if Grand Unified Theory is correct, then we would expect to see magnetic monopoles, and yet none has ever been observed. Second, the cosmic microwave background (CMB) is found to have essentially no variation in temperature in different directions in space—that is, the CMB is found to be isotropic. If we consider two regions of the CMB in opposite directions from us in space, then clearly the light from each has had time to reach us since it was produced. However, light has not yet had time to travel between the two regions: they are beyond each other's cosmic horizon. Since nothing can communicate faster than light speed, we should not expect these two regions to be in thermal equilibrium, and yet we find that they are. A third problem is that the universe is observed to be flat. General relativity tells us that space-time itself curves in the presence of matter and energy, and cosmological models show that a non-zero curvature of the universe as a whole increases in magnitude over time, quickly moving away from zero. Despite this, we appear to be in a universe with zero curvature, or at least a curvature very close to zero. To find ourselves in this situation $\sim 10^{10}$ years after the Big Bang requires incredible fine-tuning of the initial curvature.

Inflation is the idea that the universe underwent a rapid expansion early in its history—in fact growing exponentially for just the first 10^{-34} seconds, but increasing in size by a factor of around 10^{27}. Such a rapid growth is able to solve the three problems listed above. The exotic particles such as monopoles are spread so thinly across the universe by the expansion that the expected number of such particles within a given cosmic horizon is less than one. The horizon problem is solved simply by the fact that the regions of space in question really *were* once causally connected before inflation, since they were much closer together. Similarly, the expansion smooths out the wrinkles in the universe pushing the curvature toward zero, so much so that the subsequent exponential departure from zero during normal expansion has still not given the universe a detectable curvature.

So what does this have to do with particle physics? The answer is that the mechanism for inflation is commonly hypothesized to be one which we have already met—namely, symmetry breaking. The assumption is that some scalar "inflaton" field is initially in a high-energy symmetric state. During a phase transition, this field then acquires a vacuum expectation value, decreasing its energy in the process. The excess energy is what drives the sudden expansion of space. There are many details here that are being glossed over, including that the field's potential cannot be as simple as the ϕ^4 that we considered for the Higgs mechanism, but must instead decrease toward its vev at a much shallower gradient.

13.8.2 Dark Energy

Observations of distant supernovae in the late 1990s showed that we are now in another period of accelerating universal expansion. There appears to be some form of energy driving on the expansion at an ever-increasing rate. Since the exact nature of this energy is unknown, in analogy with dark matter, it has been named dark energy. There is no ordinary matter capable of producing this acceleration, since standard cosmological models tell us that all matter with positive energy also provides a positive pressure, resulting in a braking effect on the expansion. The fact that positive pressure acts to slow down the expansion of the universe can seem counter intuitive upon first encountering it. However, it is important to realize that the universe does not expand *into* anything, so we are not considering this pressure to push back on any kind of boundary. Rather, its effect is on the internal energy density of the contents of the universe, through the work done due to expansion. Recall that in thermodynamics, an expanding gas of pressure p and volume V does work pdV against its surroundings. Similarly, if the matter in the universe has a positive pressure, it does work as the universe expands. With no surroundings in this case, the energy goes instead into increasing gravitational potential energy. Before this energy was converted through the work of the expanding gas, however, it was kinetic energy. So we see that positive pressure implies a decrease in kinetic energy over time and hence a decelerating universal expansion. Therefore, we require a substance with *negative* pressure to account for an accelerating universe. Fortunately, particle physics, or more accurately quantum field theory, provides at least two possibilities.

The first possibility is that there is a "cosmological constant," Λ, representing essentially the energy density of free space. In Einstein's general theory of relativity, such a constant is an allowed but not a required term in the field equations. When it was first introduced by Einstein himself, originally as a means of ensuring a *static* universe, it did not have a physical interpretation. Instead it was simply a mathematical tool, included to make the theory behave, and later removed again when the universe was found to be expanding. Now that the constant has been revived, to account for the accelerating expansion, there is still the question of its physical meaning. As we have seen in earlier chapters, particles are described in quantum field theory as the quantized excitations of various fields. These fields, though, do not need to be in an excited state to have a nonzero energy. That is to say that the vacuum is predicted in quantum field theory to have a "zero-point energy," which appears to be the ideal candidate for the cosmological constant. However, the downside to this interpretation is that the value of the zero-point energy predicted by quantum field theory is proportional to the cutoff imposed in order to tame the infinities of the theory. If the Standard Model is the low-energy effective theory of something more fundamental, then this cutoff is in turn related to the energy scale of this underlying new physics. This leads to the prediction that the zero-point energy should be many orders of magnitude greater than the observed cosmological constant.

The second possible explanation for dark energy arising from particle physics is more closely related to the inflation model discussed in the previous section. In particular, it is possible that the universe is undergoing a phase transition at this moment, with some scalar field settling into its lowest-energy state. Such models are known as "Quintessence" models, in honor of the ancient four-element view of the universe, with quintessence being the mysterious *fifth element*. An important difference between quintessence and the cosmological constant is that the process is localized in time. That is, the amount of dark energy will vary as the phase transition begins and ends. On the other hand, the cosmological constant is just that: a constant. While energy is liberated through the inflation-like behavior of the scalar field, thus driving the universal expansion, in the case of the cosmological constant, the energy density of free space remains invariant throughout history. This does not mean that the cosmological constant implies a fixed acceleration rate, however. This is crucial, since observations show that the universe has not been accelerating since its beginning: the accelerating phase only developed later. In the case of a cosmological constant, this is a natural state of affairs. When the universe is young and dense, its expansion is dominated by the matter and radiation within it. It is only as it expands that the sheer volume of empty space, with its tiny contribution to the energy density, is able to dominate over matter and drive up the expansion rate.

13.9   THE FUTURE OF PARTICLE PHYSICS

It is of course impossible to say where the future will lead our understanding of particle physics. In the short- to mid-term, it seems likely that experiments will deepen our understanding of the exotic hadron states discussed in Section 10.6, including pentaquarks and tetraquarks. Likewise, we can expect to see a gradual increase in our knowledge of the quark-gluon plasma state of matter as well as the rest of the QCD phase diagram. More speculatively, the experimental discovery that overturns our current understanding of the field could come at any time. Some such potential discoveries would serve to steer particle physics into a particular direction that has already been theorized. For example, if the supersymmetric partner of a Standard Model particle were discovered tomorrow, then its properties would serve to narrow down the parameter space of an already well developed area of study. On the other hand, it is also entirely plausible that a new resonance may be discovered whose properties are incompatible with any of those particles yet hypothesized.

Such a discovery has the potential to alter our current understanding considerably, in any number of different ways. Depending on the interactions of such a particle, considerations such as the necessity of anomaly cancellation may immediately predict the existence of a whole family of related particles. The incomplete nature of the Standard Model, as explored in this chapter, suggests that there may well be some kind of unpredicted discovery. However, given that such a discovery could occur at essentially any energy scale, it is difficult to say whether the next big moment in particle physics will come from the current generation of accelerator experiments or from a future generation, or indeed if the Standard Model is robust to such high energy scales that there are important potential discoveries that lie beyond the energy capability of even hypothetical machines.

On the subject of new experimental searches, there are already interesting developments at various stages of planning, design, and construction. In the short term, the LHC is expected to undergo an upgrade before 2025 to increase its luminosity, and hence also its data output, by an order of magnitude. More long-term plans include the International Linear Collider (ILC) project. This collider, to be constructed in the Kitakame Mountains in northern Honshu, Japan, is planned to collide electrons and positrons at energies of 1 TeV, around five times the collision energy of the previous record holder for lepton collisions, the LEP. To achieve this, the ILC is to be around 50 km in length. Compare this with the longest existing linac, the Stanford Linear Collider (SLAC), at a length of 3.2 km, and one can appreciate the mammoth task being considered. As discussed in Chapter 5, the energy of lepton collisions is much more controlled than those of hadron collisions, in which the actual collision energy is dependent on the fraction of momentum carried by the individual parton actually involved in the interaction. However, since leptons have much smaller masses, they suffer a much greater rate of synchrotron radiation when in a circular accelerator. It is clear, therefore, why the next generation of lepton colliders will most likely be linear. Another large-scale lepton linac in the design stage is the Compact Linear Collider (CLIC), whose ingenious design will, it is hoped, allow for similar energies to the ILC to be achieved over a shorter distance. Beyond this, CERN is already conducting studies for the design of the next big thing in accelerator physics. This Future Circular Collider (FCC) project is currently in the very early stages of development, but the timing of the project is worthy of mention. The LHC and its upgrade, the high-luminosity LHC, are expected to continue to produce results until around 2035. However, history suggests that the design and construction of each new generation of experiment can take around 30 years. As such, the FCC project is likely to be the successor to the ILC and CLIC rather than the direct successor of the LHC.

All of this uncertainty does not prevent us from considering the likelihood of various experimental discoveries, though. For example, while opinion is divided on the subject of supersymmetry, it is safe to say that the simplest supersymmetric extensions of the Standard Model require the superpartner masses to be relatively low. The higher the energy that is probed without finding signatures of these particles, the less sturdy the foundations of these models. This then begs the question: how much can we extend a theory before it begins to look incredible?

While not a part of the Standard Model, arguably the most obvious "missing piece" of our current understanding of the world is the graviton. As we have seen, all other processes that are continuous in classical mechanics become discrete in quantum mechanics. The same should of course be true of the gravitational force, and so we should expect the graviton to be a real particle. So should we be looking for the graviton in our collider experiments? Unfortunately, the answer is "no." Gravity's weakness in comparison to the forces of the Standard Model, with a coupling of only around 10^{-42} times that of electromagnetism, means that individual gravitons are essentially unobservable using standard particle physics techniques. Unless a fundamentally new approach to experiment is discovered, then all we can hope to observe is the classical effect of large numbers of gravitons—which, thanks to LIGO's experimental verification of gravitational waves, we have already achieved. Now with further observations in this field, we can hope to deepen our understanding of classical gravity, and thereby indirectly develop our understanding of quantum gravity. In particular, while we almost certainly will not directly detect individual gravitons, precise analysis of gravitational waves may demonstrate quantum corrections to classical models.

Theoretical developments are somewhat easier to imagine than experimental. While the history of physics as a whole has demonstrated that there is the occasional great leap forward, such as Einstein's special theory of relativity or Planck's hypothesis of the energy quantum, more commonly progress in theoretical physics is a gradual development of the ideas that have already been laid down. So it seems reasonable that, in the near future at least, progress will mostly consist of building and refining models, as well as the improvement of existing tools and techniques. With this in mind, it seems a safe bet that lattice gauge theory will see the greatest advances in the near future. Computing power grows exponentially year upon year and lattice calculations are the area of particle physics most heavily reliant on large-scale computation. The history of lattice QCD has already effected a remarkable improvement in our understanding of the strong interactions and this trend seems set to continue.

The mathematical foundation on which particle physics is built—quantum field theory—has proven incredibly successful in describing the smallest scales of our world as we observe it. However, through predictions such as the Landau pole and the non-renormalizability of gravitational interactions, it essentially predicts its own demise at the shortest length scales and highest energies. What lies beyond this is open to speculation: whether this be string theory, loop quantum gravity, or some as-yet unconsidered and revolutionary approach. But since field theory also demonstrates the validity of the effective field theory approach, it shows itself to be the correct model to approximate whatever physics may actually lie beneath the surface. As such, in much the same way that day-to-day engineering tasks require only Newtonian mechanics, we can be sure that whatever the true underlying high-energy behavior may be, the techniques of field theory and particle physics discussed in this book will remain the standard for most applications.

EXERCISES

1. **(a)** Show that the eigenvalues of the neutrino "see-saw matrix" (Equation 13.2) are approximately m_M and $-m_D^2/m_M$.
   (b) Hence show that the low-mass neutrino is almost entirely left-chiral and the heavy neutrino is almost entirely right-chiral.
   (c) Assuming a neutrino Dirac mass of the same order as the electron, and based on current experimental limits of the physical neutrino mass, estimate the Majorana mass.

2. A neutrino mass state propagating in the x direction at time t can be expressed as

$$|\nu_1(t)\rangle = e^{-i(Et - px))} |\nu_1(0)\rangle,$$

   where $|\nu_1(0)\rangle$ is the same state at an initial time.
   (a) Since the neutrinos are close to massless, we can assume that they travel at $v \approx 1$, and that $E \gg m$. By expanding the expression for the relativistic momentum in terms of E and m, show that the above state can be written as

$$|\nu_1(t)\rangle = e^{-im_1^2 t/2E} |\nu(0)\rangle.$$

   (b) Assuming only two neutrino flavors, write the flavor states as appropriate orthonormal linear combinations of the mass states. Following a similar procedure as for the kaon mixing in Section 12.3.2, find the rate of oscillation between flavor states in this case.
   (c) How does this generalize in the case of three neutrino flavors?

3. Without performing any calculations, explain why the selectron ($\tilde{e}$) loop would cancel off the electron (e^-) loop in corrections to the Higgs mass.

4. If we wish to include a mass term in the equation for a spin-2 particle, it must be modified according to

$$\partial^2 h_{\mu\nu} + \eta_{\mu\nu}\partial^\rho\partial^\sigma h_{\rho\sigma} + \partial_\mu\partial_\nu h - \eta_{\mu\nu}\partial^2 h$$
$$- \partial^\rho \left(\partial_\mu h_{\nu\rho} + \partial_\nu h_{\mu\rho}\right) + m^2 \left(\eta_{\mu\nu} h - h_{\mu\nu}\right) = 0.$$

By acting with appropriate differential operators and the Minkowski metric, show that this equation has five degrees of freedom, as expected for a massive spin-2 particle. Hence find a simplified version of this equation obeyed by $h_{\mu\nu}$ when the appropriate constraints are imposed.

ELEMENTARY PARTICLE PROPERTIES AND OTHER USEFUL QUANTITIES

This appendix lists the properties of the fundamental particles and interactions. All quantities are given in natural units. Note that the lifetime in SI units and the decay width in natural units are related by

$$\text{lifetime in seconds} = \frac{6.58 \times 10^{-22}}{\text{decay width in MeV}}.$$

Leptons

Name	Charge	L_e	L_μ	L_τ	Mass (MeV)	Decay width (MeV)
ν_e	0	1	0	0	(~ 0)	(stable)
ν_μ	0	0	1	0	(~ 0)	(stable)
ν_τ	0	0	0	1	(~ 0)	(stable)
e^-	-1	1	0	0	0.51	(stable)
μ^-	-1	0	1	0	105.66	5.99×10^{-16}
τ^-	-1	0	0	1	1776.8	4.53×10^{-9}

Quarks

The following table lists the quark properties. Here, I and I_3 refer to isospin, rather than weak isospin. The first mass listed is the current mass, with the constituent quark (effective) mass given in brackets. Since all quarks but the top exist only in hadrons, there is no decay width to list. Likewise there is no effective mass to list for the top. The decay width of the top quark is predicted to be ~ 2632 MeV.

Name	Charge	I	I_3	S	C	$\widetilde{B}$	T	Mass (MeV)
u	$+\frac{2}{3}$	$\frac{1}{2}$	$+\frac{1}{2}$	0	0	0	0	2.3 (336)
d	$-\frac{1}{3}$	$\frac{1}{2}$	$-\frac{1}{2}$	0	0	0	0	4.8 (340)
s	$-\frac{1}{3}$	0	0	-1	0	0	0	95 (486)
c	$+\frac{2}{3}$	0	0	0	1	0	0	1290 (1550)
b	$-\frac{1}{3}$	0	0	0	0	-1	0	4180 (4730)
t	$+\frac{2}{3}$	0	0	0	0	0	1	172440

Bosons

Name	Charge	Spin	Mass (MeV)	Decay width (MeV)
γ	0	1	0	(stable)
$W^{\pm}$	± 1	1	80385	2085
Z^0	0	1	91188	2495
g	0	1	0	—
H	0	0	125090	4.218

Some Other Useful Quantities

$$e = 0.303 \qquad \theta_C \approx 13.02°$$
$$g_2 = 0.631 \qquad \theta_w \approx 28.7°,$$
$$g_3 = 1.23 \qquad G_F = 1.166 \times 10^{-11} \text{ MeV}^{-2}.$$

Note that the value of a coupling is only valid at a particular energy scale. The values of e and g_2 given here are in the low-energy limit.

The value of g_3 is given at the M_Z scale since strong interactions are confined at low energy and perturbative calculations are invalid. To lowest order, the inverse coupling $\alpha_i = g_i^2/4\pi$ depends linearly on the natural logarithm of the energy scale, with gradients

$$\alpha_1 : -25/6 \qquad \alpha_2 : 19/6 \qquad \alpha_3 : 7.$$

We also list here the magnitudes of the entries of the quark mixing (CKM) matrix:

$$\begin{pmatrix} |V_{ud}| & |V_{us}| & |V_{ub}| \\ |V_{cd}| & |V_{cs}| & |V_{cb}| \\ |V_{td}| & |V_{ts}| & |V_{tb}| \end{pmatrix} = \begin{pmatrix} 0.974 & 0.225 & 0.003 \\ 0.225 & 0.973 & 0.041 \\ 0.009 & 0.040 & 0.999 \end{pmatrix}.$$

FEYNMAN RULES

The Feynman rules have already been explored in depth in the relevant chapters. This appendix summarizes the rules for ease of use.

B.1   PROPAGATORS AND EXTERNAL PARTICLES

Each particle below is assumed to carry momentum p^μ and mass m. The additional factors in square brackets are only relevant in the QCD case.

$- - - - - \rightarrow\bullet$	$=$	incoming scalar	$=$ 1
$\bullet - - - - - -$	$=$	outgoing scalar	$=$ 1
$\bullet - - - - - \bullet$	$=$	internal scalar	$=$ $\dfrac{i}{p^2 - m^2}$
$\longrightarrow\bullet$	$=$	incoming fermion	$=$ $u(p)\,[c_A]$
$\bullet\longrightarrow$	$=$	outgoing fermion	$=$ $\bar{u}(p)\,[c_A^\dagger]$
$\longleftarrow\bullet$	$=$	incoming antifermion	$=$ $\bar{v}(p)\,[c_A^\dagger]$
$\bullet\longleftarrow$	$=$	outgoing antifermion	$=$ $v(p)\,[c_A]$
$\bullet\longrightarrow\bullet$	$=$	internal fermion	$=$ $\dfrac{i\left(\not{p}+m\right)}{p^2 - m^2}\,[\delta_{AB}]$
$\wedge\!\wedge\!\wedge\!\wedge\!\wedge\bullet$	$=$	incoming vector boson	$=$ $\varepsilon^\mu(p)\,[G_a]$
$\bullet\wedge\!\wedge\!\wedge\!\wedge\!\wedge$	$=$	outgoing vector boson	$=$ $\varepsilon^{*\mu}(p)\,[G_a^\dagger]$

•/\/\/\/\/\• $\quad=\quad$ internal vector boson $\quad=\quad \dfrac{-ig_{\mu\nu}}{p^2-m^2}\,[\delta_{ab}]$

B.2   FERMION INTERACTIONS WITH GAUGE BOSONS

Here, q is the charge of the fermion, θ_w is the weak mixing angle, and e, g_2 and g_3 are the electromagnetic, weak, and strong couplings respectively. Gauge bosons carry Lorentz indices denoted μ and color is denoted a. In the weak quark interactions, V_{ij} is the relevant component of the CKM mixing matrix. Finally, I_3 and Y_R in the last diagram are the third component of weak isospin of the fermion's left-chiral component, and the hypercharge of the right-chiral component.

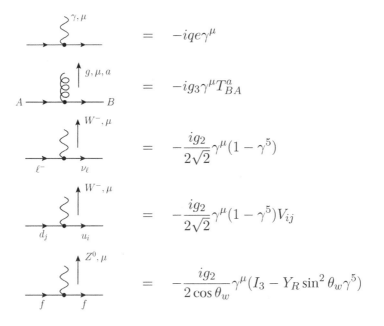

B.3 GAUGE BOSON SELF-INTERACTIONS

Here g_2 and g_3 are the weak and strong couplings respectively, θ_w is the weak mixing angle, and f^{abc} are the structure constants for $SU(3)$. Lorentz indices are denoted with Greek letters and color indices with Latin letters.

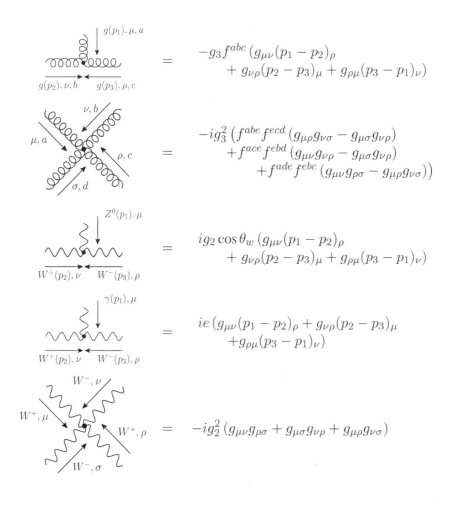

$$-g_3 f^{abc} \left(g_{\mu\nu}(p_1 - p_2)_\rho \right.$$
$$\left. + g_{\nu\rho}(p_2 - p_3)_\mu + g_{\rho\mu}(p_3 - p_1)_\nu \right)$$

$$-ig_3^2 \left(f^{abe} f^{ecd} \left(g_{\mu\rho}g_{\nu\sigma} - g_{\mu\sigma}g_{\nu\rho} \right) \right.$$
$$+ f^{ace} f^{ebd} \left(g_{\mu\nu}g_{\nu\rho} - g_{\mu\sigma}g_{\nu\rho} \right)$$
$$\left. + f^{ade} f^{ebc} \left(g_{\mu\nu}g_{\rho\sigma} - g_{\mu\rho}g_{\nu\sigma} \right) \right)$$

$$ig_2 \cos\theta_w \left(g_{\mu\nu}(p_1 - p_2)_\rho \right.$$
$$\left. + g_{\nu\rho}(p_2 - p_3)_\mu + g_{\rho\mu}(p_3 - p_1)_\nu \right)$$

$$ie \left(g_{\mu\nu}(p_1 - p_2)_\rho + g_{\nu\rho}(p_2 - p_3)_\mu \right.$$
$$\left. + g_{\rho\mu}(p_3 - p_1)_\nu \right)$$

$$-ig_2^2 \left(g_{\mu\nu}g_{\rho\sigma} + g_{\mu\sigma}g_{\nu\rho} + g_{\mu\rho}g_{\nu\sigma} \right)$$

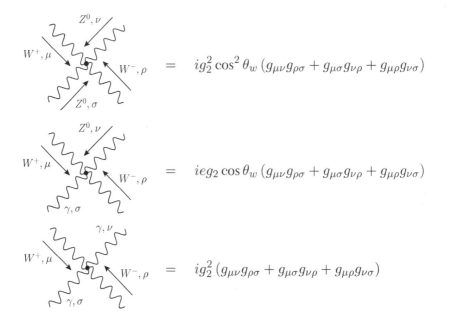

$$= ig_2^2 \cos^2 \theta_w \left(g_{\mu\nu}g_{\rho\sigma} + g_{\mu\sigma}g_{\nu\rho} + g_{\mu\rho}g_{\nu\sigma} \right)$$

$$= ieg_2 \cos \theta_w \left(g_{\mu\nu}g_{\rho\sigma} + g_{\mu\sigma}g_{\nu\rho} + g_{\mu\rho}g_{\nu\sigma} \right)$$

$$= ig_2^2 \left(g_{\mu\nu}g_{\rho\sigma} + g_{\mu\sigma}g_{\nu\rho} + g_{\mu\rho}g_{\nu\sigma} \right)$$

B.4   HIGGS BOSON INTERACTIONS

The notation here is as in the last section with the addition of m_H, M_W and M_Z which denote the masses of the Higgs, $W^\pm$ and Z^0 bosons respectively. The m in the first diagram refers to the fermion mass.

$$= -\frac{ig_2}{2}\frac{m}{M_W}$$

$$= ig_2 M_W g_{\mu\nu}$$

$$= i\frac{g_2}{\cos\theta_2}M_Z g_{\mu\nu}$$

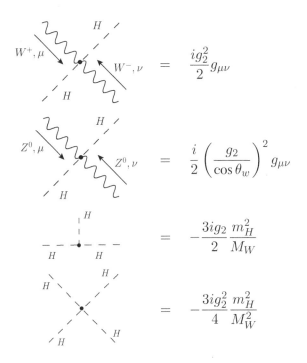

$$= \frac{ig_2^2}{2}g_{\mu\nu}$$

$$= \frac{i}{2}\left(\frac{g_2}{\cos\theta_w}\right)^2 g_{\mu\nu}$$

$$= -\frac{3ig_2}{2}\frac{m_H^2}{M_W}$$

$$= -\frac{3ig_2^2}{4}\frac{m_H^2}{M_W^2}$$

B.5   GENERAL RULES

1) Four-momentum must be conserved at each vertex.

2) Integrate over undetermined momenta in loops: $\int \mathrm{d}^4 q/(2\pi)^4$.

3) Closed fermion loops give an additional factor of -1, and the trace must be taken over the γ matrices.

4) Diagrams related by the exchange of two external-state fermions have a relative minus sign.

GAMMA MATRIX
IDENTITIES

When computing amplitudes involving fermions, the following γ-matrix identities are often useful.

$$g_{\mu\nu}\gamma^{\mu}\gamma^{\nu} = 4$$
$$g_{\mu\nu}\gamma^{\mu}\gamma^{\rho}\gamma^{\nu} = -2\gamma^{\rho}$$
$$g_{\mu\nu}\gamma^{\mu}\gamma^{\rho}\gamma^{\sigma}\gamma^{\nu} = 4g^{\rho\sigma}$$
$$g_{\mu\nu}\gamma^{\mu}\gamma^{\rho}\gamma^{\sigma}\gamma^{\tau}\gamma^{\nu} = -2\gamma^{\rho}\gamma^{\sigma}\gamma^{\tau}$$
$$\mathrm{tr}(\gamma^{\mu}\gamma^{\nu}) = 4g^{\mu\nu} \tag{C.1}$$
$$\mathrm{tr}(\gamma^{\mu}\gamma^{\nu}\gamma^{\rho}\gamma^{\sigma}) = 4(g^{\mu\nu}g^{\rho\sigma} + g^{\mu\sigma}g^{\nu\rho} - g^{\mu\rho}g^{\nu\sigma})$$
$$\mathrm{tr}(\gamma^{5}) = 0$$
$$\mathrm{tr}(\gamma^{\mu}\gamma^{\nu}\gamma^{5}) = 0$$
$$\mathrm{tr}(\gamma^{\mu}\gamma^{\nu}\gamma^{\rho}\gamma^{\sigma}\gamma^{5}) = -4i\varepsilon^{\mu\nu\rho\sigma}.$$

Note that only traces of even products of γ-matrices are listed. The trace of any product involving an odd number of γ-matrices vanishes. The last of these identities can be useful when considering weak interactions.

The Feynman parameter trick can be generalized to allow more than two factors in the denominator. The general form is

$$\frac{1}{A_1 A_2 \ldots A_n} = \int_0^1 dx_1 \ldots \int_0^1 dx_n \frac{1}{(A_1 x_1 + \ldots + A_n x_n)^n}$$
$$\delta(1 - x_1 - \ldots - x_n). \qquad (C.2)$$

BIBLIOGRAPHY

[1] I. J. R. Aitchison and A. J. G. Hey. *Gauge Theories in Particle Physics*. Institute of Physics, 2 edition, 1996.

[2] I. I. Bigi and A. I. Sanda. *CP Violation*. Cambridge University Press, 2009.

[3] F. Bissey, F-G. Cao, A. R. Kitson, A. I. Signal, D. B. Leinweber, B. G. Lasscock, and A. G. Williams. Gluon flux-tube distribution and linear confinement in baryons. *Phys. Rev.*, D76:114512, 2007.

[4] T. Bringmann, J. Hasenkamp, and J. Kersten. Tight bonds between sterile neutrinos and dark matter. *JCAP*, 1407:042, 2014.

[5] A. Broncano, M. B. Gavela, and E. E. Jenkins. The effective Lagrangian for the seesaw model of neutrino mass and leptogenesis. *Phys. Lett.*, B552:177–184, 2003. [Erratum: Phys. Lett.B636,332(2006)].

[6] J. F. Cornwell. *Group Theory in Physics*. Academic Press, 1997.

[7] A. Das and T. Ferbel. *Introduction to Nuclear and Particle Physics*. World Scientific, 2 edition, 2004.

[8] M. Drewes. The phenomenology of right handed neutrinos. *Int. J. Mod. Phys.*, E22:1330019, 2013.

[9] M. Drewes et al. A white paper on keV sterile neutrino dark matter. *JCAP*, 1701(01):025, 2017.

[10] L. D. Duffy and K. van Bibber. Axions as dark matter particles. *New J. Phys.*, 11:105008, 2009.

[11] G. Fischer and J. Plch. The high voltage read-out for multiwire proportional chambers. *Nucl. Instrum. Meth.*, 100:515–523, 1972.

[12] D. Goldbert. *The Standard Model in a Nutshell.* Princeton University Press, 2017.

[13] D. Griffiths. *Introduction to Elementary Particles.* Wiley, 1987.

[14] A. Grozin. Lectures on QED and QCD. In *3rd Dubna International Advanced School of Theoretical Physics, Dubna, Russia, January 29-February 6, 2005*, pp. 1–156, 2005.

[15] A. N. Hiller Blin, C. Fernández-Ramírez, A. Jackura, V. Mathieu, V. I. Mokeev, A. Pilloni, and A. P. Szczepaniak. Studying the $P_c(4450)$ resonance in J/ψ photoproduction off protons. *Phys. Rev.*, D94(3):034002, 2016.

[16] G. Kane. *Modern Elementary Particle Physics.* Cambridge University Press, 2 edition, 2017.

[17] I. D. Lawrie. *A Unified Grand Tour of Theoretical Physics.* Institute of Physics, 3 edition, 2002.

[18] M. S. Longair. *High Energy Astrophysics*, volume 1. Cambridge University Press, 2 edition, 1992.

[19] R. Machleidt and D. R. Entem. Chiral effective field theory and nuclear forces. *Phys. Rept.*, 503:1–75, 2011.

[20] B. R. Martin and G. Shaw. *Particle Physics.* Wiley, 4 edition, 2017.

[21] A. Merle. keV neutrino model building. *Int. J. Mod. Phys.*, D22:1330020, 2013.

[22] D. Morris. *Empty Space is Amazing Stuff.* Pantaneto Press, 2013.

[23] D. Morris. *The Nuts and Bolts of Quantum Mechanics.* Pantaneto Press, 2014.

[24] K. A. Olive et al. Review of particle physics. *Chin. Phys.*, C38:090001, 2014.

[25] R. Pasechnik and M. Šumbera. Phenomenological review on quark-gluon plasma: concepts vs. observations. *Universe*, 3(1):7, 2017.

[26] S. B. Patel. *Nuclear Physics*. Anshan, 2 edition, 2011.

[27] R. Penrose. *The Road to Reality*. Random House, 2004.

[28] C. F. Perdrisat, V. Punjabi, and M. Vanderhaeghen. Nucleon electromagnetic form factors. *Prog. Part. Nucl. Phys.*, 59:694–764, 2007.

[29] M. E. Peskin and D. V. Schroeder. *An Introduction to Quantum Field Theory*. Addison-Wesley, 1995.

[30] A. I. M. Rae. *Quantum Mechanics*. Institute of Physics, 4 edition, 2002.

[31] W. R. Rindler. *Relativity*. Oxford University Press, 2 edition, 2006.

[32] S. Scherer. Introduction to chiral perturbation theory. *Adv. Nucl. Phys.*, 27:277, 2003.

[33] M. D. Schwartz. *Quantum Field Theory and the Standard Model*. Cambridge University Press, 2014.

[34] M. Sozzi. *Discrete Symmetries and CP Violation*. Oxford University Press, 2008.

[35] A. Zee. *Quantum Field Theory in a Nutshell*. Princeton University Press, 2 edition, 2010.

INDEX

750 GeV diphoton excess, 147

A

action, 201
adjoint representation, 98
adjoint spinor, 216
ALICE, 307
α decay, 6, 9
angular momentum, 58, 90
  commutation relations, 59
  in hadrons, 174
angular momentum operator, 59
anomalous magnetic moment, 249, 275
anomaly cancellation, 349, 368
anti-fundamental representation, 97
antiparticle, 17, 189, 222, 225, 294
asymptotic freedom, 288
atomic orbitals, 16
axial vector, see pseudo-vector
axion, 357, 378, 380
Axion Dark Matter Experiment, 380

B

B meson, 305, 346, 347
bare coupling, 270

bare mass, 269, 355
bare propagator, 265, 268
baryogenesis, 353, 354
baryon, 19, 156
  number, 19, 353, 369
basis spinors, 219, 223, 253
  completeness relation, 224, 254
β decay, 6, 9, 10, 14, 15
β function, 271, 287, 352, 365, 374
betatron oscillation, 131
Bethe formula, 107
Bhabha scattering, 84
Bohr
  model of atom, 8
Bohr magneton, 69
boson, 11, 155
bottom quark, 27, 180
bottomonium, 180
branching fraction, 147
branching rules (subgroups), 367
Breit-Wigner formula, 146
Bremsstrahlung, 110
Brownian motion, 3
bubble chamber, 115
bunching, 126

C

Cabibbo angle, *see* quark mixing, angle
calorimeter, 121–122
Casimir, 59, 94
Cauchy's integral formula, 247
Čerenkov detector, 120–121
Čerenkov radiation, 111–113
charge conjugation, 81, 353
  matrix, 231
  of hadrons, 177
  of spinors, 223, 231
  violation in weak interactions, 81, 313, 340
charm quark, 27, 180, 339
charmonium, 180
chiral anomaly, 351
chiral perturbation theory, 297
chiral projection operators, 227
chiral symmetry, 297, 355
chiral symmetry breaking, 320
chirality, 226, 229, 230, 313, 327, 333
  operator, 226
chromaticity, 133
CKM matrix, 335, 343
CLIC, 134, 388
Clifford algebra, 211, 227
cloud chamber, 114–115
CMS, 144
Coleman-Mandula theorem, 371
color, 25, 161, 280, 294

color factor, 289, 292
complex coupling, *see* complex phase
complex phase, 343, 344
Compton scattering, 109, 259, 264
  amplitude, *see* scattering amplitude
confinement, 25, 156, 161, 294, 296
conservation laws, 19, 29, 75, 78, 353
conserved current, 242, 377
  for scalar in EM field, 190
  for scalar particle, 188, 198
  for spinors, 216, 235
constituent quark, 156, 172
continuity equation, 57
contravariant vector, *see* vector
coordinate transformations, 36
cosmological constant, 385
Coulomb gauge, 193
Coulomb interaction, 248
coupling constant, 12, 270
  electromagnetic, 12
  nuclear force, 14
  strong, 271, 282
  weak, 15, 311
covariant derivative, 244, 282, 300, 321
covariant vector, *see* vector
CP symmetry, 82, 314, 340, 341, 353, 356
  violation in weak interactions, 344
CPT theorem, 83

Cronin-Fitch experiment, 345
cross-section, 13, 135, 139
    classical scattering, 135
    Coulomb scattering, *see*
        Rutherford scattering
        formula
    differential, *see* differential
        cross-section, 206
    for hadron production, 162
    inelastic scattering, 138
    total, 144
crossing symmetry, 84, 189, 225
cyclotron, 126–128

D

D meson, 346
dark energy, 31, 384
dark matter, 31, 364, 379
decay constant, 5
decay mode, 147
decay rate, 145
Delbrück scattering, 250
differential cross-section, 138,
    139, 260
digamma, 147
dipole magnet, 130
Dirac equation, 17, 209
    Hamiltonian form, 222
    in momentum space, 217
    solutions, 216
Dirac mass, *see* mass, Dirac
Dirac representation, 212
double-beta decay, 233
down quark, 24
drift chamber, 118–119
dummy index, 37

E

e^+-e^- pair production, 110
effective coupling constant, 270,
    287, 352, 365
effective field theory, 297
eigenstate, 53
eigenvalue, 53
Einstein summation convention,
    37
electromagnetic current, 198
electromagnetic interactions
    of scalar particles, 198
    of spinors, 235
electromagnetism, 10
electron, 4
electron volt, xviii
electroweak symmetry breaking,
    see Higgs mechanism,
    Standard Model
electroweak theory, 28, 327, 329,
    331, 349
emission spectra, 8, 172
energy operator, 55, 186
energy-momentum relation, 45,
    185
Euclidean space, 38, 300
Euler-Lagrange equation, 66,
    67, 203
expectation value, 53, 65
exponential map, 86

F

f_0, 307
fermi (unit), xviii
Fermi interaction, 15, 311
fermion, 11, 16, 155

fermion doubling, 301, 303
Feynman diagram, 11, 29, 198
 estimation of amplitudes,
 250
 higher-order, 264
 QCD, 288
 QED, 250
 quark level, 171
 tree-level, 202
Feynman parameter, 266
Feynman rules, 199, 202, 206,
 397
 for hadron-photon
 interaction, 271
 for QCD, 285
 for QED, 251
field strength tensor, 44
fine tuning, 355
fine-structure constant, 12,
 172
flavor, 158, 165, 331, 337
 oscillation in mesons, 342
flavor mixing in mesons, 170,
 340, 342, 345
flavor-changing neutral current,
 338
flavour, 25, 27
FODO lattice, 132
form factor, 271
 pion, 275
 proton, 274
four-momentum operator,
 186
four-vector, 41, 43
free index, 37
full propagator, 265, 268
fundamental representation,
 97

G

g-factor, 70, 72, 249, 275
γ^5, *see* chirality, operator
γ matrices, 211
 identities, 255, 267, 403
 representations, 211
gauge fixing, 192
gauge invariance, 192, 195, 243,
 244, 281, 333, 355, 371,
 375
gauge theory, 285, 331
 Abelian, 243
 non-Abelian, 281, 314, 316
gauge-covariant derivative, *see*
 covariant derivative
gaugino, 374
Geiger counter, 117
Gell-Mann matrices, 94, 281
Gell-Mann–Nishijima formula,
 329
general covariance, 375
general relativity, 374
generator, *see* group generator
Georgi-Glashow model, 366,
 370
GIM mechanism, 27, 339
global phase transformation,
 242
glueball, 306–307
gluon, 26, 282
 self-interactions, 284,
 286
Goldstone boson, 320, 322
Gordon identity, 274, 275
grand unified theory, 365, 366
gravitational waves, 377
graviton, 374–378

group, 76
   Abelian, 77, 86
   non-Abelian, 77, 86, 279
group generator, 85, 281, 289,
   366
group representation, 87, 158,
   211, 232, 366, 367
   irreducible, 89, 93, 95, 158,
   168
   reducible, 88, 97
   Standard Model, 331

H

hadron, 19, 156
   multiplets, 21
   naming conventions,
   180
   properties, 21
hadron jet, 275, 296
hadronization, 296
half-life, 145
Hamilton's equations, 64
Hamiltonian, 64
   non-relativistic, 64
   of charged particle in EM
   field, 67
harmonic function, 193
Heisenberg's uncertainty
   principle, 9
helicity, 225, 229, 230, 233, 312,
   313
   spinor operator, 225
   vector operator, 196
hidden sector, 381
hierarchy problem, 354, 373
Higgs boson, 28, 32, 326, 334,
   335, 354

Higgs mechanism, 28, 316, 334,
   366
   Standard Model, 324,
   364
   toy model, 316, 322
Hilbert space, 52
hypercharge (strong), 158

I

ILC, 387
impact parameter, 104,
   135
inflation, 383–384
invariant amplitude, 142, *see
   also* transition
   amplitude, 203
invariant mass, 47, 145
ionization energy losses,
   104
irrep, *see* group representation,
   irreducible
isospin, 19, 24, 157
isotopes, 4

J, K

Jarlskog invariant, 348
J/Ψ, 27
kaon, 18, 81, 340
KATRIN experiment, 14
ket vector, 52
K_L and K_S, 345
Klein-Gordon equation, 186,
   187, 316
klystron, 129–130
Kronecker delta, 37

L

ladder operator, 61, 166
Lagrangian, 66, 75, 202, 235
    for electroweak theory, 325
    for QED, 243
    for toy Higgs model, 322
    QCD, 284
Landau pole, 352, 366
large hadron collider, *see* LHC
lattice QCD, 295, 298
Legendre polynomial, 176
Legendre transformation, 66
LEP, 312
lepton, 19, 156
    numbers, 19
Levi-Civita symbol, 39
LHC, 32, 123, 130, 144, 326, 387
LHCb, 305
Lie algebra, 87, 92
    graded, 372
Lie group, 85
    semi-simple, 366
lifetime, 145–148
lightest supersymmetric particle, 381
LIGO, 377
linac, 125–126, *see also* ILC, CLIC, SLAC
linear accelerator, *see* linac
local phase transformation, 242, 243
Lorentz group, 99, 215, 226
Lorentz transformation, 40, 42, 333
    of spinors, 213, 215, 228

Lorenz condition, 193, 195
luminosity, xviii, 126, 139

M

MACHO, 379
magnetic focusing, 132–132
magnetic moment, 70
    of electron, 69
    of hadrons, 73
    of proton, 275
magnetic monopole, 370–371
Majorana mass, *see* mass, Majorana
Mandelstam variables, 260
mass, 45, 100
    bare, *see* bare mass
    constituent quark, 173
    Dirac, 363
    fermions, 333
    Majorana, 363
    neutrino, *see* neutrino, mass
    of hadrons, 172
    of quarks, 173, 335
    physical, *see* physical mass
mass shell, 10, 205, 268
mass state, 337
Maxwell equation, 191
    classical, 44
    deformation under Higgs mechanism, 321
    quantum, 191
meson, 19, 156, 170
metric, 38, 39, 374
    Minkowski, 42
minimal coupling, 68, 71, 190
minimal substitution, *see* minimal coupling

Minkowski space, 41, 300, 374
missing mass, 48
Møller scattering, 250
    amplitude, 256
momentum cut-off, 267, 297
momentum operator, 55, 186
multi-wire proportional
    chamber, 117
muon, 18, 123, 162
MWPC, *see* multi-wire
    proportional chamber

N

natural units, xvii, xviii, 41, 52
negative-energy solutions, 187,
    189, 209, 217, 222
neutral currents, 28, 328
neutrino, 14, 15, 27, 313, 332
    as Majorana fermion, 233
    mass, 15, 332, 361, 362, 364
    oscillations, 15, 332, 361
    sterile, 364, 380
neutron, 10
Noether's theorem, 75, 242
nuclear force, 12, 26, 248, 296
    coupling constant, 14
nuclear magneton, 73
nucleon, 157, 297
nucleus, 6

O, P

off-shell, *see* mass shell
on-shell, *see* mass shell
operator, 52
orthogonal group, 89
parity, 78, 353
    of fermions, 80

of hadrons, 158, 174, 175
of photon, 79
violation in weak
    interactions, 80, 312,
    340
particle decay, 46
particle shower
    electromagnetic, 110, 121
    hadronic, 121
parton, 276, 288
path integral, 201, 298
Pauli equation, 70, 71, 230
Pauli exclusion principle, 155
Pauli matrices, 63, 70
Pauli-Lubański vector, 100
P_c^+, 305
penguin diagram, 346
pentaquark, 305
periodic table of elements, 3
perturbation theory, 202, 298
    in QCD, 288
photino, 374
photo-electric effect, 7, 108
photomultiplier, 122
photon, 7, 10, 191, 199, 244
physical mass, 269, 355
pion, 13, 18, 296
plum-pudding model, 5
PMNS matrix, 361, 362
Poincaré group, 99, 371
polarization tensor (graviton),
    376
polarization vector, 191
    basis, 192, 194
    circular polarization, 194
    completeness relation, 194,
        259
    linear polarization, 194

longitudinal, 197, 198
orthonormality, 194
positron, 17
potential between interacting
    particles, 246
principle of detailed balance, 83
probability density current
    non-relativistic, 56, 58
Proca equation, 194–195
propagator, 203, 204
    on a lattice, 301
proton, 9
    lifetime, 369
proton decay, 369
pseudo-Goldstone boson,
    320
pseudo-scalar, 79, 234
pseudo-vector, 79, 234, 235,
    312

Q

QCD, 25, 285, 349
    Lagrangian, 284
    scale-dependence, 286
QCD flux tube, see QCD string
QCD string, 295, 303
QCD vacuum, 303
QED, 10, 249, 349
    coupling constant, 12
    Lagrangian, 235, 243
quadrupole magnet, 132
quantization, 200, 349
quantum electrodynamics, see
    QED, see QED
quantum field theory, 199, 316
quantum mechanics, 7, 51
quantum state, see state vector

quark, 21, 23, 156, 158, 165, 276
    constituent, see constituent
    quark
quark mixing, 27, 335, 337, 340,
    343
    angle, 337, 344
    matrix, see CKM matrix
quark-gluon plasma, 307–308
quarkonium, 180
quintessence, 385

R

R-parity, 381
R-symmetry, 372, 381
radiation length, 110
radiative losses, 104, 110
radio-frequency acceleration,
    128–129
radioactivity, 5, 9
rapidity, 43
real particle, 10
regularization, 267, 270, 299
relativistic kinetic energy, 46
renormalization, 264, 265,
    268–270, 287, 312, 314,
    316, 354, 378
representation, see group
    representation
residual gauge freedom, 193
resonance, 48, 145, 181
RHIC, 307
RICH, see Čerenkov detector
Rutherford experiment, 6
Rutherford scattering formula,
    6, 10, 26, 137

S

scalar, 35, 37, 79
scalar particle, 186
scalar product, 37
scattering amplitude
  for Bhabha scattering, 258
  for Compton scattering, 259
  for distinguishable charged
    spinors, 252, 256
Schrödinger equation, 8, 54, 64,
    68
  for hadrons, 175
  time-dependent, 56
  time-independent, 65
scintillator, 122
see-saw mechanism, 363–365
selectron, 374
self-energy, 269
semi-conductor detector, *see*
    solid-state detector
Σ baryon, 158
significance level, 147
silicon detector, *see* solid-state
    detector
SLAC, 387
slash notation, 217
sneutrino, 374
$SO(3)$, 92
solar neutrino problem, 361
solid-state detector, 119–120
$SO(n)$, 89, 232
space-time, 40
spark chamber, 115–116
special orthogonal group, *see*
    $SO(n)$
special unitary group, *see* $SU(n)$
spectrometer, 113

sphaleron, 353
spherical harmonics, 176
spin, 16, 60, 100, 153
  commutation relations, 60
  of Dirac spinors, 220
  of hadrons, 156, 158, 165
  quantum number, 153
spin operators, 60
  for spinors, 221
  for vectors, 195
spin-averaged amplitude, 253,
    258
spin-statistics theorem, 16, 153,
    155
spinor
  color space, 280, 289
  Dirac, 211, 232
  Majorana, 232
  Pauli (non-relativistic), 63
  Weyl, 232
spontaneous symmetry
    breaking, 315, 316
squark, 374
Standard Model, 28, 29, 331,
    349
state vector, 52
  normalization, 52, 140, 220,
    319
static quark model, 172
sterile neutrino, *see* neutrino,
    sterlie
Stern-Gerlach experiment, 16
stopping power, 107
strange quark, 24
strangeness, 19, 20, 158
straw chamber, 118
stress-energy tensor, 374, 378
strong $\mathbb{CP}$ problem, 356, 378

strong nuclear force, 15, 25, 155,
   156, 163, 279
   coupling constant, 282
   structure function, 275–276
$SU(2)$, 92
$SU(3)$, 94
$SU(3)$ color symmetry, 281
$SU(5)$, *see* Georgi-Glashow
   model
sum over histories, 201
$SU(n)$, 89, 90, 157, 178, 331
supercurrent, 378
supersymmetry, 371–374
synchrocyclotron, 128
synchrotron, 130–133, *see also*
   LHC, RHIC, LEP
synchrotron radiation, 133–134

T

τ particle, 27, 123
τ-θ problem, 80
tensor, 40, 97
tetraquark, 304–306
time reversal symmetry, 82
top quark, 27, 180
TOTEM, 144
Townsend avalanche, 117
transition amplitude, 84, 142,
   see also invariant
   amplitude, 205
transition radiation, 113

U, V

up quark, 24
vacuum expectation value, 317,
   319, 325, 334
vector, 35, 36, 97

contra- and covariant, 38, 44
virtual particle, 10, 205, 287

W

$W^{\pm}$ boson, 26, 312, 326, 327
Ward identity, 274
wave function, 8
wave-particle duality, 9, 51, 53,
   199, 241
wavefunction, 53
   static quark model, 164, 167
wavefunction collapse, 53
waveguide, 128
weak focusing, 131
weak hypercharge, 324, 328,
   329, 331, 333
weak isospin, 328, 329
weak mixing angle, 326
weak nuclear force, 15, 26
   coupling constant, 311, 324,
   328
   mixing angle, 326
Weyl representation, 212, 227,
   228
Wick rotation, 266, 300
WIMP, 380
wino, 374

X, Y, Z

X boson, 369
Yukawa interaction, 13, 236,
   296, 332, 333, 345
Z^0 boson, 28, 312, 325, 327
Zeeman effect, 68, 172
   anomalous, 70
zero-point energy, 385
$Z^-(4330)$, 305